Penny Farmer

Wind Energy 1975 – 1985
A Bibliography

Springer-Verlag
Berlin Heidelberg New York
London Paris Tokyo

Penny Farmer
Prisk House
Prisk
Nr. Cowbridge
South Glamorgan
Wales, UK

ISBN-13:978-3-642-82662-7   e-ISBN-13:978-3-642-82660-3
DOI: 10.1007/978-3-642-82660-3

Library of Congress Cataloging-in-Publication Data
Farmer, Penny
Wind energy, 1975–1985.
Includes index.
1. Wind power--Bibliography.
I. Title.
Z5853.W55F37 1986   [TJ825]   016.33379'2   86-1918
ISBN-13:978-3-642-82662-7(U.S.)

2161/3020-543210

Contents

1 Research and Development

1.1 International (Including Developing Countries)

1.1(1) UTILIZATION OF SOLAR AND WIND ENERGY
TO IMPROVE THE LIVING ENVIRONMENT IN LESS
DEVELOPED COUNTRIES.
Ramakumar, R. (Oklahoma State Univ.,
Stillwater, USA.) In Proc. Institute of
Environmental Science 22nd Annual Technical
Meeting, Environmental Technology 1976
Philadelphia, Pa., 26-28 April 1976.
Mount Prospect, Ill., USA. Institute of
Environmental Science 1976 pp. 314-318.
   Discusses the prospects of utilizing
renewable energy sources such as solar and
wind energy and an approach to develop viable
rural economic units in less developed
countries. (16 refs.)

1.1(2) ELECTRICITY FROM THE WIND A GUIDE
FOR THE GENERALIST.
Bendick, P. (Solar Solution, Washington, DC.,
USA.) NTIS Report PB80-139751, 1978 23 pp.
   The possibility of using wind energy to
generate small amounts of electricity is
discussed for the generalist working in
developing countries. Basic information about
wind energy is presented, but the guide is
neither comprehensive nor technical.

1.1(3) NEW VILLAGE USES OF RENEWABLE ENERGY
SOURCES.
Howe, J.W. (Overseas Development Council, DC.)
Paper presented at U.N. Environment Programme/
Et Al Energy & Environment in East Africa
Conf., Nairobi, 7-10 May 1979 pp. 93-108.
   Renewable energy resources are suitable for
displacing conventional and costly energy
systems at the village level in rural areas.
Solar energy is one such energy source; it is
plentiful, well distributed, and associated
technologies are proven. Biogas and wind
energy systems are also evaluated. Factors
such as cost and socioeconomic impact are
identified. (16 refs.)

1.1(4) ENERGY FUTURES IN DEVELOPING
COUNTRIES.
Olende, S.A. (U.N. Centre for Natural
Resources Energy & Transport, New York)
Paper presented at U.N. Environment
Programme/Et Al Energy Environment in East
Africa Conf., Nairobi, 7-10 May 1979
pp. 14-45.

   Traces trends in global energy consumption
with particular reference to developing
nations. The energy resource base of these
areas is analysed to determine if future
energy demands will be satisfied. Coal,
natural gas, oil shale, tar sand, geothermal,
and hydroelectric resources are surveyed.
Potential contributions of wave, tidal, wind,
and solar energy are also considered.
(18 refs.)

1.1(5) ENERGY RESOURCES AND THE IMPACT
OF NEW TECHNOLOGIES
Ibrahim Said, M.A. (Al-Azhar Univ., Cairo)
Long Range Planning, 13 (8) August 1980,
pp. 63-70.
   Examines global energy resources and reserves
including tidal power, geothermal resources,
hydropower, wind power, solar emergy fusion,
fission, and fossil fuels. Energy conversion
technologies and their prospects are discussed
and associated environmental problems
summarised. Energy growth and development is
considered for future energy planning. (6 refs.)

1.1(6) ANNUAL REPORT ON ENERGY RESEARCH,
DEVELOPMENT AND DEMONSTRATION ACTIVITIES
OF THE IEA 1979-1980.
International Energy Agency/OECD Report,
1980 89 pp.
   Describes IEA Energy R & D Projects conducted
by the nine member countries during 1979 which
focused on energy conservation, coal technology,
enhanced oil recovery, nuclear reactor safety
experiments, geothermal energy systems, solar
energy, biomass conversion, ocean energy, wind
energy, fusion energy, and the production of
hydrogen from water. Organisation arrangements
and financing of each project are reported.

1.1(7) WORLD ENERGY IN THE 21ST CENTURY.
Kiely, F. Chartered Mechanical Engineer,
May 1980 pp. 26-30.
   Attempts to evaluate the worldwide availability of various energy resources in the
long term. The ultimate availability of
each resource, the rate at which it could be
produced, the extent to which conservation
measures could prolong the availability of
each resource, and the potential for
substituting renewable resources for resources

of limited availability. Includes oil, natural gas, coal, hydroelectric power, nuclear, biomass, oil shale, tar sands, and solar, geothermal, wind, and ocean thermal.

1.1(8) LOAD CASES FOR WIND ENERGY CONVERSION SYSTEMS.

Tornkvist, G. (Saab-Scania AB) In Proc. 3rd International Symposium on Wind Energy Systems, Lyngby, Denmark 26-29 August 1980 Cranfield, U.K., BHRA Fluid Engng., 1980 Paper D2, pp. 183-192.

Describes the international programme for research and development on wind energy conversion systems (WECS) organised by the International Energy Agency (IEA). One project concerning load case recommendations is presented and results on wind data for WECS and on main load cases given.

1.1(9) THE ECWA REGIONAL REPORT TO THE UNITED NATIONS CONFERENCE ON NEW AND RENEWABLE SOURCES OF ENERGY.

Paper presented at U.N. Economic and Social Council New & Renewable Sources of Energy Preparatory Conference, Beirut 12-16 January 1981 56 pp.

Describes the General Assembly's decision to convene an International Conference on New and Renewable Energy Sources. Identification and analysis of new technologies, potentials for utilisation of new and renewable forms of energy in developing countries, assessment of the economic viability of both new and old technologies, and promotion of adequate information on the subject. Includes wind energy.

1.1(10) ENERGY AND AID.

OECD Observer, No. 111, July 1981 pp. 23-26.

Increased energy production is a key priority for oil importing developing nations. Describes the investment of aid from OECD member nations France, W. Germany, Japan, The Netherlands, and the U.S. in the development of wind energy, biofuels, and other energy systems.

1.1(11) ENERGY FOR RURAL DEVELOPMENT RENEWABLE RESOURCES AND ALTERNATIVE TECHNOLOGIES FOR DEVELOPING COUNTRIES.

U.S. National Research Council Report, 1981 244 pp.

Small, low-cost, locally operated energy systems are described for application in rural areas of developing nations. Rural applications of photovoltaics are cited and other alternative technologies addressed including solar heating and cooling, wind energy, small-scale hydroelectric power, bioconversion, geothermal energy, batteries, flywheels and stirling engines.

1.1(12) ENERGY RESEARCH, DEVELOPMENT, AND DEMONSTRATION IN THE IEA COUNTRIES 1980 REVIEW OF NATIONAL PROGRAMMES.

International Energy Agency Report 1700/ES(61 81 06 1) 1981 169 pp.

The IEA Committee on Energy R & D (CRD) supports the definition and implementation of energy technology that will enable member countries to reduce their dependence on imported energy and improve the efficiency of energy use. Research programmes in the following member countries Australia, Austria, W. Germany, Ireland, Japan, and Norway; and brief reviews of programmes in Belgium, Canada, Denmark, Greece, Italy, Netherlands, New Zealand, Spain, Sweden, Switzerland, the U.K., and the U.S. are reviewed. These assessed the feasibility of solar, hydro-electric, geothermal and wind energy, and examined the potential for energy conservation in automobiles, building equipment, and industry. The amount of money spent in energy R & D is measured against the real growth of each country's budget, and each is evaluated for the status of its energy R & D programmes.

1.1(13) GLOBAL ENERGY IN TRANSITION, ENVIRONMENTAL ASPECTS OF NEW AND RENEWABLE SOURCES FOR DEVELOPMENT.

Sierra Club International Earthcare Centre Report 5, 1981 125 pp.

Over-reliance on fossil fuels, fuelwoods, and charcoal fuel sources has led to a disruption of the natural resource base, heightened expenses, and social disruption from a lack of productivity, lack of access to essential resources, and changes in cultural patterns. The U.N. Conference on New and Renewable Sources of Energy focused global attention on the contributions that new and renewable sources of energy can offer. Technical, development, and environment aspects of alternative energy sources; policy issues for energy planning and development, and selected national reports are some topics included which also incorporates wind energy assessments.

1.1(14) IEA LARGE-SCALE WECS ANNUAL REPORT 1980.

National Swedish Board for Energy Source Development Report NE 1981, 1981 49 pp.

IEA Wind Energy Conversion System research and development is designed to further develop such systems by means of cooperative efforts. Large-scale Wind Energy Conversion Systems in Denmark, W. Germany, Sweden, and the U.S. made considerable progress during 1980 and design, construction, testing, and operation of these systems are detailed in this report.

1.1(15) NEW AND RENEWABLE SOURCES OF ENERGY CONFERENCE SUMMARY.

Paper presented at U.N New and Renewable Sources of Energy Conference, Nairobi 10-21 August 1981 126 pp.

The major challenge to both industrialised and developing countries in the 80's is finding a safe, clean, and renewable source of energy. This U.N. Conference on New and Renewable Sources of Energy examines economic and political aspects of new energy sources and discusses the transition from a world dependent on depletable fuels to one reliant on ecologically sound renewable fuels. Solar, wind, hydroelectric, tidal, wood, biomass, and other energies are studied in detail for their applicability and feasibility to various regions of the world.

1.1(16) OVERVIEW OF ACTIVITIES RELEVANT TO THE INCREASED USE OF NEW AND RENEWABLE SOURCES OF ENERGY WITHIN UNITED NATIONS SYSTEM. REPORT OF THE SECRETARY-GENERAL.
Paper presented at U.N. New and Renewable Sources of Energy Conference, Nairobi 10-21 August 1981 23 pp.
Gives a general overview of technical cooperation projects included in the U.N. Development Programme dealing with new and renewable energy sources. Data on resource allocations to each of the projects and the Programme activities are presented together with the relative degree of involvement of each project in various activities related to renewable energy development. Each of the renewable energy technologies are studied for their relative importance to the various U.N. Programmes.

1.1(17) REPORT OF THE AD HOC EXPERT GROUP ON EDUCATION AND TRAINING.
Paper presented at U.N. New and Renewable Sources of Energy Conference, Nairobi 10-21 August 1981 41 pp.
This group was established by the U.N. to determine the needs of education in the development and use of new and renewable energy sources. Training needs and existing facilities for renewable energy technologies are given for solar energy, biomass, hydropower, ocean power, oil shales, geothermal energy, and wind energy and recommendations made for policy structuring and education of the general public. (10 refs.)

1.1(18) REPORT OF THE SYNTHESIS GROUP RECOMMENDATIONS MADE BY THE TECHNICAL PANELS AND THE AD HOC GROUPS OF EXPERTS.
Paper presented at U.N. New and Renewable Sources of Energy 3rd Preparatory Conference, New York City, U.S.A. 30 March-17 April 1981 97 pp.
Reports on a variety of energy forms - geothermal, wind, ocean, oil shale, solar, biomass, hydropower, fuelwood and charcoal, peat, draught animal power and other new and renewable sources of energy. Information flows, research, development, and transfer of technology, education and training, and utilisation of energy in agriculture, transportation, and allied sectors are considered.

1.1(19) REPORT OF THE TECHNICAL PANEL ON WIND ENERGY ON ITS SECOND SESSION.
Paper presented at U.N. New and Renewable Sources of Energy Conference, Nairobi 10-21 August 1981 34 pp.
Describes the current status and potential of wind energy and its relevance to developing countries. A brief bibliography of key existing studies on wind energy is included, and basic characteristics of wind energy and wind resources are determined and presented. Economics of both large- and small-scale wind energy production are calculated and the constraints limiting development and utilisation of wind energy and measures to overcome these are discussed. (21 refs.)

1.1(20) REPORT SUBMITTED BY THE FOOD AND AGRICULTURE ORGANISATION OF THE UNITED NATIONS.
Paper presented at U.N. New and Renewable Sources of Energy Conference, Nairobi 10-21 August 1981 14 pp.
FAO maintains an interest in the development of new and renewable energy sources because of the dominant role of agriculture in the energy systems of rural areas. Energy in the form of waste materials can be used for biogas production. Use of biofertilisers based on nitrogen-fixing organisms, applications of wind and solar energy, integrated pest control, and the improved utilisation of draught animal power are discussed.

1.1(21) REPORT SUBMITTED BY THE UNITED NATIONS DEVELOPMENT PROGRAMME.
Paper presented at U.N. New and Renewable Sources of Energy Conference, Nairobi 10-21 August 1981 15 pp.
Countries throughout the world are assessed for their eligibility for various types of alternative energies including hydropower, geothermal, solar, biogas, nuclear, and wind, and then awarded assistance when and where appropriate by the UNDP, budget activities are detailed, and all UNDP projects in new and renewable energy sources listed.

1.1(22) SYNTHESIS OF TECHNICAL INFORMATION ON NEW AND RENEWABLE SOURCES OF ENERGY.
Paper presented at U.N. New and Renewable Sources of Energy Conference, Nairobi 10-21 August 1981 38 pp.
Summarises available material on new and renewable sources of energy and their applications and highlights areas where information is lacking. The increasing importance of renewable energy to developing countries is investigated in a global context. Ongoing R & D in renewable resources are summarised, both generally and by country. Peat, oil shale, tidal power, wood energy, wind power, solar energy, geothermal power, and biomass are included.

1.1(23) RESEARCH AND TECHNOLOGICAL CAPACITY FOR THE USE OF RENEWABLE ENERGY RESOURCES IN DEVELOPING COUNTRIES.
Dosik, R.S. and Weiss, C. (World Bank, DC.)
Paper presented at UNESCO Non-Technical Obstacles to the Use of New Energies in Developing Countries Conference, Bellagio, Italy 25-29 May 1981 14 pp.
Reviews the capabilities of developing countries for assessing national energy needs

and resources, choosing appropriate technologies, and establishing effective institutions
for energy manufacture and distribution.
Direct assistance programmes for strengthening
these capabilities are discussed. International
programmes to build technological capacity in
biomass production, and solar, wind, and small
hydropower systems need to be implemented.

1.1(24) NON-TECHNICAL PROBLEMS IN THE
IMPLEMENTATION OF ALTERNATIVE ENERGY PROGRAMMES
IN DEVELOPING COUNTRIES.
Dunn, P.D. (University of Reading, U.K.) Paper
presented at UNESCO Non-Technical Obstacles to
the Use of New Energies in Developing Countries
Conference, Bellagio, Italy 25-29 May 1981 5 pp.
   Criteria for choosing new energy sources for
developing countries are listed and barriers to
implementation of new energy forms, such as
lack of information, financing, and expertise.
Progress has been made in the last few decades,
improving information flow, developing field
projects, and encouraging small energy-related
industry. A programme to develop wind power
for water pumping in the Sudan and a study of
the need for local renewable energy sources to
maintain Thai agriculture illustrate general
requirements for introducing new energy forms
to a developing country.

1.1(25) REFLECTIONS ON INSTITUTIONAL
ORGANISATIONAL STRUCTURES FAVOURING RENEWABLE
ENERGIES IN DEVELOPING COUNTRIES.
Durand, H. Paper presented at UNESCO Non-
Technical Obstacles to the Use of New Energies
in Developing Countries, Bellagio, Italy
25-29 May 1981 6 pp.
   Summarises problems of applying solar, wind,
small hydro, and biomass energy in developing
countries and concludes that resolution of
these problems must be carried out through an
interministerial agency with appropriate
authority and decision-making powers.

1.1(26) ENERGY FOR RURAL DEVELOPMENT: A TIME
TO ACT.
El-Hinnawi, E. (National Research Centre,
Cairo) Mazingira, 5 (1) 1981 pp. 70-79.
   A wide variety of renewable resources can be
harnessed to meet the essential energy needs of
the rural poor in developing nations. Examples
of such efforts initiated by United Nations
Energy Programme are presented. Small-scale
wind, solar, and bioconversion energy systems
were installed in a small village in
Sri Lanka to provide electricity and heat.
Similar projects conducted in the Philippines
are also surveyed.

1.1(27) THE FUTURE OF RENEWABLE ENERGY IN
DEVELOPING COUNTRIES.
Foley, G. (International Institute for
Environment & Development, London, U.K.)
Ambio, 10 (5) 1981 pp. 200-206.
   Investigates the lack of widespread
applications of advanced renewable energy
technologies - wind, solar, biogas, ethanol,

and improved charcoal kilns - to low-subsis-
tence developing countries. (17 refs.)

1.1(28) THE POSSIBLE SHARE OF SOFT/DECENTRALISED
RENEWABLES IN MEETING THE FUTURE ENERGY DEMANDS
OF DEVELOPING REGIONS.
Khan, A.M. IIASA Report RR-81-18, September 1981
29 pp.
   Developing countries can be expected to
increase their energy consumption by a factor
of about 10 over the next 50 years. The
maximum share of renewable energy sources in
meeting the future commercial energy demand of
the three market-economy developing world
regions - Latin America, Africa and Southeast
Asia - is determined. Windmills, small hydropower units, charcoal, biogas, and solar heat
are considered.

1.1(29) ALTERNATIVE SOURCES OF ENERGY INDIGENOUS
RENEWABLE RESOURCES.
Omo-Fadaka, J. (Third World Outlook, London)
Alternatives-J, 6 (3) Winter 1980-81
pp. 409-418.
   Describes ways third world countries can
benefit from their geographical location by the
use of indigenous biomass and other resources
such as sun, wind and water for their energy
needs.

1.1(30) WIND ENERGY FOR DEVELOPING COUNTRIES.
Smulders, P.T. Energiespectrum, 5 (7/8)
July-August 1981 pp. 193-200. In Dutch.
   Taking the wind's mean velocity as an
indication of wind power potential, Curaçao
and Cape Verde Islands are comparable with
Den Helder in the Netherlands, with 8 m/s at an
altitude of 40 m. Most developing countries
show only half this speed, i.e. a wind energy
availability eight times lower, but the consumer-
side aspects of the problem could be more
favourable. Considers commercial factors and
comparative energy consumption rates. (8 refs.)

1.1(31) WHO'LL WIN THE RACE?
Swift-Hook, D.W. (Central Electr. Res. Labs.)
Consulting Engineer 45 (8) August 1981
pp. 10-11, 13 and 15.
   Discusses the present position in wind energy,
with summaries of the progress in America,
Sweden, Britain and Denmark.

1.1(32) EVALUATION OF IEA WIND ENERGY RESEARCH
1977-1981.
Pershagen, B. National Swedish Board for Energy
Development Report NE 1982, March 1982 52 pp.
   Activities of the International Energy Agency's
Programme of R & D on Wind Energy Conversion
Systems (WECS) are described. Includes environmental and meteorological aspects of WECS,
integration of wind power into national
electricity supply systems Phase 1, and rotor
stressing and smoothness of operation of large-
scale WECS. An evaluation of wind models for
wind energy siting and the integration of wind
power into national electricity supply systems

Phase 2 are nearing completion and were finished during 1982. Ongoing projects planned to continue through 1983 and 1985 are the study of wake effects behind single-turbines and in wind turbine parks and the study of local wind flow at potential WECS hill sites.

1.1(33) LAYMAN'S GUIDE TO NEW AND RENEWABLE SOURCES OF ENERGY. PART 1 : BIOMASS, HYDROPOWER, OCEAN ENERGY AND TIDAL POWER.
Ponomaryov, B. and Ward, R. (U.N. Natural Resources & Energy Div., New York, U.S.A.)
Intl. J. Sol. Energy, 1 (4) April 1983
pp. 293-310.

Discusses the reasons behind increased interest in the use of traditional renewable sources of energy such as biomass, hydro and wind energy. An awareness has developed that there are other options for the older forms of renewable energy which have not been widely used plus the possibility of exploring and exploiting heretofore unused forms of ocean and solar energy and this paper attempts to define and review the spectrum of new and renewable sources of energy, their products, options and their estimated costs, and is included in the bibliography to complete the reference to this U.N. research although Part 2 concentrates on wind energy.

1.1(34) LAYMAN'S GUIDE TO NEW AND RENEWABLE SOURCES OF ENERGY. PART 2 : SOLAR ENERGY, WIND ENERGY, PRODUCTS AND OPTIONS OF ENERGY USAGE.
Ponomaryov, B. and Ward, R. (U.N. Natural Resources & Energy Div., New York, U.S.A.)
Intl. J. Sol. Energy, 1 (5) April 1983
pp. 379-397.

The authors focus on various available solar and wind energy technologies with appropriate applications and attempt to define and review the spectrum of these renewable sources of energy, their products, options and their estimated costs. (155 refs.)

1.1(35) RENEWABLE ENERGY SOURCES AND DEVELOPING COUNTRIES.
Ramakumar, R. (Oklahoma State Univ.)
IEEE Trans Power Apparatus & Systems, 102 (2) February 1983 pp. 502-510.

Considers the use of renewable energy sources in developing countries as an alternative to present methods of energy generation, causing resource degradation and economic problems. Information on the absolute rural needs and the priorities of use of presently generated energy is presented. The types of feasible energy systems, the benefits to be derived, and economic factors are examined. Includes wind energy. (33 refs.)

1.1(36) ENERGY FOR RURAL AND ISLAND COMMUNITIES III.
Proc. Third International Conference, Inverness, UK 12-16 Sept. 1983.
Twidell, J. et al (Eds.) Oxford, Pergamon July 1984 468 pp.

52 papers presented at the conference identify practical solutions to problems concerning the use and supply of energy to small, isolated communities who experience economic or technical difficulties in obtaining supplies from a central source. Includes technical innovations and improvements both in alternative energy and conventional supply systems, and the development planning necessary to implement these.

1.1(37) FREE AS THE WIND.
Lee-Frampton, J. Int. Power Generation 7 (7) September 1984 pp. 30-34.

Discusses two main obstacles which stand in the path of cost-effective wind power; high-cost, low-efficiency wind turbines and low-cost, high-efficiency nuclear or fossil fuels.

1.1(38) WIND POWER: A QUESTION OF SCALE.
Moore, T. (Electric Power Research Inst.)
EPRI J. 9 (4) May 1984 pp. 6-16.

Early operating experience with turbines of all sizes is highlighting the need for improved fundamental understanding of the effects of complex wind dynamics on the structural loading and fatigue life of these machines. Reducing such uncertainties will contribute toward development of optimal-size, commercially competitive wind machines for bulk power production.

1.2 United States and Canada

1.2.1 RENEWABLE ENERGY RESOURCES.
FEDERAL PROGRAMMES.

1.2.1(1) CENTRALISED POWER.
Messing, M., Friesema, H. and Morell, D.
Environmental Policy Inst. Report, March 1979
211 pp.

Effects of centralised and decentralised electric generating systems on the political structures of local governments are presented and evaluated. The electric utility industry is profiled, and several case studies are included covering small, medium, large and alternative power plant systems for various communities. Data on planning electric utility systems are provided, and federal, state, and local controls on power plant siting and licensing are detailed.

1.2.1(2) DISPERSED, DECENTRALISED AND RENEWABLE ENERGY SOURCES ALTERNATIVES TO NATIONAL VULNERABILITY AND WAR.
U.S. Federal Emergency Management Agency Report 2314-F, December 1980 340 pp.

U.S. reliance on imported fuel and centralised
energy production systems presents problems for
national security and emergency preparedness in
the event of a nuclear crisis or war. The
feasibility of using unconventional energy
resources and alternative approaches to
vulnerable centralised energy supply systems to
meet future U.S. energy requirements is
examined. Dispersed and renewable energy
systems considered include small-scale hydroelectric power, solar heating and cooling,
solar thermal electricity, cogeneration,
OTEC, solar photovoltaics, biomass energy,
geothermal energy, wind energy, wave energy,
load management and storage, energy conservation, and fuel cells.

1.2.1(3) ENERGY PLANNING POLICIES OF THE
AMERICAN PLANNING ASSOCIATION.
American Planning Association Report, April 1980
7 pp.
    The Association supports the long-range goal
of a shift to the use of renewable energy
resources, primarily solar, to the maximum
extent possible and calls for strategies that
will manage current energy production and
consumption patterns through a much stronger
emphasis on conservation. Action and planning
at all levels of government is required and a
diverse mix of energy sources, such as solar,
biomass, and wind energy, should be supported.

1.2.1(4) ENERGY STRATEGIES TOWARD A SOLAR FUTURE
(RENEWABLE ENERGY RESOURCES).
Union of Concerned Scientists Report, 1980
pp. 109-200.
    Reviews major renewable energy resources
available to the U.S. and technologies that are
either available or proposed for harnessing them.
Solar radiation, solar heating and cooling, solar
thermal electric conversion, photovoltaics, wind
energy, hydroelectric power, bioconversion, solar
satellite power stations, OTEC, tidal power, wave
energy, and ocean current energy are discussed.
Economic aspects and environmental impacts of
each of these resources and associated technologies are considered. (156 refs.)

1.2.1(5) FEDERAL CONSERVATION & RENEWABLE
ENERGY RESOURCE DIRECTORY.
(Department of Energy, U.S.) Paper presented at
DOE Local Alternative Energy Futures Conference,
Austin 11-13 December 1980 pp. 93-98.
    Gives quick access to federal departments,
agencies, congressional committees, and offices
of the President that have responsibilities in
energy conservation and renewable resource
development. Programmes are grouped into four
categories Outreach, R & D, Financial
Assistance, and Other Support Services. Bioconversion, wind energy, and solar energy
projects are listed in the Directory.

1.2.1(6) INTERNATIONAL APPLICATIONS OF RENEWABLE
ENERGY RESOURCES.
Senate Committee on Energy from Natural
Resources, Hearings 96 Con. 2 96-147, 19 August
& 5 September 1980 240 pp.

Questions the depth of the U.S. commitment in
developing international applications for
renewable energy. It is crucial that the U.S.
promote renewable energy and that developing
countries are not held back in doing so because
of a lack of affordable energy sources. The
developing world provides an enormous market for
such U.S.-developed technologies. The
devastating effects of overdeveloping certain
resources on the environment is pointed out and
problems inherent in promoting renewable technologies are discussed.

1.2.1(7) REGIONAL AND COMMUNITY IMPACTS OF THE
DEPARTMENT OF ENERGY ENERGY TECHNOLOGY PILOT
PROGRAMME IN THE WESTERN PACIFIC.
Case, W. (Internat. Solar Energy Society) Paper
presented at ISES-AS/Et Al 1980 Annual
Conference, Phoenix, Arizona 2-6 June 1980 V3.2
pp. 1309-1313.
    DOE has awarded 15 grants for solar energy
projects in American Samoa, the Commonwealth of
the Northern Marianas, Guam and the Trust
Territory of the Pacific. Projects include
solar water heating, solar greenhouses, solar
distillation, wind water pumping, and methane
digesters. Continuing technical assistance from
DOE and other agencies is recommended to ensure
the success of these solar ventures. (6 refs.)

1.2.1(8) NEITHER SUN NOR WIND NOR TIDE WILL
REPLACE FOSSIL FUELS QUICKLY.
Douglas, B. Canadian Petroleum, 22 (10)
October 1980 pp. 58-61.
    Describes Canada's implementation of various
R & D programmes to develop solar, wind, and
other unconventional but renewable energy
resources. However, federal and private funds
allocated for such research are insufficient to
make these forms available in the near future.
Forestry residues, solar heating and cooling,
wind energy, and hydroelectric power are being
developed to displace fossil fuels.

1.2.1(9) PROJECTING AMERICA'S ENERGY FUTURE.
Dukert, J.M. Exxon, 19 (3) July-September
1980 pp. 27-31.
    Unconventional resources, such as synthetic
methane, shale oil, solar energy, and wind
energy, are likely to provide a sizeable
proportion of U.S. energy by 2004 and Government
involvement in U.S. energy development is
expected to continue. Cooperation between the
executive and legislative branches of the
Government, world events, public attitudes to
specific energy technologies and to energy
conservation, the availability of energy
resources, and future population trends will all
affect the course of such developments.

1.2.1(10) THE RISK OF PRODUCING ENERGY.
Inhaber, H. (Oak Ridge National Laboratory)
Paper presented at Royal Society Assessment &
Perception of Risk Conference, London
12-13 November 1980 pp. 121-131.
    While studies seem to indicate that coal- and
oil-powered electricity poses more of a risk than

nuclear power, alternative sources like solar and wind power appear to be risk-free. The author argues they are not risk-free because they require more collectors per unit energy output, great amounts of steel, glass, copper and aliminium are required and their production also entails risk to workers. He examines implications of the risks involved in altern- ative energy production in terms of public policy. (12 refs.)

1.2.1(11) DEPARTMENT OF ENERGY SOLAR PROGRAMMES WITH UTILITY APPLICATIONS.
Katz, M.J. (DOE Office of Solar Power Applications) Paper presented at DOE/Univ. of Kansas Integration of Solar Energy in Utility Systems Planning Symposium, Topeka 11-13 May 1980 pp. 66-73.
Solar and wind power system development programmes currently supported by DOE are examined. Utility siting guides are being compiled as an aide in determining suitable locations for wind power plants. Wind energy prime movers are also being refined. Innovative designs for solar power plants are described. Present R & D activities extend to bio- conversion and OTEC.

1.2.1(12) APPA LAUNCHES DEED.
Leber, E. (American Public Power Association) Public Power, 38 (6) November-December 1980 pp. 30-33.
To increase the energy options available to publicly-owned utilities and their consumers, APPA, has introduced DEED (Demonstration of Energy-Efficient Developments). This helps members to identify and obtain information on new energy technologies and techniques for field testing and evaluation. Public power systems are asked to contribute small amounts of money to the programme which is then used to sponsor selected projects. Technologies considered for R & D include energy storage techniques, heat pumps, load management options, fuel cells, and wind energy.

1.2.1(13) PUBLIC POWER SYSTEMS INVOLVEMENT WITH SOLAR ENERGY.
Leber, E. (American Public Power Association) Paper presented at DOE/Univ. of Kansas Integration of Solar Energy in Utility Systems Planning Symposium, Topeka 11-13 May 1980 pp. 228-242.
Alternative energy schemes being developed by publicly-owned utility systems include solar heating/cooling, photovoltaics, wind energy, small hydro, wood energy, and animal waste energy. Specific R & D projects from around the U.S. are surveyed, including retrofitting and new power plant construction programmes.

1.2.1(14) SOCIAL ISSUES AND ENERGY ALTERNATIVES. THE CONTEXT OF CONFLICT OVER NUCLEAR WASTE. FINAL REPORT.
Lindell, M.K., Earle, T.C. and Perry, R.W. (Battelle Human Affairs Research Centres, Washington, DC, U.S.A.) NTIS Report PNL-3401 June 1980 67 pp.

Investigates attitudes toward nuclear power through the perception of risks and benefits of electric power alternatives. Opponents and supporters of nuclear power answer questions about five categories of the nuclear power problem the production potential of electric power generation technologies, energy conserv- ation, comparisons of risks among technologies, and the risks of those technologies. Nuclear supporters believe that solar, wind, and hydro- electric power generation can contribute to the energy supply.

1.2.1(15) THE AUDUBON ENERGY PLAN TECHNICAL REPORT.
National Audubon Society Report, April 1981 95 pp.
A practical, low-cost strategy to ensure that the U.S. obtains adequate energy while protecting the environment. The plan's total energy budget for the U.S. in 2000 is the same as the current budget; but the mix of energy resources is considerably different. Oil will provide only 20% of the total energy, as opposed to the current 50%. Oil imports will be reduced to about 25% of the current level. Solar energy will contribute about 25% of the total energy budget and wind energy will also contribute substantially to the total energy mix. Only limited use of coal, nuclear power, natural gas, and synthetic fuels to help ease the transition from oil to renewable resources is envisaged and these resources will be strictly regulated. Industry is seen as investing heavily in energy-efficient technology. (8 refs.)

1.2.1(16) ENERGY AND THE ENVIRONMENT THE CHALLENGE OF THE '80S.
Energy Consumer, January 1981 44 pp.
To reduce U.S. foreign oil dependence and satisfy public health and safety standards, research in fuel resources that offers energy alternatives to petroleum has begun. Current and future R & D in solar energy, nuclear power, renewable resources, clean coal technologies, and wind, biomass, and hydropower energy production attempt to rectify present dwindling U.S. energy supplies. The Synthetic Fuels Corporation aims to supply an additional demand of 150-200 mm tons of coal by the 1980's, and 300 mm tons by 2000. Emphasis placed on the development and use of energy resources that do not endanger public health or the environment.

1.2.1(17) ENERGY AUDUBON'S ANSWER.
Audubon, 83 (4) July 1981 pp. 94-103.
The Audubon Energy Plan offers a practical approach for assuring sufficient national energy supply to produce more goods and services than currently generated and for providing a higher standard of living for a larger population. Economic growth need not be linked to energy growth and its resulting environmental degradation. Energy demand will be held constant by increasing energy efficiency and using existing technology. Solar and wind energy will be exploited to supplement existing resources.

1.2.1(18) NATIONAL REPORT SUBMITTED BY CANADA.
Paper presented at U.N. New and Renewable
Sources of Energy Conference, Nairobi
10-21 August 1981 38 pp.

Canada imports oil to supply about 43% of its
energy needs, but a plan is under way to
eliminate oil imports by 1990 and reduce oil
consumption to about 10% of the total demand
in each of the economic sectors. Oil sands,
oil shales, hydroelectric power, small hydropower plants, geothermal energy, wind power,
solar energy, and ocean energy are among the
many renewable resources currently under
development.

1.2.1(19) NEW AND RENEWABLE ENERGY IN THE
UNITED STATES OF AMERICA.
Paper presented at U.N. New and Renewable
Sources of Energy Conference, Nairobi
10-21 August 1981 110 pp.

Gives a history of U.S. utilisation of and
dependency upon renewable energy sources. Past
and current uses of renewable energy sources
are analysed with focus on high and low
temperature solar collectors, solar cells,
biomass energy systems, wind energy systems,
ocean energy systems, hydropower, geothermal
systems, oil shale, tar sands. Economics,
private and government activities, and
barriers and impacts for all these energy
sources are described.

1.2.1(20) PRELIMINARY DRAFT OF A PROGRAMME OF
ACTION. REVISED TEXT.
Paper presented at U.N. New and Renewable
Sources of Energy Preparatory 4th Conference,
New York City 7 July 1981 84 pp.

Discusses a framework for national action on
the development and utilisation of new and
renewable energy resources. Suggestions include
integrating existing programmes for development
of these resources; planning national energy
strategies which define the role of these
resources in the economy; establishing adequate
research and development programmes; and
integrating both public and private sectors in
this effort. The fundamental goal of the
programme of action is to promote concerned
action in terms of energy transition. Implementation and monitoring techniques are summarised.
(23 refs.)

1.2.1(21) GETTING ON THE RIGHT PATH: CONSTITUENCY
BUILDING THROUGH ENERGY PLANNING.
Benson, J. (Institute for Ecological Policies,
Virginia, U.S.A.) Alternative Sources of Energy,
(49) May-June 1981 pp. 20-25.

U.S. politics presently favours major reliance
on nuclear power, increased coal use, and huge
taxpayer subsidies for expensive and unproved
synthetic fuels experiments. Various approaches
to community energy planning that could help
build a strong national base for energy conservation and community-based renewable energy
systems are explored.

1.2.1(22) PUBLIC UTILITY PARTICIPATION IN
DECENTRALISED POWER PRODUCTION.
Gentry, B.S. (Harvard University) Harvard
Environmental Law Review, 5 (2) 1981
pp. 297-345.

The implications of a programme of federal
incentives designed to increase utility
participation in such options as solar, wind,
hydroelectric, and cogeneration technology are
explored. The current state of the electricity
industry and benefits that can be derived from
greater use of these energy forms is discussed.
(References.)

1.2.1(23) THE RCS PROGRAMME, A NEW NATIONAL PLAY.
Johnson, B. (Mid-American Solar Energy Complex,
Minnesota, U.S.A.) Paper presented at ISES-AS/
Et Al Wind Power Energy Alternative for MidWest
2nd Conference, Minnesota 3-4 April 1981
pp. 45-48.

The aim of the Residential Conservation
Service (RCS) Programme authorised by the
National Energy Act of 1977, is to encourage
the use of energy-saving devices and solar and
wind energy techniques in the residential
sector. Electricity suppliers are required to
provide energy audits on request for all
residential customers and to assist in the
financing, supply, and installation of energy-
saving and solar and wind energy equipment.
Regulations relating to the installation and
operation of small wind energy conversion
systems are reviewed.

1.2.1(24) OUR ENERGY OPTIONS.
Meyers, P. and Witt, F. Rockford Public Schools,
Illinois, U.S.A. 1981 47 pp.

Looks at various U.S. energy options and their
advantages and disadvantages. Undeveloped
energy sources, such as geothermal, solar, and
tidal energy, could provide limitless and clean
energy. Oil shale, coal gasification, wind
energy, solid waste energy, and fusion are also
considered. Nuclear energy is an energy source
that has proved itself worthwhile. Coal is a
plentiful resource but has many drawbacks,
including air pollution, land disturbance, and
miner respiratory diseases. (58 refs.)

1.2.1(25) A DYNAMIC INPUT-OUTPUT ANALYSIS OF
NET ENERGY EFFECTS IN SINGLE-FUEL ECONOMIES.
Penner, P.S. (University of Illinois, U.S.A.)
Energy Systems & Policy, 5 (2) 1981 pp. 89-117.

Describes the application of a partially
dynamic input-output model of the U.S. economy
to determine the net energy costs of constructing and operating a single-fuel energy-supply
system over time. Three energy resources were
considered oil, nuclear, and wind. The model
provides an accurate representation of the net
energy effects of declining ore resource quality
and accelerating production. Results show that
new energy effects can reduce the effective life
of a resource by as much as 40%. (26 refs.)

1.2.1(26) NCAT'S BEST.
Rucker, C., Young, J. and Melcher, J. (National
Centre for Appropriate Technology) National

Centre for Appropriate Technology Report, 1981
71 pp.

The National Centre for Appropriate Technology
(NCAT) provides a programme of financial assistance to local projects that seek workable
solutions to the problems of poverty through the
use of appropriate technology including wind
energy. The grants stress research and development, demonstration, and transfer of
technologies for each project. The projects are
successful when they are simple, easily
implemented, adaptable, lasting, and foster
community development. Projects from all areas
of the U.S. are described and a contact
provided.

1.2.1(27) THE ROLE OF ENGINEERING-CONSULTING
FIRMS.
Scarborough, J.C. (NUS Corp., Maryland, U.S.A.)
Paper presented at ANS Alternative Energy
Sources for Electrical Power Conference,
Maryland 4-7 October 1981 37 pp.

Calculates U.S. electrical generation requirements from 1980-2000 based on world oil prices
and U.S. economic growth. The role of
engineering-consulting firms in the development
of alternative energy sources for electrical
generation is outlined. Technologies examined
include hydroelectric, geothermal, biomass,
wind, solar thermal, and photovoltaics.

1.2.1(28) ALTERNATIVE ENERGY DATA SUMMARY FOR
THE UNITED STATES 1981-1985 - NEAR TERM
PROJECTIONS ENERGY CONSUMPTION BY SECTOR.
Quarterly Alternative Energy Data Summary,
March 1982 3 75 pp.

Presents data on projected U.S. consumption of
biomass, geothermal, hydroelectric, solar, and
wind energies for 1981-85. Decreased overall
energy consumption is expected for the residential, commercial, industrial, and transportation sectors, with demand being related to
increased energy prices. Biomass R & D is
expected to begin paying out, with an estimated
consumption of about 3.75 million tons by 1985;
in full-design passive solar systems, the trend
is expected to continue upward because of
improved technical performance and increasing
cost effectiveness.

1.2.1(29) ENERGY RESOURCES AND TECHNOLOGIES
IN CANADA.
Canada Today, March 1982 11 pp.

Overview of research in potential energy
sources and currently applied technologies in
Canada. Solar energy, conservation, biomass
and sawdust, tidal power, nuclear power, hydropower, wind energy, and hydrogen are described.
Natural energy resources are expected to gain
in popularity as fossil fuel supplies dwindle.

1.2.1(30) PROGRESS ON ALTERNATIVE ENERGY
RESOURCES.
Couch, H.T. Astronaut. & Aeronaut., 20 (3)
March 1982 pp. 42-47.

Outlines the Department of Energy's
predictions for energy demand in the U.S.A.,

and briefly reviews present developments in
energy resources. Topics include fluidized
bed combustion and coal gasification/lique-
faction, magnetohydrodynamic (MHD) power,
solar energy (solar/thermal devices and
solar photovoltaic devices), wind energy,
ocean thermal energy conversion, geothermal
and nuclear fusion.

1.2.1(31) ENERGY RISK ASSESSMENT
Inhaber, H. (Atomic Energy Control Board)
Gordon & Breach, 1982 345 pp.

A comprehensive review of available energy
options and their attendant risks. The energy
cycle is examined in depth and alternative
energy sources such as solar energy and wind
power, coal, oil, natural gas, nuclear power,
methanol, ocean hydroelectric systems are also
considered with comment upon the materials and
labout statistics in relation to Canada and the
United States.

1.2.1(32) RENEWABLE ENERGY TARGET FOR 2050.
Rowe, W. (American University) IEEE Spectrum,
19 (2) February 1982 pp. 58-63.

Opportunities for supplying the world's
future energy needs already exist, primarily
solar, wind, and thermal gradient resources.
Prospects for the commercial development and
use of these technologies by 2050 are examined.
International cooperation, technological break-
throughs, and massive capital investment will
be needed to fully develop renewable energy
technology.

1.2.1(33) A PERSPECTIVE ON OUR ENERGY OPTIONS.
Scott, J.H. (U.S. Sandia Labs., New Mexico)
Paper presented at 27th Symposium of Society
for the Advancement of Material & Process
Engineering, San Diego 4-6 May 1982 pp. 73-82.

America's energy future will be characterised
by an increasing dependence on energy recovery
and conversion techniques. The problems
associated with these methods are discussed
from the perspective of a national laboratory.
The number of energy sources available has been
increasing, making the selection process
difficult. The future of enhanced oil recovery,
liquid fuels, natural gas, solar energy, wind
energy, and geothermal energy is examined.

1.2.1(34) THE ALTERNATIVE ENERGY SCENE.
Wyatt, A. (Montreal Engineering) Ascent, 3 (3)
1982 pp. 20-27.

Alternative energy resource development in
Canada is described. Utilities encounter
planning problems with hydro, wind and solar
power; these three and other energy sources
including tidal power are nevertheless being
investigated in research projects by Utilities
and Provincial Governments.

1.2.1(35) ENERGY-RELATED MANPOWER REQUIREMENTS
IN THE 1980S.
Energy Economics Policy & Management, 2 (3)
Winter 1983 pp. 43-60.

Since the mid-1970s, the number of scientific, engineering, technical, and blue collar workers employed in energy-related fields such as R & D, exploration, conservation, and production has increased steadily. Prospects for future energy-related employment are examined. Demand for scientists and engineers involved in energy R & D will probably decrease in the future. Federal funding for R & D in all technologies except nuclear energy has decreased in recent budgets and is expected to continue to decline in the coming years.

1.2.1(36) RENEWABLE ENERGY HAS SOME BIG MINUSES TOO.
Mannon, J.H. Chemical Business, 2 May 1983 pp. 40-44.
Reductions in federal budgets and international oil prices have created an unfavourable climate for renewable energy R & D in the U.S.. Advocates of alternative energy are seeking to implement such technology in developing nations instead. However, the potential side effects and environmental impacts of massive exploitation of renewables in these nations have not been considered. Adverse and varied impacts associated with wind energy, hydropower, solar energy, and geothermal energy are discussed.

1.2.2 RENEWABLE ENERGY RESOURCES.
STATE PROGRAMMES.

1.2.2(1) SOLAR AND WIND ENERGY APPLICATIONS IN HAWAII.
Koide, G.T. and Takahashi, P.K. (Univ. of Hawaii, Hilo) Solar Energy, 21 (4) 1978 pp. 297-305.
Describes a project to stimulate the development of solar and wind energy by assisting in the first crucial steps of the planning process. Projects which had potential for immediate implementation in the State of Hawaii were selected. Variety in the form of application type of engineering analysis and location within the State was sought, and applications range from agriculture to aquaculture to tourism to education. (9 refs.)

1 1.2(2) CALIFORNIA ENERGY COMMISSION 1979 BIENNIAL REPORT.
California Energy Commission Report, 1979 67 pp.
California cannot meet its energy needs in the next decade, if energy supplies are limited to conventional sources. State energy policies must encourage electricity conservation, and promote development of preferred alternatives, including hydroelectric power, geothermal power plants, synthetic fuels and wind energy.

1.2.2(3) ALTERNATIVE TECHNOLOGY IN LOW INCOME NEW YORK ENERGY TASK FORCE'S FIRST FOUR YEARS
Christianson, M. (New York City Task Force) Paper presented at DOE/SERI Community Energy

Self-Reliance Conference, Boulder 20-21 August 1979 pp. 80-91.
New York City Energy Task Force is a group of designers, builders, and educators which advises low income grass roots organisations on energy matters. Energy conservation, solar energy, and wind energy systems to demonstrate the potential of alternate technology for urban communities have been built. Efforts to develop, finance, and carry out this work have shown that energy projects in low income communities must be integrated into other neighbourhood redevelopment plans.

1.2.2(4) LOCAL ENERGY PLANNING THROUGH COMMUNITY LEVEL TECHNOLOGY ASSESSMENT IN THE SOUTHERN TIER CENTRAL REGION OF NEW YORK STATE.
Miller, C.G. (Southern Tier Central Regional Planning & Development Board, New York) Paper presented at DOE/SERI Community Energy Self-Reliance Conference, Boulder 20-21 August 1979 pp. 214-221.
Describes a year long project by regional planners, volunteers and public officials in the Southern Tier Region of New York to create a local policy for energy conservation and the development of renewable energy resources. Scenarios were developed to envision the use of solar energy, wind energy, and biofuels and implementation and financing considerations also surveyed.

1.2.2(5) ELECTRICAL ENERGY SELF-SUFFICIENCY FOR HAWAII BY 1990.
Neill, D.R. (University of Hawaii, U.S.A.) Paper presented at DOE/SERI Community Energy Self-Reliance Conference, Boulder 20-21 August 1979 pp. 282-289.
As Hawaii is heavily dependent on imported oil, schemes are being developed to realise electrical energy self-sufficiency by 1990. Hawaii's abundance of renewable natural resources including wind energy, solar energy, biomass, OTEC, geothermal energy, and hydroelectric power are to be exploited.

1.2.2(6) GRASSROOTS ENERGY PLANNING THE WYOMING COMMUNITY GRANTS PROGRAMME.
Welch, J. (Wyoming Energy Conservation Office) Paper presented at DOE/SERI Community Energy Self-Reliance Conference, Boulder 20-21 August 1979 pp. 185-190.
The Wyoming Energy Conservation Office has completed the first year of a programme to fund community-based energy conservation and renewable energy projects. Each participating community forms an energy council which is responsible for developing a first year energy programme that meets the specific need of the community. Grants of up to $10,000 are available. Includes wind energy.

1.2.2(7) ELECTRICITY TOMORROW 1980 PRELIMINARY REPORT.
California Energy Commission Report, 30 October 1980 354 pp.
Discusses critical issues in the electricity sector for California and projections

of electric demand to 2000. Individual utility forecasts of electric demand are given, and various forecasting models are presented and criticised. A general overview of alternative fuel options for the State is provided, and case studies illustrating the need for new power supplies are also included, including wind energy.

1.2.2(8) ENERGY IN NEVADA. A SUMMARY OF HISTORICAL AND PROJECTED ENERGY USES.
Nevada Dept. Energy Report, May 1980 166 pp.
Electricity and fuel usage data for Nevada are tabulated from 1961, where available, to the present, and projected to 2000. Coal, electricity, natural gas, and alternative energy source consumption data are tabulated for the commercial, residential and industrial sectors. Includes wind energy.

1.2.2(9) ENERGY IN PUERTO RICO'S FUTURE.
NRC/NAS Report, 1980 192 pp.
Current and future energy demands through 2000 are discussed and solutions to Puerto Rico's energy problems are suggested. No single low-cost energy source can solve the area's energy dilemma, but modest contributions can be made from a variety of domestic energy resources including small hydroelectric projects, biomass fuels, solar water heating, and wind energy. (30 refs.)

1.2.2(10) ENERGY SOURCES. THERE'S MORE TO ALASKA THAN OIL AND GAS.
Alaska Construction & Oil, 21 (12) December 1980 pp. 17-24.
Alaska's alternative energy reserves are only beginning to elicit serious interest. These energy resources can be developed to displace petroleum fuels and other conventional energy forms. Programmes are currently under way to mine and market the low sulphur coal and to harness hydro resources. Geothermal hot water resources are being developed for heating, cooling, and electric power generating applications. Peat, wind energy, and heat recovery are also discussed.

1.2.2(11) TALL TREES, STRONG WINDS, HOT ROCKS AND CHEAP POWER.
Reynolds, J.S. (University of Oregon) Paper presented at ISES-AS/Et Al 5th Passive Solar National Conference, Amherst 19-26 October 1980 V1 pp. 574-577.
Citizens of Oregon were appointed to the alternative energy development commission and six associated task forces charged with planning for the development of solar, geothermal, wind, biomass, and hydropower energy technologies. Energy policy monitoring and decision-making activities of the solar task force are outlined.

1.2.2(12) THE BIG SWITCH.
Rodberg, L. and Stokes, G. (Public Resource Centre, New York City) Village Voice, 25 (7) 18 February 1980 pp 1-10.

This article argues for a major governmental programme designed to break the oil suppliers' stranglehold on the City. Includes conservation for apartment dwellers, alternate energy sources, including solar, wind, and garbage power, and waste energy. Financial support for the plan is outlined, and the immediate and long-term results of the plan are depicted.

1.2.2(13) THE HAWAII INTEGRATED ENERGY ASSESSMENT.
Weingart, J.(LBNL) LBNL Energy & Environmental Division 1979 Report, LBL-11650 October 1980 5 pp.
The Hawaii Integrated Energy Assessment is designed to evaluate the opportunities for and impacts of imported petroleum displacement by use of indigenous renewable resources. These include solar energy, geothermal energy, wind energy, and bioconversion. Several scenarios of electricity and fuel demand by sector were developed. Conservation and improved efficiency, with photovoltaics and OTEC plants, can help to make this State virtually energy self-sufficient. (7 refs.)

1.2.2(14) ENERGY PLANNING GUIDE FOR MINNESOTA COMMUNITIES.
Minnesota Dept. of Economic Development/Energy Agency Report, undated 134 pp.
States which import large quantities of energy producing materials - oil, natural gas, and coal - are particularly vulnerable to energy crises and economic disruptions; Minnesota is such a State. Suggestions and guides designed to aid local Minnesotan governments in implementing programmes which will decrease their towns' dependence upon centralised energy supply and distribution systems are detailed, including renewable energy systems such as wind energy. (refs.)

1.2.2(15) EXECUTIVE SUMMARY VERMONT STATE ENERGY CONSERVATION PLAN.
Vermont State Energy Office Report, January 1981 58 pp.
Outlines major sections of the Vermont State Energy Conservation Plan. Vermont - a rural State consuming a low amount of energy per capita - is highly dependent on imported oil. Conservation programmes are examined through audits and standards, transportation, government activities, and industrial/commercial/agricultural/utility sectors. Alternative energy sources such as wood, solar, hydro, wind, and barriers are discussed.

1.2.2(16) NEW ENGLAND CAN REDUCE ITS OIL DEPENDENCE THROUGH CONSERVATION AND RENEWABLE RESOURCE DEVELOPMENT, VOLUME 2.
GAO Report, EMD-81-58A 11 June 1981 349 pp.
Presents an energy consultant's analysis of the impact that conservation and renewable energy resources could have on New England's high oil dependence. New England's energy needs through 2000 are projected under three policy options business as usual, vigorous

conservation, and increased use of renewable resources. Topics discussed include the alternative energy supply strategy scenario, and the impact of energy conservation on employment. (Refs.)

1.2.2(17) NEW YORK STATE ENERGY RESEARCH AND DEVELOPMENT AUTHORITY QUARTERLY REPORT.
New York State R & D Authority Report, 1 April 1981 238 pp.
Reports on the expenditures and activities of the New York State Energy R & D Authority for the first quarter of 1981. Surveys various energy conservation and R & D programmes and summarises projects designed to increase energy efficiency in the residential, commercial, utility, industrial, and transportation sectors. Includes hydropower, coal usage, solar technologies, wind energy, bioconversion, and advanced energy systems.

1.2.2(18) REGIONAL CONSTITUENCY SUPPORT PROGRAMME FOR 1981.
NTIS Report MASEC-R-81-075, September 1981 65 pp.
Outlines State programmes in the Mid-American Solar Energy Complex (MASEC) region to augment use of solar and other renewable energy sources and energy conservation. The economics of solar, wind, hydro, biomass, and more efficient energy usage are superior to traditional fuels; however institutional barriers such as local ordinances, lack of experience, and public uncertainty obstruct the widespread implementation of these strategies.

1.2.2(19) REPORT TO THE GOVERNOR FROM THE GOVERNOR'S TASK FORCE ON WISCONSIN ENERGY POLICY.
State of Wisconsin Governor's Task Force Report, January 1981 90 pp.
State reserves of coal, uranium, and other resources are surveyed and conservation measures suggested for implementation, including building standards, transportation speed limits, encouragement of mass transit patronage, and use of solid and agricultural wastes for fuel. Solar space heating and industrial applications should be developed, prospects of developing hydro, wind and biomass resources are considered. (153 refs.)

1.2.2(20) SCOPE AND PURPOSE OF SPONSORING ORGANISATIONS.
Paper presented at ISES-AS/Et Al Wind Power Energy Alternative for MidWest 2nd Conference, Minnesota, 3-4 April 1981 pp. 93-99.
Four organisations that sponsored the Wind Power Energy Alternative for the MidWest Conference are examined. ISES-AS provides a meeting ground for stimulating the exchange of ideas and advances in the field of solar energy conversion. Alternative Sources of Energy Inc., aims to publicise and share information pertaining to the development of renewable energy sources. Mid-American Solar Energy Complex aims to increase the understanding and use of solar

energy in the MidWestern U.S. Rochester Adult Vocational Technical Institute coordinates efforts to provide information, education, and technical assistance to energy consumers in the MidWestern U.S.

1.2.2(21) SOLAR AND WIND ENERGY IN NEW MEXICO.
Paper presented at LASL/Et Al Showcase for Technology Symposium, Albuquerque, 28-30 October 1981 pp. 123-131.
Describes solar and wind energy development. New Mexico is currently leading all other states in the number of solar installations. Wind energy's potential as a renewable energy source is due to the simplicity of the windmills and the availability of wind.

1.2.2(22) EXPLORING NEW ENERGY CHOICES FOR CALIFORNIA THE 1981/82 REPORT TO THE LEGISLATURE.
Deller, N.J. and others. California Energy Commission Report, March 1981 88 pp.
The California Energy Commission is responsible for a R & D programme in energy supply, consumption, and conservation. Energy-related matters such as expansion and acceleration of alternative energy sources are also addressed by the Commission. Significant programme changes that have occurred are outlined. Future programme directions are revealed, including wind energy developments.

1.2.2(23) FRANKLIN COUNTY AN INSIDE VIEW.
Gery, M.E.C. (Ashfield Energy Conservation Committee) Alternative Sources of Energy (49) May-June 1981 pp. 6-10.
Since its inception in 1976, the Franklin County (Mass.) Energy Planning System has significantly changed the energy consumption habits of both businesses and households within the County. The Franklin County Energy Project, Inc., is involved in numerous activities to help guide county energy development. These include building a demonstration trombe wall for passive solar heating and cooling, organising a woodstove safety programme, conducting tours of local renewable energy installations, and testing wind sites and loaning anemometers.

1.2.2(24) RENEWABLE ENERGY IN COMMUNITIES SIX CASE STUDIES.
Green, B., Levine, A. and Perkins-Smith, D. (Solar Energy Research Institute) DOE Report SERI/SP-744-1276 November 1981 30 pp.
Describes six case studies of communities that have administered conservation and renewable energy programmes to reduce dependence on fossil fuels. The projects are a fairly representative mix of renewable energy technologies, financing mechanisms, and other owner/management approaches: a hydroelectric project in Lawrence, MA; a wind turbine generator on Cuttyhunk Island, MA; an active solar water heating project in New York City; a solar pond in Miamisburg, OH; Kaplan Industries biogas facility in Barton, FL; and the Lihue Plantation cogeneration plant in Kauai, HI.

1.2.2(25) RENEWABLES PROMINENT IN CALIFORNIA'S
ENERGY PLANS.
James, P. Electrical Review, 208 (21) 5 June
1981 pp. 24-26.
  Southern California Edison has built the first
solar electric central receiver power plant in
the U.S. - Solar One, near Barstow in the
Mohave Desert. The utility has also abandoned
all future plans for nuclear and coal power
plants. Wind power, fuel cells, and cogeneration are also discussed, with corporate
support of the new technologies examined. A
new system of load management developed by the
utility is presented.

1.2.2(26) HAWAII'S ENERGY SELF-SUFFICIENT
PROGRAMME FROM RENEWABLE ENERGY SOURCES.
Neill, D. (University of Hawaii) Paper presented
at ISES-AS/Et Al 1981 Annual Conference,
Philadelphia, 26-30 May 1981 V2 pp. 1452-1456.
  Hawaii is currently dependent on imported oil
for over 90% of its energy requirements but it is
expected to be self-sufficient in terms of
electrical power generation within the next 20
years. Wind energy, bioconversion, hydroelectric power, and solar photovoltaic systems
are being developed to meet Hawaii's electricity
needs and each island's activities in these
areas are described.

1.2.2(27) ENERGY TOMORROW: CHALLENGES AND
OPPORTUNITIES FOR CALIFORNIA.
Schwieckart, R.L. and others. (California
Energy Commission) California Energy Commission
1981 Biennial Report, 1981 212 pp.
  California's official electricity demand is
projected through two long-range energy
scenarios and a review of the State's past and
future energy trends. Through these projections, a comprehensive, analytically viable
long-range plan ensuring a secure energy future
for California is developed, and includes
scenarios of solar and wind energy substit-
utions.

1.2.2(28) ENERGY: HOW HAWAII USES THE WIND,
THE SUN, AND THE SEA.
Trumbull, R. New York Times, 29 March 1981
Sec. 3 PF-9.
  Hawaiian experiments in biomass conversion,
utilisation of volcanic heat, wind energy
conversion, ocean thermal conversion, and solar
energy utilisation are described. Rapid
development of alternative energy resources
could help Hawaii achieve self-sufficiency in
electricity generation by 2000.

1.2.2(29) ENERGY MOVING UNCERTAINLY INTO A
NEW ERA.
Washburn, R.P. Cry California, 16 (3)
Summer 1981 pp. 29-35.
  Although the Governor, the State Legislature,
and the Utilities agree that embracing renewable
resource technology has become a necessity, the
Utilities appear to be lagging behind in making
a firm commitment. Many executives feel that
solar, wind, cogeneration, geothermal, and

other alternative technologies are not the
answer.

1.2.2(30) HAWAII NEW DYNASTY IN RENEWABLE
ENERGY.
Wholey, J. Solar Age, 6 (5) May 1981
pp. 22-29.
  Describes numerous large- and small-scale
projects either underway or under study to
develop the Island's renewable resources solar,
wind, geothermal, ocean thermal, and biomass.

1.2.2(31) NEW YORK STATE ENERGY RESEARCH AND
DEVELOPMENT AUTHORITY SEMI-ANNUAL REPORT.
New York State Energy R & D Authority Report,
1 April 1982 259 pp.
  Activities of the Authority from 1st October-
1st April, 1982 are outlined. Topics discussed
include the following: increasing energy usage
efficiency in the residential, commercial,
utility, industrial and transportation sectors;
increasing new energy supplies through hydropower, solar, wind, geothermal, shale gas,
methane and solid fuels; coal use; advanced
energy systems.

1.2.2(32) OUR ENERGY PREDICAMENT - ABUNDANT
RESOURCES, DEPRESSED MARKETS.
Stubbs, M.F. New Mexico Business, 6 (11)
November 1982 pp. 19-23.
  Energy-rich Mexico ranks sixth nationally in
production and reserves of oil, coal, and natural gas. Production trends and consumption
of these fossil fuel resources are surveyed.
Depressed markets for such fuels have resulted
in a loss of revenue for the State. However,
the growing solar energy industry in New Mexico
is creating employment and bringing in profits,
and participation in wind and geothermal
energy developments is also discussed.

1.2.3. WIND ENERGY PROGRAMMES.

1.2.3(1) WIND ENERGY - A REVITALISED PURSUIT.
Blackwell, B.F. and Feltz, L.V. (Sandia Lab.,
Albuquerque, NM.) In proc. ASME 15th Annual
Symposium Albuquerque, NM. 6-7 March 1975
pp. 41-54.
  From an estimate of the wind energy available
in the Great Plains area it was concluded that
this energy capacity is large in comparison to
the United States electrical energy consumption.
The status of the Darrieus-type vertical-axis
wind turbine being investigated by Sandia
Laboratories is reviewed. (17 refs.)

1.2.3(2) PLANS AND STATUS OF THE NASA-LEWIS
RESEARCH CENTRE WIND ENERGY PROJECT.
Thomas, R. and others. (NASA, Lewis Research
Centre, Cleveland, Ohio) In proc. 2nd Energy
Technology Conference, Washington, DC.
12-14 May 1975 pp. 290-314. Washington DC.
Gov. Inst. Inc. 1975.
  Describes the national five-year wind-energy
programme managed by the NASA-Lewis Research

Centre and the Lewis Research Centre's Wind Power Office, its organisation and plans and status. An experimental 100 kW wind-turbine generator; the first generation industry-built and user-operated wind-turbine generators; and the supporting research and technology task are reported on. Data in tabular and graphical form are included. (8 refs.)

1.2.3(3) HIGH POTENTIAL OF WIND AS AN ENERGY SOURCE.

Coty, U. and Dubey, M. (Lockheed-Calif. Co., Burbank) In Los Angeles Council of Engineering & Science, Proc. Ser. V2: Energy LA; Tackling the Crisis, Greater Los Angeles Area Energy Symposium, California 19 May 1976. North Hollywood, California, West Period Co., 1976 pp. 181-187.

Wind energy is an abundant resource in the United States. It can be captured economically by large wind turbine generators and used to supplement electric energy provided from fossil fuel and hydroelectric resources. A study sponsored by the Energy Research and Development Administration shows that the full potential of the wind is far greater than previously estimated. Full implementation would produce clean energy without pollution, and would make a significant contribution to the conservation of fossil fuel reserves. (3 refs.)

1.2.3(4) WIND ENERGY UTILISATION - AN OVERVIEW.

Ramakumar, R. and Hughes, W.L. (Okla. State University, Stillwater) In proc. of the 16th Annual ASME Symposium: Energy Alternatives, Albuquerque, NM. 26-27 February 1976. New York, NY. ASME, NM. Sect. 1976 pp. 143-147.

Overview of the potential of wind as a natural clean and renewable source, covering aeroturbines and the Savonius rotor. (24 refs.)

1.2.3(5) ERDA-NASA WIND ENERGY PROJECT READY TO INVOLVE USERS.

Thomas, R. and others. (NASA, Lewis Research Centre, Cleveland, Ohio) Energy (Stamford, Connecticut) 1 (2) Winter 1976 pp. 27-30.

Wind energy, being a clean renewable source of energy that has proven technically feasible is now being investigated as an alternative source of energy. In 1973 the National Science Foundation (NSF) was given the responsibility for planning and executing a sustained wind energy programme. The objective of this programme is to develop the technology needed to build reliable and cost-effective wind-energy conversion systems that have the potential for early and rapid commercial implementation. In January 1975 the wind energy programme was transferred from NSF to the newly formed Energy Research and Development Administration (ERDA). (8 refs.)

1.2.3(6) U.S. WIND ENERGY - 1978.

Cliff, W.C. (Battelle, Pac. Northwest Lab., Richland, Washington) In proc. of Annual Meeting of American Sect. of Intl. Solar Energy Society, Inc.: Solar Diversification, Denver, Colorado 28-31 August 1978. Univ. of Del., Newark AS/ISES 1978 Tutorials vol. pp. 23-32.

Presents an overall view of the state of the art in wind energy in the United States for 1978. The national assessment of wind energy is presented including a discussion of current wind turbines suitable for extraction of wind energy for the generation of electricity. (12 refs.)

1.2.3(7) WIND ENERGY ACTIVITIES AT THE SOLAR ENERGY RESEARCH INSTITUTE.

Hardy, D. (Solar Energy Research Institute, Golden, Colorado, USA) In proc. of Annual Meeting of American Sect. of Intl. Solar Energy Society, Inc.: Solar Diversification, Denver, Colorado 28-31 August 1978, Univ. of Del., Newark AS/ISES 1978 Symposium vol. pp. 99-104.

Reviews wind energy project activities at the Solar Energy Research Institute. Overall objectives are discussed in relation to the broader objectives of the Institute in furthering wind energy utilisation. SERI's role in supporting the Department of Energy programme interactions with the private sector in wind energy development, and efforts to promote information transfer throughout the wind energy sector are also covered. (18 refs.)

1.2.3(8) NRC'S WIND ENERGY PROGRAMME.

Rangi, R.S., South, P. and Templin, R.J. (Natl. Research Council of Canada, Ottawa, Ontario) In Renewable Alternatives: proc. of 4th Annual Conference of Solar Energy Society of Canada, Univ. of West Ontario, London 20-24 August 1978. Winnipeg, Manitoba, Solar Energy Society of Canada, 1978 V 1, pap. 4. 3. 2. 15 pp.

Describes current work on wind power assessment and on the Vertical Axis Wind Turbines at the National Research Council (NRC) Ottawa. A map showing the annual average wind power density is presented and the electrical wind power potential for all Canada and individual provinces has been calculated. Wind power potential is also assessed for all land area; existing electrical network +300 km; existing electrical network +150 km. The theoretical and development work on the VAWT and demonstration projects sponsored by National Research Council are described.

1.2.3(9) PRINCE EDWARD ISLAND WIND ENERGY PROGRAMME.

Lodge, M. (Institute of Man & Resources) In Renewable Alternatives: proc. of 4th Annual Conference of Solar Energy Society of Canada, Univ. of West Ontario, London 20-24 August 1978. Winnipeg, Manitoba, Solar Energy Society of Canada 1978 V 1, pap. 4. 3. 5. 13 pp.

Describes the Wind Energy Programme of the Institute of Man and Resources which commenced June 1977. Composed of four sub-programmes, the first is an analysis of six years of historical wind data available from eight monitoring sites operated for the Atmospheric Environment Service (A.E.S.) of Canada by the

P.E.I. Department of Agriculture, subsequent spatial correlation of data from pairs of these sites is performed. Application of a Darrieus VAWT for direct heat generation for space heating of a greenhouse and residence is described. Preliminary objectives and description of the Atlantic Wind Test Site are outlined. (5 refs.)

1.2.3(10) WIND ENERGY PROSPECTING IN PRINCE EDWARD ISLAND: A PROGRAMME OVERVIEW AND STATUS REPORT.
Lodge, M. (Institute of Man & Resources, Canada) In 2nd Intl. Symposium on Wind Energy Systems, Amsterdam, Netherlands
3-6 October 1978. Cranfield, Bedford, U.K., BHRA Fluid Engineering 1978 pp. B3. 21-B3. 32.
Research and development of wind as a renewable and sustainable energy source is a major programme of the Institute of Man and Resources (IMR). The Institute's wind energy programme commenced in July 1977 and completed in 1978, covers wind monitoring, data analysis and siting, large wind energy conversion systems (WECS) integration with the utility grid and rural and farm application of small WECS. Three IMR wind monitoring stations have been established and data from these stations are being correlated with five years of recent historical data from eight meteorological stations to develop a siting technology. Assessment of large WECS integrated into the utility is being performed and audits of farm energy uses and costs have been performed and are being evaluated. (6 refs.)

1.2.3(11) PROGRAMME OVERVIEW FOR THE WIND CHARACTERISTICS PROGRAMME ELEMENT OF THE UNITED STATES FEDERAL WIND ENERGY PROGRAMME.
Wendell, L.L. and Elderkin, C.E. (Battelle Pac. Northwest Lab., Richland, Wash.) In 2nd Intl. Symposium on Wind Energy Systems, Amsterdam, Neth. 3-6 October 1978. Cranfield, Bedford, U.K., BHRA Fluid Engineering 1978 pp. B2. 11-B2. 20.
The U.S. Department of Energy runs a management/research programme balancing the expertise in government laboratories, private industry and universities. The needs for wind characteristics information have been divided into four major categories: reliable wind and turbulence descriptions pertinent to WECS design and performance evaluation; effective analyses and methods for the determination of wind energy potential over large areas; dependable and cost-effective methodologies for the siting of WECS; and descriptions of day-to-day wind variability and predictability for WECS operations. (18 refs.)

1.2.3(12) ENVIRONMENTAL DEVELOPMENT PLAN WIND ENERGY CONVERSION.
Washington DC., USA. U.S. Dept. of Energy July 1979 34 pp. (DOE/EDP-0030)
Presents projects or sub-programmes pertaining to windpower utilisation that are likely to result in a demonstration or commercialisation within the near term. Projects that are expec-

ted to be considered in greater detail in the Environmental Development Plan update are also identified.

1.2.3(13) OVERSIGHT: WIND ENERGY PROGRAMME.
Hearing before the Subcommittee on Energy Development and Application of the Committee on Science and Technology, 96th Congress, 1st Session, No. 35, 30 July 1979. Washington, USA, Washington GPO 1979 190 pp.
Efforts to develop numerous sizes of wind machines to serve various applications and to meet differing market requirements are highlighted. Wind energy is identified as having the highest potential of all the solar electric technologies to contribute sizeable amounts of energy by the year 2000. The capability of the wind energy programme to realise this potential is discussed.

1.2.3(14) WIND ENERGY SYSTEMS.
QUARTERLY REVIEW, 1 JULY-30 SEPTEMBER 1979.
(Solar Energy Research Institute, Golden, CO.) Washington, DC. US Department of Energy, December 1979 176 pp.(Solar Energy Research Inst. Report No. 351-480.)
A visual presentation prepared by the Solar Energy Research Institute (SERI) as an overview of the efforts in the programme. This quarterly review fulfills SERI's Annual Operating Plan (AOP) reporting requirements. Presents the objectives, accomplishments, activities, and outputs of each of the tasks in the WES programme. Concentrates on wind turbines.

1.2.3(15) WIND POWER, RECENT DEVELOPMENTS.
De Renzo, D.J. Park Ridge, USA, Noyes Data Corp. 1979 365 pp. (Energy Technol. Rev. No. 46)
Presents the most important developments in wind power technology in the United States. Gives an overview, synthesis of national wind energy assessments; rotor development; innovative wind turbines; application to electric utilities; farming and rural use systems; and legal, social and environmental issues.

1.2.3(16) ENERGY FROM THE WINDS.
Washington, DC., USA. US Department of Energy, February 1980 11 pp. (DOE/PA-DO13)
A leaflet giving a short historical summary of the US Department of Energy's involvement in wind energy and its exploitation.

1.2.3(17) GOING WITH THE WIND.
Epri, J. 5 (2) March 1980 pp. 6-17.
The concepts behind wind-generated electricity, including the fundamentals of wind energy and the market requirements are presented. The federal research programme into wind energy in the United States is included, as is the development of wind turbines in the private sector.

1.2.3(18) WIND POWER AN INDUSTRY IN THE MAKING.
Energy Forum 1 (3) Autumn 1980 pp. 1-8.

Sixteen New England companies currently produce turbines ranging in size from those with blades 125 feet long to small residential machines. Site selection for wind turbines, the role of Utilities in wind power, wind projects around the US, and DOE's actions to promote wind power are reviewed.

1.2.3(19) NATIONAL WIND ENERGY CONSTRUCTION PROGRAMME: ITS ENERGY AND ECONOMIC IMPACT.
Curtis, E.H.T. (Department of the Interior, Washington, DC., USA.) In Energy Technol. Proc. 7th Energy Technol. Conference, V 2, Expanding Supplies and Conservation, Washington, DC., USA. Govt. Inst. Inc. 1980 pp. 1504-1523.
Wind energy is described as economical solar-electric technology. With an aggressive development programme electricity could be produced at costs below that of oil-fired electric plants. Important facts which support this optimism are discussed. (12 refs.)

1.2.3(20) A PROPOSED LARGE-SCALE WIND ENERGY PROGRAMME FOR CALIFORNIA.
Ginosar, M. (California Energy Commission) Energy Sources, 5 (2) 1980 pp. 141-171.
Describes a programme to generate at least 10 percent of California's projected electrical power with wind-electric systems by the year 2000. A six-year, $105 million programme would be financed and run by California State Government and Utilities. About 100 Utility-owned wind-electric farms with about 3300 3-MW wind-electric conversion systems would provide the electrical power. Wind resource development and prototype installation and testing of wind-electric systems would be necessary. (20 refs.)

1.2.3(21) CURRENT DEVELOPMENTS IN SMALL WIND ENERGY CONVERSION SYSTEMS.
Hansen, A.C. and Butterfield, C.P. (Rockwell Intl., Golden, CO.) J. Ind. Aerodyn. 5 (3-4) May 1980 pp. 337-356.
Presents a survey and review of research and development for Small Wind Energy Conversion Systems (SWECS). Current technical and economic status of SWECS is given, including some advantages and disadvantages of various technical approaches to wind energy conversion. Programmes managed by Rockwell International, for the US Department of Energy, to advance the commercialisation of SWECS are described. (23 refs.)

1.2.3(22) WIND ENERGY.
Kovarik, T., Pipher, C. and Hurst, J. Chalmington, U.K., Prism Press 1980 157 pp.
Gives a history of wind power utilisation concentrating on recent projects in the USA. Properties of wind, the generation of electricity by wind power and energy storage are discussed. Contains a bibliography, lists of suppliers and equipment and an index.

1.2.3(23) INTERIM STATUS REPORT ON DOE PROTO-TYPE DEVELOPMENT SWECS.
Moment, R.L. and Trenka, A.R. (Rockwell Intl., Golden, CO., USA.) In IECEC '80, Energy to the 21st Century, proc. 15th Intersoc. Energy Conversion Engng. Conference, Seattle, USA. 18-22 August 1980, V 1, New York, USA. American Institute Aeronaut. & Astronaut. 1980 No. 809163 pp. 815-820.
Describes development of several small wind energy conversion systems (SWECS) underway for over two years as part of the Department of Energy's (DOE) small wind systems programme, directed by Rockwell International, Rocky Flats Plant, as part of its commercialisation and use of SWECS. Design and fabrication efforts are complete on prototype systems in three sizes: 1 - 2 kW for remote applications requiring high reliability, 8 kW and 40 kW, these latter two to be linked. (From Author's abstract.)

1.2.3(24) WIND ENERGY IN MARYLAND.
Tompkins, D. Friendly, USA., Daryl Tompkins July 1980 37 pp.
A state-of-the-art review of applications, economic and environmental issues, the results of a preliminary wind energy resource assessment completed in April 1979 and recommendations for further resource definition.

1.2.3(25) REVIEW OF THE SERI WIND ENERGY INNOVATIVE SYSTEMS PROGRAMME.
Vas, I.E. and South, P. (Solar Energy Research Inst., Golden, CO., USA.) In 3rd Internat. Symposium on Wind Energy Systems, Lyngby, Copenhagen, Denmark 26-29 August 1980. Cranfield, Bedford, U.K., BHRA Fluid Engng. 1980 pap. A4 pp. 61-74.
Presents the major efforts of the sub-contracted research studies of the Wind Energy Innovative Systems Programme supported by the Solar Energy Research Institute. Eleven of the 17 studies funded consider potentially cost-effective innovative concepts and these are highlighted.

1.2.3(26) WIND ENERGY SYSTEMS QUARTERLY REVIEW, 1 JANUARY-31 MARCH 1981.
(Solar Energy Research Inst., Golden, CO., USA.) Washington, DC. US. Department of Energy May 1981 190 pp. (NTIS Report No. SERI/PR-635-1239)
Presents summaries of research programmes concerning wind turbine markets; economics of WECS owned by the end user; WECS-utility interface; selected utilities value analysis; WECS application in nongenerating utilities; WECS utility guide; small wind turbine production evaluation and cost analysis; WECS storage assessment and options; WECS legal issues; environmental assessment of wind systems; television interference and WECS; WECS noise studies; WECS incentives; advanced and innovative wind energy concepts; and engineering and cost analyses of AIWEC.

1.2.3(27) WIND ENERGY ACTIVITIES WITHIN THE
DEPARTMENT OF DEFENSE.
Barattino, W.J. (DOE Sandia Labs., New Mexico,
USA.) In ISES-AS/Et Al 1981 Annual Conference,
Philadelphia 26-30 May 1981 V 2 pp. 1549-1553.

The Department of Defense is the largest
single consumer of energy in the US and has
numerous bases in good wind resource areas.
R & D has focused primarily on determination of
the adaptability of commercially available
machines for military bases. Future activities
will include development of a wind energy data
base, identification of the electromagnetic
effects of wind turbines on communication-
electronics equipment, and development of
siting procedures.

1.2.3(28) WIND ENERGY IN NEW MEXICO.
Barnett, K. and Schoenmackers, R. (New Mexico
State University) In ISES-AS/Et Al Wind Power
Energy Alternative For MidWest 2nd Conference,
Minnesota 3-4 April 1981 pp. 87-92.

An estimated 40 electricity generating wind
turbines were operating throughout New Mexico,
and a large increase in the utilisation of wind
energy is expected during the next decade.
Twelve separate wind energy R & D projects
underway in New Mexico are described and
reasons for growth examined including cost-
effectiveness, reliability and government
incentives.

1.2.3(29) WIND ENERGY SYSTEMS INFORMATION USER
STUDY.
Belew, W.N. and others. (Solar Energy Research
Inst., Golden CO.) Washington, DC. US. Depart-
ment of Energy January 1981 278 pp. (Solar
Energy Res. Inst. Report No. SERI/TR-751-749)

Describes the results of a series of telephone
interviews with potential users of information on
wind energy conversion. Part of a larger study
covering nine different solar technologies, it
attempted to identify: the type of information
each distinctive group of information users
needed, and the best way of getting information
to that group. These include: wind energy
conversion system researchers; conversion
system manufacturer representatives; conversion
system distributors; wind turbine engineers;
utility representatives; educators, county
agents and extension service agents; and wind
turbine owners.

1.2.3(30) NAVY - NEW HAMPSHIRE WIND ENERGY
PROGRAMME.
Bortz, S.A. and others. Chicago, USA, IIT
Research Inst. November 1979 251 pp. (Report
No. IITRI M6052) (AD A086 506/3)

1.2.3(31) FEDERAL ROLE IN WIND ENERGY
COMMERCIALISATION.
Boyd, D.W., Buckley, O.E. and Haas, S.M.
(Decis Focus Inc., Palo Alto, California, USA.)
Energy Syst. Policy 5 (4) 1981 pp. 271-302.

Large wind energy conversion systems (WECS)
are among the most promising solar energy

technologies now under development. A decision-
analytic framework for evaluating federal
incentives designed to accelerate the
commercialisation of large WECS is described.
The framework is applied to a variety of hypothetical programmes similar to those currently
under consideration by the federal government.
(18 refs.)

1.2.3(32) THE USDA AGRICULTURAL WIND ENERGY
RESEARCH PROGRAMME.
Clark, R.N. (US Department of Agriculture)
In proc. 5th Biennial Wind Energy Conference
& Workshop, Washington, DC., USA.
5-7 October 1981, I.E. Vas (ed.) V 1, Palo Alto,
USA, Solar Energy Research Inst. 1981,
Session 1 pp. 85-92. (SERI/CP-635-1340)
(CONF-811043)

The Programme encompasses uses for both
general on-farm applications and special purpose
applications. Uses for a single application
need to include a storage system. Economic
studies show that systems must be used all year
round, rather than for only part of the year if
the investment is to be cost-effective.

1.2.3(33) WISCONSIN POWER AND LIGHT COMPANY'S
WIND ENERGY RESEARCH DEVELOPMENT PROGRAMME.
Dewinkel, C.C. (Wisconsin Power & Light Co.)
In 2nd ISES-AS/Et Al Wind Power Energy Altern-
ative For MidWest Conference, Minnesota
3-4 April 1981 pp. 83-85.

Discusses the major objectives of the Pro-
ramme which are to evaluate the economics and
reliability of small wind energy conversion
systems (SWECS), and to determine the electrical
safety of interconnecting SWECS with the utility
grid.

1.2.3(34) THE FEDERAL WIND ENERGY PROGRAMME.
Divone, L.V. (US Department of Energy,
Washington, DC.) In proc. 5th Biennial Wind
Energy Conference & Workshop, Washington,
DC., USA. 5-7 October 1981 I.E. Vas (ed.) V 1,
Palo Alto, USA, Solar Energy Research Inst. 1981,
Session 1 pp. 3-26. (SERI/CP-635-1340)
(CONF-811043)

Presents an overview of the state of wind
energy technology in the United States up to
1981.

1.2.3(35) WIND ENERGY DEVELOPMENT IN NORTH
AMERICA.
Divone, L.V. (US Department of Energy,
Washington, DC.) In proc. Internat. Colloquium
on Wind Energy, Brighton, U.K. 27-28 August
1981. Cranfield, U.K., BHRA Fluid Engng. 1981,
Session 2 pp. 68-90.

Discusses recent advances in wind energy
conversion systems of all sizes in the US,
including multimegawatt systems and progress on
the Darrieus system, on which the Canadian
effort is based. In both countries, wind
machines were built and are under test in a
variety of applications.

1.2.3(36) WIND POWER A TURNING POINT.
Flavin, C. (Worldwatch Inst.) Worldwatch Paper,
N45 July 1981 56 pp.
  Wind energy may provide a substantial renewable source of energy for the 1980's. A
growing market for wind machines exists in
remote areas where diesel engines and small
electricity grids provide expensive pumping
power and electricity. Wind machines are also
being developed for the provision of heat.
Principles of wind energy and options for its
future use are surveyed. (References.)

1.2.3(37) THE VELE PROJECT: WIND ENERGY
R & D PROGRAMMES AT ENEL.
Naccioni, C. and Sesto, E. (ENEL) In proc.
5th Biennial Wind Energy Conference & Workshop
Washington, DC., USA. 5-7 October 1981
I.E. Vas (ed.) V 2, Palo Alto, USA, Solar
Energy Research Inst. 1981, Session 2
pp. 87-110. (SERI/CP-635-1340) (CONF-811043)
  Discusses wind energy plant siting, the
design and use of wind turbine generators and
their components, and design, construction and
experimental operation of a 500 kW wind power
plant.

1.2.3(38) WIND POWER.
Macintyre, L. Ascent 3 (1) 1981 pp. 8-14.
  Reviews wind energy conversion systems
attracting increased interest in Canada.
Newly designed windmill systems are being
developed for application in remote areas and
Darrieus-type, vertical axis wind systems have
been sited on the Magdalen Islands to enhance
rural electrification. Limits to wind energy
potential and operating problems experienced
with these systems are examined.

1.2.3(39) THE WIND ENERGY INDUSTRY CONCERNS AND
PROMISES.
Marier, D.L. (Alternative Sources of Energy)
In 2nd ISES-AS/Et Al Wind Power Energy Alternative For MidWest Conference, Minnesota
3-4 April 1981 pp. 59-62.
  Challenges and potentials of the wind energy
industry are explored. Large-scale development
of the wind power alternative will require an
economic incentive to attract capital and a
suitable infrastructure to build and maintain
wind energy systems. Examines both domestic
and commercial applications. (5 refs.)

1.2.3(40) WIND POWER'S FUTURE DIMMED BY OIL
GLUT, BUDGET CUTBACKS.
Massey, S.R. Wall Street Journal, 21 August
1981 p. 21.
  Discusses problems which face the wind power
industry, among them cutbacks in federal
subsidies and a waning interest in alternative
energy caused by the current oil surplus.
Utilities involved in wind power are presented
and examined, and wind farms are studied.
The feasibility of large and small wind
turbines is assessed, and the MOD-2 wind power
generator is studied.

1.2.3(41) GUIDE TO WIND ENERGY IN TEXAS.
Nelson, V., Gipe, P. and McGaughey, D.
(West Texas State University, Canyon. Alternative Energy Inst.) Report No. TENRAC/EDF-016B
1 January 1981 51 pp.
  Parameters affecting wind generator performance, i.e., output power, are described,
including wind speed and turbine area. Wind
turbines are then described according to class-
ifications depending upon whether the rotor is
driven by drag or lift, and whether the axis
of the rotor is horizontal or vertical. Uses
of wind turbines are then described, including
pumping water and generating electricity, and
the operating characteristics of a wind
generator are discussed. Some utility programmes in the field are reviewed. Economics
of wind power are then examined, considering
initial cost, tax credits, the value of the
energy produced, carrying charges, and the
calculation of payback. Siting advice is
given regarding obstructions, terrain, and
tower sizing. Appended are tables of regional
wind speeds and of installed costs and energy
output for several wind energy conversion
systems. Also appended are monograms for
estimating payback periods for the different
wind regions. A very useful report.

1.2.3(42) SUMMARY OF PROCEEDINGS: OKLAHOMA AND
TEXAS WIND ENERGY FORUM, 2-3 APRIL 1981.
Nelson, S.C. and Ball, D.E. (eds.) (Southern
Solar Energy Centre, Atlanta, GA. Onyx Corp.,
Atlanta, GA.) Washington, DC. Department of
Energy. Report No. SSEC/SP-31230 CONF-8104147-
Sum. June 1981 34 pp. (Oklahoma and Texas Wind
Energy Forum, Amarillo, TX., USA. 2 April 1981)
(NTIS Report DE 82009474)
  The Wind Energy Forum for Oklahoma and Texas
was held to bring together the diverse groups
involved in wind energy development in the
Oklahoma and Texas region to explore the
future commercial potential and current barriers
to achieving this potential. Major topics of
discussion included utility interconnection of
wind machines and the buy-back rate for excess
power, wind system reliability and maintenance
concerns, machine performance standards and
state governmental incentives.

1.2.3(43) PROSPECTS FOR WIND ENERGY IN CANADA.
Ostroni, A.R. (Canada Dept. Energy, Minerals &
Resources, Ottawa) In proc. Intl. Colloquium
on Wind Energy, Brighton, U.K. 27-28 August
1981. Cranfield, U.K., BHRA Fluid Engng. 1981,
Session 1 pp. 45-50.
  Discusses four areas for wind energy opportunities in Canada: utility-connected; remote
communities; special-purpose; and agriculture/
household applications. Recognising the large
potential for grid-coupled wind systems, the
federal government has recently broadened its
research programme in wind energy to include
the design and development of a 4000 square
metre vertical axis prototype. Off-grid
remote communities are identified as an
emerging opportunity for wind turbines to
substitute for diesel-generated electricity.

1.2.3(44) WIND ENERGY.
Sissine, F., Luxemberg, B. and Moore, G.
(US Library of Congress Science Policy Research
Div.) US Library of Congress Congressional
Research Service Report IB80091 12 February
1981 13 pp.

Wind energy conversion technology, equipment,
and associated federal support programmes are
surveyed. Issues include incentives for wind
energy users and manufacturers; safety, reliability and durability of conversion systems;
and the potential contribution wind energy
may make in meeting US energy needs in the
next decade.

1.2.3(45) SMALL WIND SYSTEMS AN UPDATE.
Trenka, A.R. and Dodge, D.M. (Rockwell Inter-
national) Mineral & Energy Resources 24 (3)
May 1981 pp. 1-20.

An overview of wind energy R & D activities
conducted by Rockwell International since 1976.
Several prototype wind energy conversion systems have been developed and a field evaluation programme has applied institutional barriers to
small wind energy systems. National awareness
of small wind systems potential has been
increased and valuable information on
applications, financial and product liability
issues, and regulatory aspects has been
gained.

1.2.3(46) THE SERI WIND ENERGY PROGRAMME.
Vas, I.E. (Solar Energy Research Institute,
Golden, CO., USA.) In proc. 5th Biennial Wind
Energy Conference & Workshop, Washington,
DC., USA 5-7 October 1981 (I.E. Vas (ed.)) V 1,
Palo Alto, USA, Solar Energy Research Inst. 1981,
Session 1 pp. 93-106. (SERI/CP-635-1340)
(CONF-811043)

Describes the Solar Energy Research Institute
research and development studies relevant
to the widespread application of wind energy
systems. Three general areas: planning, management and analysis; advance and innovative
concepts; and information development are outlined.

1.2.3(47) THE WIND CHARACTERISTIC PROGRAMME.
Wendell, L.L. (Battelle Pacific Northwest Lab.)
In proc. 5th Biennial Wind Energy Conference &
Workshop, Washington, DC., USA 5-7 October
1981 (I.E. Vas (ed.)) V 1, Palo Alto, USA,
Solar Energy Research Inst. 1981, Session 1
pp. 27-38. (SERI/CP-635-1340) (CONF-811043)

An overview of the United States Wind
Characteristic Programme investigating the
following topics as its major areas of research:
Resource assessment techniques; Site selection
and evaluation techniques; meteorological
characteristics for design and performance
evaluation, and wind characteristics for wind
turbine operation. Highlights both recent
accomplishments and areas where further information is needed before wind energy programmes
can successfully be implemented.

1.2.3(48) WIND ENERGY.
Canadian Workshop, February 1982 pp. 26-30.

Gives an overview of wind energy as an important resource for the present and the future.
Recent oil crises have increased the demand for
wind energy. Questions as to how much of available wind energy can be appropriately used must
be answered. A description of the operation of
a wind turbine is provided and the cost-benefit
of wind energy briefly discussed.

1.2.3(49) THE WIND ENERGY RESEARCH AND DEVELOP-
MENT PROGRAMME IN CANADA.
Chappell, S.M. In proc. 4th BWEA Wind Energy
Conference, Cranfield, U.K. 24-26 March 1982
P.J. Musgrove (ed.) Cranfield, U.K., BHRA Fluid
Engng. 1982 pp. 12-28.

1.2.3(50) HARNESSING THE WINDS.
Flavin, C. (Worldwatch Inst., DC.)
Weatherwise, 35 (5) October 1982 pp. 211-218.

Summarises recent developments in wind power.
Windpumps - long a standby of the rural farmer -
are valuable because they free the user from
dependence on utilities. The Federal Government has investigated several wind turbines
for large-scale production of electricity,
including one proposal for grouping such
turbines offshore on floating platforms to
harness the permanent wind forces at sea.

1.2.3(51) OVERVIEW: THE DEVELOPING WIND
INDUSTRY.
Marier, D. Alternative Sources of Energy,
58 November-December 1982 pp. 8-15.

Although the wind energy industry has gained
a considerable amount of experience and growth
in 1982, 17,000 wind energy-related companies
went out of business in 1982, more companies
have entered the field than have left it. The
real growth in the industry has been in the
number of new companies with machines that
have ratings up to 100 kW for wind farm
applications. Wind energy conversion equipment companies that have recently entered the
market are identified and discussed.

1.2.3(52) WIND ENERGY IN THE USA - 1.
Taylor, D. (Open University, Milton Keynes,
Bucks., England) Energy J. (Auckland NZ) 55 (1)
January 1982 pp. 2-7.

Describes the Federal Wind Energy Programme.
Aspects such as wind resources, development of
large horizontal axis wind turbines (HAWTs),
small wind turbines, vertical axis wind
turbines (VAWTs), innovative wind systems,
agricultural applications of wind turbines,
institutional, economic and environmental
analyses, commercialisation and market development have been covered.

1.2.3(53) WIND ENERGY INFORMATION: ENERGY OUT
OF THIN AIR.
(Tennessee Valley Authority, Chattanooga. Div.
of Energy Conservation and Rates) Report No.
TVA/OP/ECR-82/29, 1983 20 pp. (NTIS Report No.
DE 83902759)

Presents a brief non-technical discussion of
wind power, mostly for on-site, private

application, which includes the amount of power available in wind according to the cube law, factors affecting wind flow, wind prospecting and monitoring, the user's power needs, system selection, and economics.

1.2.3(54) STATUS OF WIND TURBINE WAKES RESEARCH IN THE FEDERAL WIND ENERGY PROGRAMME.
Hadley, D.L. and Renne, D.S. (Battelle Pacific Northwest Labs., Richland, WA.) Washington, DC., USA. US Department of Energy Report No. PNL-SA-10999, CONF-830622-16 May 1983 22 pp. (American Solar Energy Society meeting, Minneapolis, MN., USA 1 June 1983) (NTIS Report No. DE 830 13828)
    Report of a series of wake studies conducted at Goodnoe Hills, sponsored by the Department of Energy's Wind Energy Technology Division. They represent the first in a series of field experiments designed to assess the impact of wind turbine wakes on downwind machines operating in clusters or arrays. Based on the results of the field measurement programmes at Goodnoe Hills, a much better understanding of the instantaneous structure of a wake in the near-wake region behind a large wind turbine, and of the distribution of the wake-induced velocity deficits out to 10 diameters downwind of a single machine is evolving.

1.2.3(55) DOE LARGE HORIZONTAL AXIS WIND TURBINE DEVELOPMENT AT NASA LEWIS RESEARCH CENTRE.
Linscott, B.S. (NASA, Lewis Research Centre, Cleveland, Ohio, USA) NASA Tech. Memo 83444 1983 15 pp.
    The large wind turbine programme is a major segment of the Federal Wind Energy Programme sponsored by the Department of Energy (DOE). The NASA Lewis Research Centre manages the large horizontal axis wind turbine programme for DOE, which is directed toward development of the technology for safe, reliable, environmentally acceptable large wind turbines that have the potential to generate a significant amount of electricity at costs competitive with conventional electric generation systems. In addition, they must be fully compatible with electric utility operations and interface requirements. (Includes a useful literative review of 77 references.)

1.2.3(56) WIND ENERGY PROGRAMME. 1983 FIRST QUARTERLY REVIEW.
Mitchell, R.I. and others. (Solar Energy Research Inst., Golden, CO.) Washington, DC. Department of Energy Report No. SERI/PR-211-1863 January 1983 30 pp. (NTIS Report No. DE 83005580)
    After the FY 1983 Wind Programme budget, the objectives, accomplishments, planned activities, and outputs of each task in the wind energy programme, which includes: programme management and planning, wind energy conversion systems (WECS) applications in nongenerating utilities, technical feasibility of stand-alone small WECS (SWECS), WECS performance/value analysis, wind energy industry analysis, wind systems co-ordination, wind workshops, noise and television interference studies, and advanced and innovative wind energy concepts. Also listed are subcontracted projects, and current publications. Previous quarterly reviews have the following numbers: 1982 3rd Quart. Rev. SERI/PR-211-1672. 1982 2nd Quart Rev. SERI/PR-211-1589.

1.2.3(57) SERI ADVANCED AND INNOVATIVE WIND-ENERGY-CONCEPTS PROGRAMME.
Mitchell, R.I. and Jacobs, E.W. (Solar Energy Research Inst., Golden, CO.) Washington, DC. US Department of Energy Report No. SERI/TP-211-1984 CONF-830622-13 June 1983 13 pp. (NTIS Report No. DE 83011960)
    The objective has been to determine the technical and economic potential of advanced wind energy concepts. Assessment and R & D efforts have included theoretical performance analyses, wind tunnel testing, and costing studies. Several concepts, such as the Dynamic Inducer, the Diffuser Augmented Wind Turbine, the Electrofluid Dynamic Wind-Driven Generator, the Passive Cyclic Pitch concept, and higher performance airfoil configurations for vertical axis wind turbines, have recently made significant progress.

1.2.3(58) WIND-BEHAVIOUR RESEARCH IN THE WIND-ENERGY-TECHNOLOGY PROGRAMME.
Wendell, L.L. (Battelle Pacific Northwest Labs., Richland, WA.) Washington, DC. US Department of Energy Report No. PNL-SA-11172, CONF-830631-9 May 1983 8 pp. (NTIS Report No. DE 83013823)
    The priorities of the section of Wind-Energy-Technology Programme devoted to research on wind behaviour have undergone a significant shift in emphasis to research focusing on the direct interaction of the turbine and the winds. Because of the activity in the design and testing of large wind turbines over the past two years, a high priority has been given to experiments and studies of microscale turbulence affecting wind-turbine stresses. Emphasis has also been placed on wind variability as it relates to turbine operating strategy and energy capture. The spacing between turbines in a cluster is an important factor in the effectiveness of the cluster in extracting maximum energy from the wind without significantly affecting the turbine design life.

1.2.3(59) SOME CURRENT ACTIVITIES IN THE CANADIAN WIND ENERGY PROGRAM.
Chappell, M.S. (National Research Council of Canada) In American Wind Energy Assn. Wind Energy Expo Natl. Conference, San Francisco 17-19 October 1983 pp. 379-411.
    The results of approximately 25,000 hours of field operation by five VAWT, coupled to utility grids, are reported. Recent activities have also addressed dynamic stall wind tunnel data and its effects on stall-limited performance prediction. Resource assessment, safety and performance tests, utility interconnection standards, and development of wind systems for electrical and pumping applications are surveyed. (16 refs.)

1.3 Europe

1.3(1) ENERGY: THE BURNING QUESTION FOR EUROPE.
European Democratic Group EDG, October 1981
30 pp.
   Discusses the need for a coordinated European
Communities Energy Policy. The long term oil
supply problems, and some of the possible
alternative energy sources such as coal and
liquefied coal, combined heat and power, heat
pumps, geothermal energy, solar energy, gas,
biomass, wind power, tidal power, and nuclear
power are also discussed. Fuel for transport
and industry, taxation, and the possibility of
energy bonds are considered.

1.3(2) REPORT SUBMITTED BY THE ECONOMIC
COMMISSION FOR EUROPE.
Paper presented at UN New & Renewable Sources
of Energy Conference, Nairobi, 10-21 August 1981
15 pp.
   Main trends, developments, and prospects for
the new and renewable energy sources of the
Economic Commission for Europe (ECE) are outlined. Biomass, geothermal, wind, hydropower
and solar sources are analysed as regards past
and current uses and programmes.

1.3(3) THE SELECTION OF AN ENERGY R & D PORT-
FOLIO FOR THE EUROPEAN COMMUNITY.
Love, P.E. and Michel, J. (Energy Management
Centre, NY, USA) CEC Report Eur. 8049 EN, 1982
62 pp.
   Discusses an energy analysis for EEC
countries that will not require frequent revising
of significant policy recommendations. The time
scale of energy development requires that
serious consideration of any new technology
cover at least 30 - 50 years. It is accepted
that various distributed parameters are independent and that the analysis of each competing
technology is made individually. The costs and
benefits associated with 30 technologies -
among them wind power - are identified. They
are used to provide a ranking procedure for the
development of a suitable technological port-
folio which will meet the energy goals of CEC
given the limits of the R & D budget.
(References.)

1.3(4) LARGE WIND TURBINE SYSTEMS SEEN FROM
THE EUROPEAN VIEWPOINT.
Selzer, H. (Erno-Raumfahrtechnik, W. Germany)
ASES Advances in Solar Energy, 1982 pp. 188-207.
   Looks at criteria for improving the design
and operation of wind energy systems. Areas
for improvement include profiles with increased
flap-bending stability and data for thick profiles and high angle of attack. Dynamic and
aeroelastic problems discussed include natural
vibrations, impulse loads, and periodic loads.
Load calculations, system dimensioning,
materials, environmental impacts, wind resource
assessment, and plant safety are also considered. Elements of national and international wind energy R & D programmes in Europe
are summarised. (10 refs.)

1.3.1 UNITED KINGDOM

1.3.1.1 RENEWABLE ENERGY RESOURCES

1.3.1.1(1) ENERGY OPTIONS IN THE UNITED
KINGDOM: A SYMPOSIUM.
Evans, S.C. Latimer New Dimensions 1975 128 pp.
   Considers geothermal energy, solar energy,
wind power, water power in relation to energy
policy in the U.K.

1.3.1.1(2) ENERGY AND THE ENVIRONMENT.
Hutton, M. and Nicoll, R.E. (Royal Institution
of Chartered Surveyors) RICS 1975 41 pp.
   The Gold Medal Address from the Annual
Conference held in Edinburgh, 16th July 1975.
Considers electricity production from nuclear,
gas, coal, peat, wind and tidal energy sources
with a review of pollution and conservation
issues.

1.3.1.1(3) RENEWABLE SOURCES OF ENERGY - HOW
FAR THEY MIGHT BE MADE TO MEET BRITAIN'S
ENERGY NEEDS.
Royal Society Arts J, 124 (5244) November 1976
pp. 723-737.
   Summary of a Symposium held on 16th June 1976,
its proceedings and accounts of six technical
papers on solar energy, tidal power, wave
power, the future of wind generation in
Britain, geothermal energy from the geophysical
and geological aspects and the technical and
economic aspects.

1.3.1.1(4) ENERGY AD INFINITUM.
Hardy, A.C. Building, 230 (7) 13 February 1976
pp. 93, 95.
   The alternative energy sources of sun and
wind appear attractive, but climatic conditions
in Britain require a great deal more research
to be done into the precise economic benefits
to be expected.

1.3.1.1(5) ALTERNATIVE ENERGY SOURCES.
Keynote Publications Ltd. 1979 9 pp.
   Reviews the market for four sources of
alternative energy that might be viable
propositions for domestic or commercial use
in the United Kingdom under certain conditions
- solar heating, solar electricity, wind power,
and water power.

1.3.1.1(6) ENVIRONMENTAL IMPACT OF RENEWABLE
ENERGY SOURCES - A REVIEW PAPER PREPARED FOR
THE COMMISSION OF ENERGY AND THE ENVIRONMENT.
Department of Energy, DOE March 1979 20 pp.
Revised Edition.
   The likely roles of the renewable energy
sources in the U.K.'s energy future are outlined and a brief account is given of present
views on the main environmental and related
issues which they raise. Includes consideration of wind energy.

22

1.3.1.1(7) THE STATE OF PLAY ON ALTERNATIVE
SOURCES OF ENERGY.
Environmental Data Services Report 29 July 1979
pp. 21-22.
   Brief look at research on solar, wind, wave,
tidal and geothermal power sources.

1.3.1.1(8) TOMORROW'S WORLD LOOKS TO THE
EIGHTIES.
Blakstad, M. BBC 1979 214 pp.
   Series of articles on technological developments expected during the decade. Among many
topics it includes development of alternative
energy systems such as solar, wind, tidal and
wave energy and geothermal energy.

1.3.1.1(9) THE 'RENEWABLES' AND THE
ENVIRONMENT.
Dawson, J.K. (United Kingdom Atomic Energy
Authority) UKAEA 1979 8 pp. (Reprint from
Atom 273 July 1979).
   Preliminary review of the environmental issues
which would be raised by the introduction of
renewable energy sources into the United Kingdom
energy supply, summarises evidence given to the
Commission on Energy and the Environment.

1.3.1.1(10) ENERGY OPTIONS AND EMPLOYMENT.
Elliott, D. North East London Polytechnic
(CAITS) NELP March 1979 146 pp.
   Compares the cost of and employment generated
by nuclear power programme to the year AD 2000
with that of a non-nuclear programme based on
conservation, the more efficient use of fossil
fuel and the exploitation of renewable sources,
including wind energy. Gives a detailed breakdown by technology, industry and skill of the
alternative programme.

1.3.1.1(11) THE POWER GUIDE: A CATALOGUE OF
SMALL SCALE POWER EQUIPMENT.
Fraenkel, P. Intermediate Technology Publications Ltd. 1979 240 pp.
   Series of short papers each followed by a
trade directory of manufacturers of relevant
products. Solar energy, wind power, biomass
and thermal energy, internal and external
combustion engines, the Stirling engine, and
electricity generation are covered.

1.3.1.1(12) THROUGH THE LOOKING GLASS - AND
INTO TOMORROW'S ENERGY.
Hill, L.M. Gas Engineering Management, 2 (9)
September 1980 pp. 347-360.
   Discusses energy policy issues including coal,
gas, oil, nuclear, wind, tidal and solar energy
sources in relation to energy conservation.

1.3.1.1(13) FACTORS DETERMINING ENERGY COSTS
AND AN INTRODUCTION TO THE INFLUENCE OF
ELECTRONICS.
Watt Committee on Energy. Watt Committee on
Energy September 1981 77 pp. (Report No. 10)
   Two reports:- the factors influencing the
basic costs and prices of primary energy; and

the impact of the use of electronics on methods
of procuring energy and on energy utilisation.
Forms of primary energy considered are oil,
coal, gas, uranium and electricity. The
report on electronics includes papers on wind
power applied to domestic requirements,
electronic controls for refrigeration plant,
and the use of computers in air traffic control.

1.3.1.1(14) NATIONAL REPORT SUBMITTED BY THE
UNITED KINGDOM OF GREAT BRITAIN AND NORTHERN
IRELAND.
Paper presented at UN New & Renewable Sources
of Energy Conference, Nairobi 10-21 August 1981
9 pp.
   Recent research efforts on renewable resources
development and the future of such energy
indicate similarities between the U.K. and
other island communities in energy needs and
availability. Political, environmental, and
economic constraints on the development of
renewable resources in the U.K. are studied.
Geothermal, wind, biomass, wave, tidal, and
hydroelectric energy data are given with
emphasis on their state of development and
constancy of supply.

1.3.1.1(15) RENEWABLE ENERGY SOURCES IN THE U.K.
British Business, 24 September 1981 pp. 140-143.
   Paper prepared for the United Nations
Conference on New and Renewable Sources of
Energy, Nairobi, August 1981. Deals with some
of the renewable energy systems in use or under
development in Britain - geothermal energy,
solar hot water systems, windmills generating
electricity and tidal energy schemes

1.3.1.1(16) THE U.K. ENERGY SCENE.
Conway, A. Energy J 54 (4) April 1981
pp. 3-7.
   Surveys energy supplies, consumption, and
resources in the U.K. and looks at what policies
are being implemented to encourage sectoral
energy conservation and the development of
solar, geothermal, wind, wave, tidal and other
renewable resources.

1.3.1.1(17) ALTERNATIVE ENERGY OPTIONS:
ENVIRONMENTAL AND PLANNING IMPLICATIONS.
Guscott, J.B. (Polytechnic of Central London,
School of Environment) PCL 1981 90 pp.
(Planning Studies No. 10).
   A brief survey of finite fuels and outline of
the energy problem; highlights issues in the
nuclear power debate, then covers alternative
renewable sources of energy research such as
wave power, tidal power, wind power, solar
energy, and geothermal energy.

1.3.1.1(18) ENERGY CONSERVATION.
Public Serv. Local Gov. 12 (9) July/August
1982 pp. 28-62.
   Special feature issue on central government's
role in promoting conservation; cost effective
conservation methods in housing; implementation
of an energy audit; use of wind power; energy

conservation consultancy services; heat pump technology; cavity wall and external wall insulation; automatic lighting control systems; and energy conservation in swimming pools and other local authority facilities. Includes details of new products.

1.3.1.1(19) WHITEHALL CHILL OVER RENEWABLE SOURCES.
Dafter, R. Financial Times 9 June 1982 Sec. 1 p. 12.
United Kingdom government research and development spending on renewable energy sources includes wind, wave, geothermal, solar, tide, biomass, conservation, and coal lique-faction technologies. It has risen slowly because of ample supplies of conventional fuels. New PWR stations will be needed to replace the first generation of nuclear power plants - alternative energies are not sufficiently cost-effective or technologically advanced to compete with nuclear generation for large-scale electricity generation.

1.3.1.1(20) ENERGY OPTIONS TO 2030.
Fells I. (University of Newcastle Department of Chemical Engineering) Long Range Planning, 15 (4) August 1982 pp. 76-85.
Evaluates three recent world energy demand forecasts and considers the constraints which may retard technological progress in energy supply. Considers whether nuclear power and coal supply will increase fast enough to provide a substitute for oil, and concludes that demand for energy will continue to grow in the next fifty years, with continued heavy dependence on fossil fuels while solar power, wave power and wind power will be incapable of providing adequate substitutes.

1.3.1.1(21) NEED FOR ALTERNATIVE ENERGY RECEDES.
Holmes, B.A. Financial Times, 6 July 1982 Sec. 1 p. 9.
Renewable energy development is being left in the hands of private enterprise and energy market forces, as energy officials have emphasised that abundant reserves of coal, other hydrocarbons, and nuclear power will satisfy long-term energy needs.

1.3.1.1(22) STRATEGIC REVIEW OF THE RENEWABLE ENERGY TECHNOLOGIES: AN ECONOMIC ASSESSMENT VOLUME I EXECUTIVE SUMMARY AND MAIN REPORT.
Department of Energy, Energy Technology Support Unit, London, HMSO 1982 110 pp.
Presents cost benefit analyses of renewable energy sources, whose value is shown to be set largely by the amount of conventional fuels they replace. Considers how the economies of each source combine with other factors to influence their potential contribution to future United Kingdom energy supplies.

1.3.1.1(23) STRATEGIC REVIEW OF THE RENEWABLE ENERGY TECHNOLOGIES: AN ECONOMIC ASSESSMENT VOLUME II APPENDICES CONTAINING DETAILED ANALYSIS.
Roberts, T. (Atomic Energy Research Establishment) Department of Energy, Energy Technology Support Unit, London, HMSO November 1982.
Various pagings, diagrams, tables (ETSU R13).

1.3.1.1(24) ALTERNATIVE ENERGY SOURCES.
Todd, R.W. (U.K. Cent. Altern. Technol.) Intl. J. Ambient Energy, 3 (2) April 1982 pp. 69-80.
Surveys renewable energy sources and considers how and on what scale they can contribute at both local and national level. The present state of development and recent cost estimates are given in most cases, together with brief mention of the current research effort on each source. A useful review of the United Kingdom situation. (75 refs.)

1.3.1.1(25) THE RENEWABLE SOURCES OF ENERGY IN THE UNITED KINGDOM.
Dawson, J.K. (Department of Energy, Energy Technology Support Unit) Institution of Civil Engineers Proceedings, Part 1 February 1983 (74) pp. 47-60.
Reviews recent progress in the Department of Energy research programme on the renewable energy sources with particular emphasis on economic prospects.

1.3.1.1(26) CONTRIBUTION OF RENEWABLE ENERGY TECHNOLOGIES TO FUTURE ENERGY REQUIREMENTS.
Buckley-Golder, D.H. and others (AERE Energy Tech. Div.) Statistician, 33 (1) March 1984 pp. 111-132.
Contributions from various renewable technologies including onshore and offshore wind power were determined for a series of snapshot years up to 2025 using a set of scenarios in which various assumptions were made about economic growth and conventional fuel prices. Essentially a U.K. analysis.

1.3.1.2 WIND ENERGY

1.3.1.2(1) POSSIBLE MEANS OF LARGE-SCALE USE OF WIND AS A SOURCE OF ENERGY.
Bockris, J.O. Environmental Conservation, Winter 1975 2 (4) pp. 283-288.
Looks at technical possibilities of using wind energy where the mean annual wind exceeds about 16 mph. Concludes that large sea-borne rotors in high-velocity wind-belts with long-distance hydrogen transmission offer a more readily attainable and more environmentally acceptable prospect than atomic or solar sources of energy.

1.3.1.2(2) THE PROSPECTS FOR THE GENERATION OF ELECTRICITY FROM WIND ENERGY IN THE UNITED KINGDOM.
Bird, R.A. and Allen, J. (Department of Energy) London, HMSO 1977 66 pp.

Examines wind as an energy resource; the technology necessary; sites, installation and interconnection; costs: small aerogenerators; past, present and future developments; and environmental and planning aspects.

1.3.1.2(3) VISITING WINDMILLS IN WALES.
Hanlon, J. New Scientist, 76 (1075) 27 October 1977 pp. 216-218.
Describes the work and achievements of the Centre for Alternative Technology in Machynlleth, Powys, Wales.

1.3.1.2(4) WIND ENERGY PROSPECTS IN THE U.K.
Musgrove, P. Coal Energy, (16) Spring 1978 pp. 15-21.
Claims that converting wind energy to electricity on a large scale is a much more practicable proposition than the large-scale conversion of solar radiation to electricity, and explores wind energy's potential to contribute very significantly to U.K. energy needs. The economics of such a conversion are found to be encouraging, eg. if environmental impact reasons precluded hill top windmills they could be deployed in shallow waters off the coast, providing at least 25% of the country's electricity needs. (5 refs.)

1.3.1.2(5) LECTURE TO BRITISH WIND ENERGY ASSOCIATION.
Bedford, L.A.W. In proc. Second BWEA Wind Energy Workshop, Cranfield, U.K. 17 April 1980. London, U.K. Multi-Sci. Publishing Co. Ltd. 1980 pp. 1-4.
Reviews the U.K. Wind Energy Programme from its start in 1976. Feasibility of both land-based and offshore sited aerogenerators and the contribution they might make to the U.K. energy supply have been considered in detail, and the cost advantages of using horizontal axis wind turbines in the 100-500 kW range investigated.

1.3.1.2(6) THE FUTURE FOR THE WIND ENERGY IN THE U.K.
R.E. News, No. 5 June 1981 pp. 2-4.
Describes the Department of Energy's R & D programme on wind energy, managed by ETSU, begun in 1976, It is aimed at determining the feasibility of both land-based and offshore aero-generators and the size and nature of the contributions they might make to the U.K. energy supply. Through a combination of paper and design studies, small-scale experiments, and prototype construction, it is hoped to establish how much energy can be extracted from the wind in the U.K., what types of machine should be employed, what the price of electricity is likely to be from this source, and how soon wind energy could be made available if needed. The Central Electricity Generating Board has its own programme of research and development into wind energy in order to improve its position as an informed buyer, which is co-ordinated with the Department's programme and provides a valuable input to it. CEGB has selected a site at Carmarthen Bay power station

on which to build a medium-sized aerogenerator to provide experience and research information.

1.3.1.2(7) WIND ENERGY REPORT.
Int. Power Generation, 4 (4) May 1981 pp. 41, 43, 46-48.
Summary of the Third British Wind Energy Association Conference 1981 held at Cranfield Institute of Technology, Bedford, U.K. The conference acted as an informative clearing house for news and views on current developments and future possibilities.

1.3.1.2(8) REVIEW OF THE U.K. WIND ENERGY PROGRAMME.
Clarke, F.J.P. In proc. Third BWEA Wind Energy Conference, Cranfield, U.K. 9-10 April 1981. Cranfield, U.K. BHRA Fluid Engng. 1981 pp. 1-6.
Brief review of the aims and implications of the Department of Energy's programme on wind energy.

1.3.1.2(9) WIND ENERGY R & D IN THE UNITED KINGDOM.
Lindley, D., Swift-Hook, D. and Stevenson, W.G. (British Wind Energy Assoc.; North Scotland Hydro Electr. Board) In proc. 5th Biennial Wind Energy Conference & Workshop, Washington, D.C., USA 5-7 October 1981 (I.E. Vas (ed.)) V. 2, Palo Alto, USA, Solar Energy Research Institute 1981 Session 2 pp. 43-54. (Report No. SERI/CP-635-1340 CONF 811043).
Summarises current wind energy activities in the U.K. including those sponsored by Government agencies, the CEGB and private companies. Studies and measurements are described which will assist in the evaluation of the whole U.K. wind resource. There are plans to erect two horizontal axis machines of 20 m and 60 m diameter on Orkney; a 25 m vertical axis machine is at the design stage: a coordinated programme of small machine development is underway. Field tests are supported by studies in areas such as wind tunnel wake measurements. Studies of offshore siting have shown economic and environmental attractions, and work has been going on into the effects on public supply networks of both onshore and offshore wind turbine clusters.

1.3.1.2(10) WIND ENERGY SYSTEMS INTEGRATION STUDIES BY THE READING UNIVERSITY & RUTHERFORD AND APPLETON LABORATORIES GROUP.
Lipman, N.H. and others. (Reading Univ.) In proc. Int. Colloquium on Wind Energy, Brighton, U.K. 27-28 August 1981. Cranfield, U.K. BHRA Fluid Engng. 1981 Session 2 pp. 91-96.
Work carried out by the Reading University/ Rutherford and Appleton Laboratories team into integration of wind energy into the National Grid is reviewed. There are three main areas: understanding and describing the behaviour of the wind; describing the wind-windmill interactions and examining the range of possible operating options for windmills; and modelling a national grid which includes substantial wind power inputs.

1.3.1.2(11) WIND ENERGY RESEARCH EXPERIENCE
OF THE READING UNIVERSITY ENERGY GROUP.
Lipman, N.H. and others. In Energy for Rural
and Island Communities, proc. Conference
Inverness, U.K. 22-24 September 1980.
(J. Twidell (ed.)) Oxford, U.K. Pergamon Press
Ltd. 1981 Topic D, Paper Dl pp. 123-136.
   Overview of wind energy, using, as illus-
tration, the results obtained by the Reading
Energy Group. Wind turbines, large and small,
wind potential and power output are topics
included.

1.3.1.2(12) THE WIND ENERGY RESEARCH AND
DEVELOPMENT PROGRAMME OF THE U.K. DEPARTMENT
OF ENERGY.
Pooley, D. (Atomic Energy Research Establishment, Harwell) In proc. Int. Colloquium on Wind
Energy, Brighton, U.K. 27-28 August 1981.
Cranfield, U.K. BHRA Fluid Engng. 1981 Session 1
pp. 2-10.
   Wind energy research and development funded
by the Department of Energy has three major
parts: demonstration of fairly conventional
horizontal-axis machines on sites where they
might become cost effective soon; seeking to
reduce machine costs and explore advanced
designs; and establishment of the feasibility
of all aspects of offshore siting for wind
generators.

1.3.1.2(13) WIND ENERGY FOR THE EIGHTIES.
Lipman, N.H., Musgrove, P.J. and Pontin, G.W.W.
Stevenage, U.K. Peter Peregrinus Ltd. 1982
388 pp.
   Review by members of the British Wind Energy
Association of wind resources available, particularly in the U.K. Considers wind turbines,
including aerodynamics, materials of construction, structural dynamics, maintenance,
inspection and instrumentation, design and
application of small, medium and large wind
turbines, integration of wind turbine generators into an electricity grid, offshore siting
of wind turbine and legal institutional and
environmental effects. Considers vertical and
horizontal axis wind turbines, the Darrieus
wind turbine, and includes comparisons of
other countries' experience.

1.3.1.2(14) A REVIEW OF UNITED KINGDOM
WINDPOWER.
Millar, M.N. (Napier College, Scotland)
Intl. J. Ambient Energy, 3 (1) January 1982
pp. 35-47.
   Describes many existing installations and
machines that are part of current research
efforts. Different turbines being
developed are discussed, including the three-
blade upwind horizontal-axis Aldborough
generator.

1.3.1.2(15) REVIEW OF U.K. WIND ENERGY
ACTIVITIES.
Musgrove, P.J. (University of Reading, Depart-
ment of Engineering, Reading, U.K.) Int. J.
Solar Energy, 1 (2) October 1982 pp. 145-160.

   Wind energy activities in the U.K. have concentrated on comprehensive studies of system
integration aspects and a detailed assessment
of offshore wind energy systems. Many small
wind turbines have been built and tested, but
design and construction of large wind turbines
is slow. However, two industrial groups are
developing large wind turbines, one horizontal
axis and one a novel vertical axis design.
Electricity utilities in the U.K. are showing
considerable interest in the use of wind energy
and are participating actively in developments.
Includes a valuable bibliography of 300
references.

1.3.1.2(16) WIND ENERGY DEVELOPMENT FOR
THE U.K.
Pooley, D. In proc. Fourth BWEA Wind Energy
Conference, Cranfield, U.K. 24-26 March 1982.
P.J. Musgrove (ed.) Cranfield, U.K. BHRA
Fluid Engng. 1982 pp. 1-11.
   Outlines how the Energy Technology Support
Unit (ETSU) and the Department of Energy see
wind energy research and development fitting
into their renewable energy programme, and
discusses factors thought to be important for
the development of wind energy in the U.K.

1.3.1.2(17) WIND POWER IN THE DOLDRUMS.
Engineer, 256 (6623) 3 March 1983 pp. 48-49.
   The U.K.'s largest wind energy machine, rated
at 3MW has been designed and is to be erected
in the Orkneys to start generating electricity
in Spring 1985, and a 200kw wind turbine has
been erected in Camarthen Bay, Wales. Claims
that more funding is still needed if major
engineering problems associated with translating
designs for large machines into hardware are to
be solved.

1.3.1.2(18) BRITISH WIND ENERGY - A DEVELOP-
MENT STRATEGY.
London, U.K., British Wind Energy Association
March 1984 23 pp.
   Present projects include the horizontal axis
wind turbines under study at the Bergar Hill
Site, Orkney and the vertical axis wind turbine
at Carmathen Bay. Costs and funding of these
medium and larger wind turbines are discussed,
and recommendations made for certain government
support in U.K. manufacturers. Examines the
potential for small wind power (up to 100 kW)
development, particular applications, and
suggests ways of informing consumers and expanding the market for such machines.

1.3.1.2(19) THE U.K. PROGRAMME OF WIND ENERGY.
Skipper, R.O.S. and Bedford, L.A.W. (U.K. Dept.
Energy and A.E.R.E. Harwell) In proc. Wind
Energy Conversion, 5th BWEA Wind Energy Conf.,
Reading, U.K. 23-25 March 1983. P. Musgrove (ed.)
pp. 1-9 Cambridge, Cambridge University Press
1984 375 pp. (Isbn: 0-521-26250-X)
   Describes the aims and objectives of the
Department of Energy programme for wind power
research and development. Onshore and offshore
wind power have been studied, and direct support

and collaborative projects cover a wide variety of areas from wind speed data collection and analysis to assessment of horizontal axis and vertical axis wind turbines.

1.3.1.2(20) WIND POWER ON LUNDY ISLAND.
Somerville, W.M. and Puddy, J. (Int. Res. & Dev. Co. Ltd.) In proc. Wind Energy Conversion, 5th BWEA Wind Energy Conf., Reading, U.K. 23-25 March 1983. P. Musgrove (ed.) pp. 185-197 Cambridge, Cambridge University Press 1984 375 pp.

Discusses design of the wind turbine generator system. Describes access for equipment, site selection and estimation of fuel savings over the present diesel generator system operating alone. Foundations for the wind turbine tower, transport and erection of the 55 kW wind turbine are described. All dwellings are supplied with frequency sensitive load control consumer units and wind turbine performance is estimated.

1.3.2 NETHERLANDS

1.3.2(1) NETHERLANDS RESEARCH PROGRAMME ON WIND ENERGY.
Piepers, C.G. and Sens, P.F. (Netherlands Energy Research Foundation, Petten, Netherlands) In 2nd Intl. Symposium on Wind Energy Systems, Amsterdam, Neth. 3-6 October 1978. Cranfield, Bedford, U.K. BHRA Fluid Engng. 1978 pp. A1.1-A1.6.

Presents an overview of the Dutch wind energy programme. The first phase of the programme, from March 1976 lasting until 1st March 1977, had the nature of orientation, collection and evaluation of existing information. Proposals for research and development were specified in more detail for the second phase, that ran from March 1977 until January 1979. Two basic designs were taken into consideration, a wind turbine with horizontal axis and the vertical axis turbine. Computer models have been developed by the National Aerospace Laboratory to describe the aeroelastic and dynamic behaviour of a Darrieus turbine.

1.3.2(2) PHYSICAL PLANNING ASPECTS OF LARGE-SCALE WIND ENERGY EXPLOITATION IN THE NETHERLANDS.
van Essen, A.A. and others. In 2nd Intl. Symposium on Wind Energy Systems, paper presented at Amsterdam, Netherlands 3-6 October 1978. Cranfield, Bedford, U.K. BHRA Fluid Engng. 1978 pp. B1.1-B1.10.

Reports on an attempt to study the physical planning aspects of a large-scale exploitation of wind energy. The coastal area of the Netherlands is subdivided into 52 regions with an average yearly wind velocity of more that 6 m/s at 40 m height, each sufficiently large to contain one or more wind energy parks of more than 100 WECS. The siting of WECS is compared for each region, with a number of physical planning aspects such as: nature, influence on the landscape, agriculture and the present-day physical planning policy of the government.

1.3.2(3) WHAT ARE THE PROSPECTS FOR WIND ENERGY.
van Holten, T. (Technische Hogeschool, Delft, Netherlands) Land Water Int., (41) 1980 pp. 6-11.

Presents the argument that the potential of wind power is greater than was previously thought. The National Research Programme of the Netherlands into wind energy is described. The use of wind turbines, and the market for wind energy are included.

1.3.2(4) WIND ENERGY PLANNING CONSIDERATIONS (WINDENERGIE IN PLANOGISCH PERSPEKTIEF).
Vansetten, A. and Voogd, H. (Technische Hogeschool, Delft, Netherlands. Department of Urban and Regional Planning.) Presented at Planning Conference, Amsterdam April 1980. Report No. PM-80-4 April 1980 pp. 19. (In Dutch)

Windpowered utilisation in the Netherlands and the search for windy areas is discussed. It is suggested that large water areas, like the North Sea and Market Lake, are preferred areas for wind powered energy production. Storage of wind energy in large water basins is suggested. Government wind energy production programmes like the installation of wind parks, and difficulties because of opposing local interests are discussed. Windmill exploration by small private industry, rather than by government administered programmes are outlined.

1.3.2(5) WATER PUMP RESEARCH AND PROTOTYPE WORK FROM THE NETHERLANDS.
Windirections, (6) May 1981 pp. 8-9.

Two independent organisations in the Netherlands are working in the field of wind energy for developing countries, SWD (Steering Committee Wind Energy Developing Countries) and WOT (Working Group on Development Technology). The main emphasis at present is on applying wind energy for pumping water and the development of windmills is aimed at adaptation to local conditions, available materials and skills. Research activities are devoted to horizontal axis wind turbines. The development of the Wind Energy Utilisation Project in Sri Lanka is described.

1.3.2(6) WIND AVAILABILITY IN THE NETHERLANDS.
Builtjes, P.J.H. Energiespectrum, 5 (7-8) July-August 1981 pp. 177-181.

The National Research Programme on Wind Energy has yielded data on wind sources and availability. However, the actual location of the wind turbine determines the extent to which these resources can be utilised. The Apeldoorn MT-TNO establishment is now studying this location aspect of the problem. (14 refs.)

1.3.2(7) THE AUTHORITIES AND WIND ENERGY.
Campen, J.P. Energiespectrum, 5 (7-8) July-August 1981 pp. 212-214.

Address given at the 'Wind Energy in the Netherlands' conference summarising the attitude of the Netherlands Ministry of Economic Affairs to the subject. Potential and practical contributions to the Netherlands energy balance, the cost/profit ratio of wind energy, the required technological developments, the innovation possibilities for industry, specific fields of application relative to marketing, work opportunities, safety and environment factors and technical knowhow, are all considered and recommended as potentially favourable for further research.

1.3.2(8) IS WIND AN IMPORTANT FUTURE ENERGY SOURCE FOR THE NETHERLANDS?
de Zeeuw, W.J. Ned. Electron. & Radiogenootschap, 46 (2-3) 1981 pp. 55-59. (In Dutch)
A shortened version of a paper read at a NERG meeting at the Dutch Royal Institute of Engineers in The Hague on 20 November 1980, considers the capacity, efficiency, location and social aspects of wind energy. In addition to reducing chemical pollution, claims employment is created (possibly 100,000 jobs in the Netherlands in 30 years), oil, coal and uranium imports are cut, and there is also an export potential.

1.3.2(9) ASSESSMENT OF THE POTENTIAL OF LWECS IN THE NETHERLANDS.
Dub, W. (Regensburg University) In, Implementing Agreement for Programme of Research & Development on Wind Energy Conversion Systems, 1981 Meeting of Experts of Annex III & IIIa: Integration of Wind Power into National Electricity Supply Systems (Regensburg, Federal Republic of Germany 29-30 January 1981), Julich, Fed. Rep. Germany, Kernforschungsanlage Julich GmbH, April 1981 Session B, Paper B1 pp. 41-62. (Jul/Spez-108)
Considers meteorological data, wind turbine design, wind energy production and electricity consumption, and wind turbine data on numbers and specifications.

1.3.2(10) PROPOSALS FOR A CONTINUING NATIONAL RESEARCH PROGRAMME ON WIND ENERGY.
Piepers, G.G. Energiespectrum, 5 (7-8) July-August 1981 pp. 201-204. (In Dutch)
Summarises the final report of the five-year National Research Programme on Wind Energy (NOW) wound up on 1st March 1981. Suggests that wind energy is sufficiently promising to justify its active development, but many aspects of the problem have to be resolved, as regards both industrial and domestic installations. Research planning and financing are also considered.

1.3.2(11) THE POTENTIAL OF WIND ENERGY.
Piepers, G.G. (Netherlands Research Foundation) Resour. & Conserv., 7 November 1981 pp. 37-47.
Discusses three aspects of wind energy exploitation: problems of design concerning reliability and lifetime when faced with metal fatigue, and the need to control the device for

the production of electricity at constant voltage and frequency; economic difficulties with structures of large capital cost and uncertain operation and maintenance costs; and the need for windy sites which have no environmental impact, and for Holland this could mean establishing offshore sites in shallow coastal waters.

1.3.2(12) THE WIND PROGRAMME IN THE NETHERLANDS.
Piepers, G.G. and Sens, P.F. In proc. 5th Biennial Wind Energy Conference & Workshop, (Washington, D.C., USA. 5-7 October 1981) (I.E. Vas (ed.)) V. 2 Palo Alto, USA, Solar Energy Research Institute 1981 Session 2 pp. 111-130. Report No. SERI/CP-635-1340 CONF-811043.
Summarises the conclusions and recommendations given in the final report on the programme to investigate the possibilities of using wind power in the Netherlands. Prospects for wind energy conversions systems (WECS) are good but a number of technical, environmental and legal problems have to be solved, probably by executing an adequate follow-up research, development and demonstration programme.

1.3.2(13) WIND ENERGY IN THE NETHERLANDS: PLANNING ASPECTS.
van Essen, A.A. Energiespectrum, 5 (7-8) July-August 1981 pp. 189-193. (In Dutch)
Surveys the Netherlands national Research Programme on Wind Energy's initial stocktaking in 1978, of the large-scale centralised application of wind energy. Further possible action in this field is examined with regard to the nature of the relation between the energy and national economy, and also the legal and control aspects of this relation. Tabulated regional wind energy data are presented. See also pp. 172-177 of the same journal for article on the National Research Programme.

1.3.2(14) THE POTENTIAL OF RENEWABLE ENERGY SOURCES.
van Koppen, C.W. (University of Technology, Netherlands) Resources & Conservation, December 1981 V. 7 pp. 17-36.
Discusses the potential for renewable energy resources to displace conventional fossil fuels. The solar light radiation intercepted by the earth represents an energy flow more than four orders of magnitude larger than the present global energy demand. This flow is also available in the form of wind energy, wave energy, hydroelectric power, and biomass. A renewable energy supply model is constructed for application to energy conditions in the Netherlands. (20 refs.)

1.3.2(15) WIND ENERGY IN THE NETHERLANDS.
Irrig. & Power., 39 (4) October 1982 pp. 409-414.
Includes details of large wind turbines (e.g. the 300 kw turbine), glass fibre re-

inforced polyester components, and ideas for
wind energy storage. Small wind turbines and
use in alternative applications are mentioned.
Use of tip vanes to increase capacity is noted.
Comments on work on wind energy in remote
areas, and the development of systems for
developing countries.

1.3.2(16) THE DUTCH WIND ENERGY PROGRAMME.
Beurskens, H.J.M. and Piepers, G.G. In proceedings of the Fourth Intl. Symposium on Wind
Energy Systems, Stockholm, Sweden 21-24 September 1982 V. 1, Cranfield, U.K. BHRA Fluid Engng.
1982, Session B, Paper Bl pp. 93-111.

Report on the five years research programme
on wind energy completed in 1981. The aim of
the programme was to investigate the possibilities of the utilisation of wind power in the
Netherlands as a substitute for fossil and
nuclear fuel. The research concluded that wind
power offers sufficient prospects to justify
encouragement of the application of wind
energy conversion systems, and it was recommended to execute a follow-up research,
development and demonstration programme. (See
also Windirections 3 (1) July 1983 pp. 12-13.)

1.3.2(17) ENERGY CONSUMPTION AND CONSERVATION
IN THE NETHERLANDS.
Tammes, E. (Bouwcentrum, Rotterdam) Paper
presented at An Foras Forbartha, Energy Conservation in the Built Environment 3rd Intl.
CIB Symposium, Dublin 31 March,
2 April 1982 V. 1 PA-63 10 pp.

Establishes energy conservation goals in the
Netherlands for 1985, 1990 and 2000. Policy
consists of price settings and taxation,
information campaigns, special legislation,
energy R & D, and financial incentives to
encourage reduced consumption. Government
regulations specifying conservation measures
are directed toward both residential and non-
domestic structures. Includes wind energy.

1.3.2(18) SOCIOECONOMIC ASPECTS OF A VALUE
ANALYSIS OF WIND ENERGY ILLUSTRATED FOR THE
NETHERLANDS.
Dub, W. and Pape, H. (Regensberg University
W. Germany) Energy Sources, 6 (3) 1983
pp. 245-259.

The socioeconomic aspects of a value analysis
of wind power as a source of electricity generation are discussed and an economic assessment
of large scale wind turbines in the Netherlands
is outlined, noting value determination based
on fuel displacement and capacity displacement
capability. Examines normal independence of
energy sources, scarcity of fossil fuels,
safety considerations and costs of nuclear
power stations with reference to the
Netherlands.

1.3.2(19) A BREATH OF LIFE FOR WIND POWER.
van Kasteren, J. Financial Times, 1 September
1983.

Reports that Holland's Government has developed
a national plan to promote wind energy that
includes the development of medium-sized
turbines of up to 0.5 mw. and the creation of
a massive prototype wind turbine with capacity
of 3 mw. The programme represents an investment of $7 million, with Government providing
about 40% of the costs. Most of the power will
be sold for agricultural uses, but the electricity market for wind energy is hampered by
generating companies who charge $.20/kwh for
back-up electricity but buy spare wind-
generated power at half that price.

1.3.2(20) RECENT DEVELOPMENTS OF WIND ENERGY
IN THE NETHERLANDS: R, D AND D, COMMERCIAL-
ISATION AND INSTITUTIONAL PROBLEMS.
Beurskens, H.J.M. and Piepers, G.G. (Nether-
lands Energy Res. Found.) In 2nd ASME Wind
Energy Symp. 6th Annual Energy-Sources
Technol. Conf. & Exhib., Houston, USA
30 January-3 February 1983 New York, USA,
American Soc. Mech. Engrs. 1983 pp. 559-577
(See also paper in Wind Energy Conversion
proc. 5th BWEA Wind Energy Conf. 23-25 March
1983 Reading, U.K. pp. 10-18.)

The National Programme is divided into
centralised applications; decentralised applications and general research. The research,
development and demonstration of the two types
of applications are described and commercial
developments are discussed. The two medium
sized Dutch made wind turbines, commercially
available are briefly described, while
the applications of wind turbines in the 10 kW
to 60 kW range are discussed.

1.3.3 SCANDINAVIA

1.3.3(1) RENEWABLE ENERGY RESOURCES: WIND
ENERGY FROM A SWEDISH VIEWPOINT.
Engstrom, S. (Swedish Board for Technical
Development, Stockholm) Ambio, 4 (2) 1975
pp. 75-79.

Interest in wind power as a renewable source
of energy has revived. Looks at the historical
perspective, new technical developments,
questions of economy, ecological considerations, and the future possibilities. (15 refs.)

1.3.3(2) FRESH BREEZE FOR DENMARK'S WINDMILLS.
Cawood, P. and Hinrichsen, D. New Scientist,
70 (1004) 10 June 1976 pp. 567-570.

Wind power in Denmark has received considerable support in recent years as an alternative
energy source to oil. Denmark is one of the
windiest areas in Europe, and government
reports indicate that wind power is economically competitive with other energy sources for
the generation of electricity. Previous
estimates indicated that wind power was 2 to 3
times more expensive than oil or nuclear power.
The TVIND Windmill which will produce 3.6
million kw hours of electricity per year is
described.

1.3.3(3) ENERGY RESEARCH AND DEVELOPMENT IN
SWEDEN 1978/81.
Swedish Energy R & D Commission Report 13 1978
37 pp.

Survey of the objectives, structure, budget and organisation of government-supported energy R & D programmes in Sweden. Includes energy use in industrial processes, fuel usage in transportation, energy use for buildings, energy supply, and general system studies. Sub-programmes entail synthetic fuels, fusion, coal and wind energy.

1.3.3(4) METEOROLOGICAL FIELD PROJECT AT THE WIND ENERGY TEST SITE KALKUGNEN, SWEDEN.
Faxen, T. and others. (Uppsala University, Sweden) NTIS Report UUIM-51 1978 27 pp.
A meteorological field station has been set up in Sweden near a 63 kw wind energy test station. The wind energy facility has a hub height of 24 m. and a turbine diameter of 18 m. The station consists of two 42 m. masts, each equipped with cup anemometers. Wind speeds, temperature, humidity, and solar radiation are monitored. An analysis of wind data from October-December 1977 is included.

1.3.3(5) SWEDISH WIND ENERGY PROGRAMME.
Hugosson, S. (National Swedish Board for Energy Source Development, Sweden) In 2nd Intl. Symposium on Wind Energy Systems, Amsterdam, Netherlands 3-6 October 1978. Cranfield, Bedford, U.K. BHRA Fluid Engng. 1978 pp. A2.7-A2.15.
The Swedish Wind Energy Programme was presently leaving its "Studies and Experiments" phase and going into the "Prototype" phase. Conclusions from the first phase have resulted in a specification for full-scale prototypes of wind power units planned to enter operation and evaluation in early 1981. The first phase had shown that the use of small-scale units is technically simple, but that existing units need development. In parallel a comprehensive measurement project to map the winds over Sweden at 50-150 metres level was started, together with development of wind forecasting methods.

1.3.3(6) A SWEDISH ENERGY PROGRAMME FOR THE 1980'S.
Human Env. in Sweden, N.10 April 1979 pp. 1-14.
On 9 March 1979 Sweden's Government presented a bill containing guidelines for an energy programme for the period ending in 1990. Designed to reduce dependence on foreign oil imports and maximise energy self-sufficiency, hydroelectric and nuclear power capacity will be increased and renewable resources such as wind energy, solar energy and biomass will be extensively utilised.

1.3.3(7) SOME METEOROLOGICAL ACTIVITIES IN THE NATIONAL SWEDISH WIND ENERGY PROGRAMME.
Faxen, T. (University of Uppsala, Sweden) In Pac. Northwest Lab. Tech. Report PNL 3214, proc. of the Conference and Workshop on Wind Energy Charact. and Wind Energy Siting, Portland, Oregon 19-21 June 1979. Richland, Washington, Pac. Northwest Lab. 1979 pp. 113-130.

Two projects conducted by the University's Department of Meteorology on behalf of the National Swedish Board for Energy Source Development are discussed. First, the meteorological field station at Sweden's 60 kw windmill is presented. Then, the Gotland-Skane project, which was to determine sites for two 1 mw to 2 mw prototype windmills is described. (10 refs.)

1.3.3(8) RESEARCH AND DEVELOPMENT ON ENERGY IN SWEDEN.
Hormander, O. Paper presented at University of Auckland/Et Al 4th New Zealand Energy Conference, Auckland May 1979 V.1, pp. 113-135.
Examines Sweden's energy policy and research programmes geared to the development and commercialisation of wind energy and bio-conversion, amongst others. Energy supplies and policies are compared with those of New Zealand.

1.3.3(9) SOLAR OR NUCLEAR - ON THE CHOICE OF ENERGY FUTURE.
Johansson, T.B., Lonnroth, M. and Steen, P. (University of Lund, Sweden) In CEC Energy Systems Analysis International Conference, Dublin 9-11 October 1979 pp. 343-357.
In the nuclear scenario, LWR's supply commercial, industrial and domestic energy demands, but uncertainties include environmental issues, land use, and political controversy. The solar scenario envisions energy from biomass, solar cells, and wind energy, but uncertainties include high costs of solar technology and environmental effects of energy plantations. The economic and technical feasibilities are compared.

1.3.3(10) WIND ENERGY RESEARCH.
Nat. Swedish Board for Energy Source Develop. Report NE 1980, December 1980 209 pp. (In Swedish)
Examines recent wind energy research and development projects in Sweden and prospecting and measurement studies are detailed. Experimental wind energy conversion units, employing horizontal axis turbines, have been constructed and operated, and design features of several prototype full-scale power plants are summarised. Wind energy costs were computed for three of these facilities, revealing a cost of 25/kwh. Surveys future research needs.

1.3.3(11) RECENT DANISH DEVELOPMENTS IN WIND ENERGY CONVERSION SYSTEMS.
Johansson, M. In proc. 11th World Energy Conference in Munich, Germany 8-12 September 1980 V. 1A, London, U.K. World Energy Conference, 1980 Section 1.2 pp. 569-586.
A report on the four year development programme of the Ministry of Trade and Electric Utilities in Denmark for the extension of existing supply systems by wind power plants.

1.3.3(12) WIND POWER IN SWEDEN.
Summerton, J. (National Swedish Board for
Energy Source Development) Human Env. in
Sweden, (15) March 1980 pp. 1-7.

Sweden is heavily dependent on imported oil,
but favourable conditions exist for the
exploitation of wind energy, which can serve to
displace costly oil imports. Strong winds are
found in many regions suitable for driving
turbines to produce heat and/or electric power.
Estimates of wind energy contribution to the
national energy economy are in the range of
5-30 twh/yr. by 2000 and research programmes
to develop wind energy conversion systems for
large and small applications have begun.

1.3.3(13) NEW AND RENEWABLE ENERGY RESOURCES IN
SWEDEN.
Swedish Ministry of Foreign Affairs/Ministry of
Industry Report February 1981 38 pp.

The technical and economic feasibility of
developing various renewable resources are
examined. Hydroelectric, solar, wind, bioconversion, and other resources are surveyed.

1.3.3(14) RENEWABLE ENERGY SOURCES IN SWEDEN.
Swedish Energy R & D Commission Report 37 1981
108 pp.

Three categories of renewable energy resources
are defined. 1. Sources that can be exploited
now with proven technology; direct combustion
of forest energy, peat resources, and agricultural residues. 2. Sources for which it is
possible to estimate the length of the development period and assess their future energy
contribution, and wind and solar energy fall
into this category. 3. Energy sources for
which the development potential can be assessed
only after a research effort the length of
which cannot be predicted; includes energy
plantations, solar cells for electricity
production, wave energy, and geothermal energy.

1.3.3(15) SUMMARY OF THE NATIONAL REPORT
SUBMITTED BY DENMARK.
Presented at U.N. New & Renewable Sources of
Energy Conference, Nairobi 10-21 August 1981
2 pp.

Denmark has created a national system of
energy conservation and development of cogeneration for heat and power. Financial
support from the Government is provided for the
investigation of alternative energy sources,
and wind power, solar energy and biomass are
detailed.

1.3.3(16) SUMMARY OF THE NATIONAL REPORT
SUBMITTED BY SWEDEN.
Paper presented at U.N. New & Renewable Sources
of Energy Conference, Nairobi 10-21 August 1981
1 pp.

Hydropower, wood energy, forest residues,
peat, solar energy and wind power are all
expected to play some role in energy production
by 1990. Economic incentives have been
introduced by the Government to spur growth of
these technologies, and several governmental

and non-governmental agencies are funding
research in various areas of renewable
resources, including wind power.

1.3.3(17) POTENTIAL FOR A DANISH POWER SYSTEM
USING WIND ENERGY GENERATORS, SOLAR CELLS AND
STORAGE.
Blegaa, S. and Christiansen, G. (Tech. Univ.
of Denmark, Lyngby) Intl. J. Ambient Energy,
2 (4) October 1981 pp. 223-232.

Based on the solar and wind conditions in
Denmark during the years 1959 to 1972 the
energy saving potential of a solar/wind power
system is at a maximum when the production
capacity is divided between solar cells and
wind energy generators in the ratio 40 : 60.
The energy saving produced using storage
systems of different capacities is calculated
for a solar/wind power system with a production
capacity of 1.25 times the annual load.
(9 refs.)

1.3.3(18) ENERGY IN SWEDEN.
Dampier, B., Summerton, J. and Hinrichsen, D.
Ambio, 10 (5) 1981 pp. 216-229.

Presents the Swedish Government's plan to
cut oil consumption up to 50% by 1991 - without severely curtailing economic growth or
overall energy consumption. Potential uses of
renewable energy sources including solar, wind,
peat, biomass, and forest wastes are outlined.
Coal and nuclear power will continue to supply
a significant amount of Sweden's energy.

1.3.3(19) THE WIND ENERGY PROGRAMME IN SWEDEN.
Engstrom, S. (National Swedish Board for
Energy Source Development) In proc. 5th
Biennial Wind Energy Conference & Workshop,
Washington, D.C., USA 5-7 October 1981
(I.E. Vas (ed.)) V. 2, Palo Alto, USA, Solar
Energy Research Institute 1981 Session 2
pp. 29-36. (Report No. SERI/CP-635-1340
CONF-811043)

Looks at wind energy conversion system
prototypes being constructed. Evaluations of
these are planned for 1982-84 when one is to
be placed at Maglarp and the other at Nasudden,
using the horizontal axis wind turbine design.

1.3.3(20) THE WIND ENERGY PROGRAMME IN NORWAY.
Nitteberg, J. (Energy Technol. Institute,
Norway) In proc. 5th Biennial Wind Energy
Conference & Workshop, Washington,
D.C., USA 5-7 October 1981 (I.E. Vas (ed.))
V. 2, Palo Alto, USA, Solar Energy Research
Institute 1981 Session 2 pp. 15-28.
(Report No. SERI/CP-635-1340 CONF-811043)

A discussion of the prospects for wind
energy production which concludes that the
future is so promising that the proposed
research and development programme should
continue with experimental and demonstrational
projects for large wind energy conversion
systems.

1.3.3(21) USERS EXPERIENCES IN DENMARK:
DEVELOPMENTS, ACHIEVEMENTS AND EXPERIENCE OF
THE DANISH ACTIVITIES IN WIND ENERGY
UTILISATION, 1974-1981.
Pedersen, B.M. (Tech. Univ. of Denmark, Lyngby)
In Lecture Series on Wind Energy Conversion
Devices, Rhode Saitn Genese, Belgium, Von
Karman Inst. Fluid Dyn. 1981 44 pp. (Lecture
Series 1981-88)

1.3.3(22) THE WIND PROGRAMME IN DENMARK.
Pederson, B.M. (Tech. Univ. of Denmark,Lyngby)
In proc. 5th Biennial Wind Energy Conference
& Workshop, Washington, D.C., USA
5-7 October 1981 (I.E. Vas (ed.)) V. 2, Palo
Alto, USA, Solar Energy Research Institute 1981
Session 2 pp. 55-69. (Report No. SERI/CP-
635-1340 CONF-811043)
 Surveys the wind energy R & D activities in
Denmark over the past five years. The experiences gained from operating two 630 kW demonstration wind turbines (the NIBE wind turbines)
and development of small scale wind turbines
(5-100 kW) in particular the activities at the
test station for small wind turbines are
described. An outline of future plans and
prospects for incorporating wind energy as
part of the Danish energy supply system are
discussed.

1.3.3(23) RENEWABLE ENERGY SOURCES IN SWEDEN.
Summerton, J. and Tyden, T.(National Swedish
Board for Energy Source Development; Swedish
Energy R & D Commission) Human Env. in Sweden,
(19) September 1981 pp. 1-14.
 Awareness of the need for transition from the
current oil-based energy system in Sweden has
led to an increased demand for development of
viable alternative sources of energy within a
short time. Current energy policies emphasise
the need to develop domestic, renewable energy
forms. Swedish R & D programmes are aimed at
the commercialisation of solar heating,
forestry waste, peat, fuel crop, biogas and
wind energy.

1.3.3(24) WIND ENERGY POTENTIAL IN SWEDEN:
THE IMPORTANCE OF NON-TECHNICAL FACTORS.
Carlinan I. (Lund University, Sweden) In proc.
4th Intl. Symposium on Wind Energy Systems
Stockholm, Sweden, 21-24 September 1982 V. 2
Cranfield, U.K., BHRA Fluid Engng. 1982
Session N, Paper N1 pp. 335-348.
 Such non-technical factors include public
opinion and land planning conflicts, and the
author discusses attitudes towards wind
power, and their effects upon the siting of
wind power systems with respect to potential
energy supply. Environmental issues are
important.

1.3.3(25) THE NORWEGIAN ENERGY PROGRAMME.
Johansen, O.S. In 4th Intl. Symposium on Wind
Energy Systems, Stockholm, Sweden 21-24 Sept-
ember 1982. V. 1 Cranfield, U.K., BHRA Fluid
Engng. 1982 Session B, Paper B2 pp. 111-117.

Describes the different projects included in
the Norwegian energy programme. A general
study of wind conditions in Norway was carried
out and a study of wind conditions in selected
areas started for which some results are given.
Problems with measuring instruments and systems
are described. VINDKOM, an advising committee
on wind power, recommends a prototype station
of 1-3 mw to be built.

1.3.3(26) SWEDEN PUTS WINDMILLS TO THE TEST.
Williams, E. Financial Times, 14 October 1982
pp. 11-12.
 Report on Sweden's intention to begin testing
wind power devices as part of a national plan
to reduce dependence on imported oil during a
phaseout of nuclear power. Two horizontal axis
design generators will begin operating in
Gotland and Maglarp. Studies show that wind-
generated electricity could provide as much
as 5% of the energy Sweden consumes.

1.3.3(27) ALTERNATIVE ENERGY IN SWEDEN.
Financial Times, 25 May 1983 pp. 6.
 Sweden has decided to halt all construction
of nuclear power plants and will phase out
existing plants, the nation must therefore
look at alternative forms of generation,
including wind energy, heat pumps, district
heating, bioconversion, and solar energy.

1.3.3(28) TEST PLANT FOR WIND POWER KALKUGNEN
1977-1980. SUMMARY OF OPERATION RESULTS.
Forsgren, M. and Back, T. Stockholm, Sweden,
Naemnden foer Energiproduktionpforskning
May 1983 54 pp. (NE-1983-15)
 The Swedish wind energy programme included
a test plant at Kalkugnen, district of Uppland.
The wind power plant was erected in 1977 and
the following measurements were carried out:
mechanical tests, stress analysis, electrical,
acoustical meteorological measurements and
television reception disturbance. The diameter of the turbine was 18 m, blades were
made of aliminium and later of glass fibre
reinforced plastics. The height of the tower
was 23 m.

1.3.3(29) WINDS IN SWEDEN. EVALUATION OF
WIND MEASUREMENT ON HIGH MASTS AND SURVEYING
OF WIND ENERGY IN SOUTHERN SWEDEN.
Kvick, T. and others Stockholm, Sweden,
Naemnden foer Energiproduktions. Report
No. NE-1983-16 1983 121 pp. (In Swedish)
(Dept. of Energy Report No. DE 84 770226)
 Vertical profiles of wind and temperature
have been recorded at seven stations, and
weekly wind speed frequency distributions
have been recorded at the other seven stations.

1.3.3(30) DANISH WIND ENERGY PROGRAMME.
Rasmussen, B. Intl. J. Solar Energy 2 (6)
July 1984 pp. 441-467.
 Initiatives have been taken on small wind-
mills to large wind turbines, from R & D in
materials and computer models and to the

problems of incorporating wind power into the public grid. Wind energy activities were supported and partly initiated by the Danish Government. A governmental and electric utility R & D programme on large-scale wind energy conversion systems (WECS) was formulated and implemented in 1977. At the same time private industry embarked on the development of small-scale wind energy conversion systems (SWECS) for the private user.

1.3.3(31) HELGOLAND TO GET M.A.N. WIND TURBINE.
Windirections 4 (1) July 1984 pp. 14-16.

Describes a new energy supply for Helgoland, 90 km off the North Sea coast of Germany. It consists of two diesel generator sets driving electrical generators and heat pumps, five small diesel generators, two boilers and a 1.2 mW wind turbine. Lists the specifications for this M.A.N. constructed 60 m diameter machine. The wind turbine will operate in parallel to the diesel engines as a fuel saver.

1.3.4 OTHER COUNTRIES

1.3.4(1) UTILISATION OF WIND ENERGY.
Massart, G. (CNEXO, Brest, France)
Onde ELectr., 55 (4) April 1975 pp. 225-230.
(In French)

Article sums up factors, meteorological as well as technological relating to wind energy utilisation, then surveys attempts to use wind energy on a more or less large scale.
(11 refs.)

1.3.4(2) EVALUATION OF WIND ENERGY CONVERSION SYSTEMS FOR ISRAEL.
Rudman, P.S.(Tech-Isr. of Technol. Haifa)
J. Assoc. Eng. Archit. Isr., 34 (4-5)
April-May 1975 pp. 78-59.

Argues that wind energy conversion (WEC) has not been competitive with fossil fuels with the promise of essentially unlimited and cheap nuclear energy. The modern steam or gas turbine with a high speed, aerodynamically profiled rotor does not leave much room for improvement in efficiency. While all of the machines of up to 200 kW have been operated successfully for years without breakdown, all of the mW range machines have broken down disastrously. Efficient, trouble free, 200 kW wind turbines can be built with the present state-of-the-art to last 20 years. (25 refs.)

1.3.4(3) WIND ENERGY POTENTIAL IN THE NORTHERN PART OF GERMANY.
Tetzlaff, G. and Beyer, R. (Univ. Hannover, Germany) in Proc. of the Conference and Workshop on Wind Energy Characteristics and Wind Energy Siting, Portland, Oregon 19-21 June 1979. Richland, Washington, Pacific Northwest Lab. 1979 pp. 253-259. (Report No. Pacific Northwest Lab. Tech. Report PNL 3214).

The German wind energy programme centres around a major wind turbine to be built in Germany. It is designed to be introduced as a power plant in the main power grid and is in particular meant to be integrated in the network of the different existing power plants. The general meteorological conditions encountered in Germany are characterised by the topography, as far as the wind is concerned. (6 refs.)

1.3.4(4) WIND ENERGY STATUS REPORT (SEMINAR)
(Bundesminist. fuer Forsch. und Tecknol., Bonn, Germany) Statusber. Windenerg. (Seminar), Hamburg, Germany 24-26 June 1980. Dusseldorf, Germany, VDI 1980 403 pp. (In German).

24 papers dealing with the design and construction of large wind power plants, their economic efficiency and their possible contribution to overall power generation, from the West German point of view. Also covers meteorological investigations of wind power potential in West Germany and research and development of small wind energy converters.

1.3.4(5) POSSIBILITIES OF USING WIND POWER IN HUNGARY.
Blaho, M. (Budapest Tec-. Univ.) Period.
Polytech.-Mech. Engng., 25 (1) 1980
pp. 57-64.

Discusses general aspects of utilising wind power including wind speed measurements and available wind energy in Hungary. Includes data on costs.

1.3.4(6) SOLAR ENERGY AND WIND ENERGY - CONTRIBUTION TO COVERING FUTURE ENERGY DEMAND.
Feustel, J.E. and Stoy, B. Brennst-Warme-Kraft 32 (9) September 1980 pp. 360-366. (In German)

An overview is given for the direct possibilities of using solar collectors and their most important developments in the lower, medium and high temperature range are described, as well as development activities in the field of the photo-electric transformation of light into electric current, with the help of solar cells. Applications of wind as a further form of indirect solar energy utilisation are discussed and the then present international state of wind energy convertors, future possibilities as well as limits of the energy utilisation.

1.3.4(7) ON THE UTILISATION OF WIND ENERGY RESOURCES.
Sidorov, V.I. and others. Power Eng. (New York) (3) 1980 pp. 61-68.

Presents a brief historical report on the development of Soviet wind power engineering. Resources of the Soviet Union in the three zones with the highest wind energy potential are evaluated. Design proposals for a 5 MW wind generator and a multi-rotor wind generator up to 40 MW are examined, with a proposal for their use as a basis for creation of a system with capacity up to 1000 MW as the first stage of utilisation of wind energy resources for electric power. (11 refs.)

1.3.4(8) ASPECTS OF THE EUROPEAN ENERGY DEBATE: EXAMPLE FEDERAL REPUBLIC OF GERMANY.
Smidt, D. (Kernforschungszentrum Karlsruhe GmbH, W. Germany) Paper presented at ANS/European Nuclear Society Thermal Reactor Safety Conf., Knoxville 7-11 April 1980 V. 1 pp. 26-34.

Briefly reviews the differences between France, W. Germany and the U.K. with respect to nuclear energy policy. Contributing factors include the degree of dependence on foreign oil, competitive pressure, and public perceptions of nuclear power. W. German government policy places great emphasis on energy conservation, development of coal resources, and development of renewable energy resources, such as solar, wind and geothermal in response to influential public opposition.

1.3.4(9) WIND ENERGY.
Goethals, R. (Poitiers Univ., France) Recherche, 11 (109) March 1980 pp. 262-271. (In French)

One hundred or so large wind generators of several megawatt capacity would make their kWh fed to the grid, competitive in price with the thermal power station kWh. In France, two thousand such generators would meet 10% 1980 power demand.

1.3.4(10) IT'S NOT SO QUIXOTIC.
Petromin Asia, December 1981 pp. 47, 51.

Discusses wind energy potential in Russia and considers one power plant in particular which has a capacity of 40,000 kW and will operate in any wind velocity. Instead of one, it has 8 rotors which divide the load equally. Energy is stored when the wind is strong and hydrogen is generated from excess energy by water electrolysis, and in windless weather a thermal generator provides electricity using this fuel.

1.3.4(11) NATIONAL PAPER OF AUSTRIA.
Paper presented at U.N. New & Renewable Sources of Energy Conference, Nairobi 10-21 August 1981 47 pp.

Hydropower supplies Austria with 70% of its electricity, but the nation imports the remainder of its energy needs in the form of oil. Research in solar energy, wind energy, geothermal energy, and other renewable energy technologies is underway. Recommendations for national and international energy policies are included.

1.3.4(12) NATIONAL REPORT SUBMITTED BY IRELAND.
Paper presented at U.N. New & Renewable Sources of Energy Conference, Nairobi 10-21 August 1981 35 pp.

Ireland produces some natural gas and is exploring for offshore oil, but its main indigenous resources are peat and hydropower. Both sources are detailed, and wave energy, wind power, biomass, tidal energy, and solar energy are also examined. Constraints on these technologies include limited information and lack of funding and trained personnel.

1.3.4(13) NATIONAL REPORT SUBMITTED BY ITALY.
Paper presented at U.N. New & Renewable Sources of Energy Conference, Nairobi, 10-21 August 1981 17 pp.

Discusses Italy's national energy conservation policy developed to reduce energy consumption. Various technologies in solar energy, solar cells, biomass, wind energy, and geothermal power are discussed. Energy consumption patterns of different economic sectors are presented and compared, and legislative action concerning renewable energy sources is reviewed.

1.3.4(14) NATIONAL REPORT SUBMITTED BY PORTUGAL.
Paper presented at U.N. New & Renewable Sources of Energy Conference, Nairobi 10-21 August 1981 16 pp.

75% of Portugal's energy consumption is dependent upon imported fuel sources. Potential energy resources and technologies are being explored, particularly solar, wind, and water sources.

1.3.4(15) NATIONAL REPORT SUBMITTED BY THE FEDERAL REPUBLIC OF GERMANY.
Paper presented at U.N. New & Renewable Sources of Energy Conference, Nairobi 10-21 August 1981 52 pp.

W. Germany is currently exploring and utilising several forms of new and renewable sources of energy, but because of its topography, geography, climate, and highly industrialised infrastructure, these energy sources will not provide much of its energy needs. Biomass wind energy, geothermal energy oil shale, and solar energy each provides 1 - 4% of its energy requirements.

1.3.4(16) NATIONAL REPORT SUBMITTED BY THE GERMAN DEMOCRATIC REPUBLIC.
Paper presented at U.N. New & Renewable Sources of Energy Conference, Nairobi 10-21 August 1981 19 pp.

East Germany has almost no oil or gas resources, but major coal reserves allow it to meet much of its energy needs. Renewable energy sources are currently being researched, but they will provide only 1% of the primary energy needs of the nation by 2000. Geothermal, hydropower, solar energy, wind energy, and other renewable resources are described.

1.3.4(17) NATIONAL REPORT SUBMITTED BY THE UNION OF SOVIET SOCIALIST REPUBLICS.
Paper presented at U.N. New & Renewable Sources of Energy Conference, Nairobi 10-21 August 1981 62 pp.

The USSR contains about 40% of the world's known fuel reserves, but only 16% of these are currently being utilised. Nuclear energy is seen as a viable alternative to oil and gas energy, but various types of renewable energy sources including solar energy, wind power, geothermal energy, and biomass are being developed in conjunction with nuclear power. Details the current status of these technologies, and the role of the USSR in international energy R & D is studied.

1.3.4(18) SUMMARY OF NATIONAL REPORT SUBMITTED BY FRANCE.
Paper presented at U.N. New & Renewable Sources of Energy Conference, Nairobi 10-21 August 1981 4 pp.
Provides energy projections for France for 1990 and 2000 to indicate the need for development of renewable energy sources. Hydropower, biomass, solar, wind, ocean, geothermal, and oil shale resource technologies are all detailed.

1.3.4(19) SUMMARY OF NATIONAL REPORT SUBMITTED BY ROMANIA.
Paper presented at U.N. New & Renewable Sources of Energy Conference, Nairobi 10-21 August 1981 6 pp.
Several programmes are underway to develop alternative energy sources. Direct solar heatir hydropower plants, biogas, and wind energy are detailed, and constraints on the development of these technologies are identified.

1.3.4(20) WIND ENERGY : AN ASSESSMENT OF THE TECHNICAL AND ECONOMIC POTENTIAL.
Jarass, L. and others. Berlin, Fed. Rep. Germany, Springer-Verlag 1981 220 pp.
Discusses the possible position of wind power as part of the future energy supply of West Germany covering wind conditions; determinants of wind power utilisation; conversion of wind power into electrical energy; a simulation model for the integration of wind power into the National Grid; fuel saving through the use of wind power plants; displacement of conventional power plant capacity by wind power plants (capacity credit) with summaries on evaluation of fuel saving and break-even-costs of investment and maintenance for wind power plants. (A summary of the full report is available in : Wind Energy, 5 (3) 1981 pp. 154-161.)

1.3.4(21) THE WIND PROGRAMME IN IRELAND.
Kinsella, E.M. (Eire Dept. Energy) In proc. 5th Biennial Wind Energy Conference & Workshop (WWV), (Washington, D.C., USA 5-7 October 1981) (I.E. Vas (ed.)) V. 2, Palo Alto, USA, Solar Energy Research Institute 1981 Session 2 pp. 37-41. (Rept. No. SERI/CR-635-1340)
A National Wind Energy Programme was launched in 1980 with the objectives of assessing the resources available, demonstrating the practicality and economic feasibility of windpower, taking the first steps in introducing windpower for utility generation, and developing Irish industry where appropriate. In 1981 the Department erected six different WECS at different locations and the Electricity Supply Board (ESB) the only electrical utility in the country, erected four WECS ranging from 120 to 10 kW, also at different locations, and all grid linked. Resource assessment including the

analysis of historic data and a developing programme of new data collection will lead to the production of a source book for designers, users and wind energy planners.

1.3.4(22) THE SIMPLE ALTERNATIVES SUNSHINE, WIND AND BIO-HEAT.
Muller, P. Austria Today, 1981 pp. 17-20.
Surveys joint Maltese-Austrian research programmes involving bioconversion, solar energy, and wind energy. Experimental wind energy converters are discussed.

1.3.4(23) THE WIND PROGRAMME IN THE FEDERAL REPUBLIC OF GERMANY AND CURRENT WIND ENERGY PROJECTS.
Windheim, R. In proc. 5th Biennial Wind Energy Conference & Workshop (WWV), (Washington, D.C., USA 5-7 October 1981) (I.E. Vas (ed.)) V. 2, Palo Alto, USA, Solar Energy Research Inst. 1981 Session 2 pp. 71-85.
The programme supports two projects: adaptation of small WECS for special applications and; electricity generation by large scale WECS for the grid, especially the GROWIAN programme.

1.3.4(24) SEMINAR AND STATUS REPORT 'WIND ENERGY'.
Bundesministerium fuer Forschung und Technologie, Bonn-Bad Godesburg (Germany, F.R.) Kernforschungsanlage Juelich G.m.b.H. (Germany, F.R.) Projekleitung Energieforschung.
Report No. CONF-8210136-, 1982 542 pp. (In German). (Seminar and Status Report on Wind Energy, Juelich, F.R. Germany 11 October 1982.)
Reports about the current projects of the Federal Government which are subsidised by the Federal Ministry for Research and Technology. The 29 lectures report on meteorological studies, fundamental research, pilot projects and profitability calculations.

1.3.4(25) THE WIND POWER PROGRAMME IN GERMANY - PRESENT STATUS.
Hau, E. and Windheim, R. In proc. of 4th International Symposium on Wind Energy Systems, Stockholm, Sweden 21-24 September 1982 V. 1, Cranfield, U.K. BHRA Fluid Engng., 1982 Session B, Paper B4 pp. 131-139.
After several years of research and development activities the German wind power programme entered into a new phase in 1982. Some large wind energy converters have started on their test programme. The progress in the development of small wind energy converters has reached the first application projects. This overview shows the status of the most important projects and research activities.

1.3.4(26) THE NORTH SEA ISLAND OF PELLWORM : WIND INSTEAD OF COAL.
Lienert, E. Konstr. Elem. Methoden, 19 (12) December 1982 pp. 66.
Describes the development and research programme on wind energy plants of the

German Federal Ministry of Research and
Technology which commenced work on Pellworm
Island six years ago. Now a single large
turbine produces 17 - 20 million kWh per year,
via its steel and plastic rotor blades, of
about 100 m. diameter and almost 5 m. root
depth. The plant, Growian I, starts to
generate at wind force 4 and reaches maximum
power, 3 MW, at a wind speed of 12 m/s. Design
and performance data, and general economic
factors are presented.

1.3.4(27) OUTLINE OF AUSTRIA'S WIND ENERGY
PROGRAMME.
Szeless, A. (Oesterr. Elektr. AG) In proc. 4th
BWEA Wind Energy Conference, Cranfield U.K.
24-26 March 1982 (P.J. Musgrove (ed.))
Cranfield, U.K. BHRA Fluid Engng. 1982
pp. 29-33.

1.3.4(28) SPAIN'S WIND ENERGY PROGRAMME TAKES
OFF.
Enrich, J.L.C. Windirections, 3 (1) July 1983
pp. 10-11.
  Describes a study of wind power prospects in
Spain which includes details of a 100 kW proto-
type UNESA-INI turbine with 18 m. diameter rotor,
and fibreglass blades. The five small machines
under development include a 22 kW prototype,
PEUI turbine.

1.3.4(29) AN ANALYSIS OF WIND POWER POTENTIAL
IN GREECE.
Lalas, D.P. and Theoharatos, G. (Univ. of
Athens, Greece) Solar Energy, 30 (6) 1983
pp. 497-505.
  Available wind data for the Aegean Sea are
analysed to determine the area's potential for
wind energy development. Wind energy density
is highest in this region and the local
economics of wind power make such utilisation
advantageous. Mean wind speed and annual
wind energy estimates are provided.
(24 refs.)

1.3.4(30) WIND ENERGY EVALUATION FOR THE
EUROPEAN COMMUNITIES.
Musgrove, P. (Commission of the European
Communities, Luxembourg) Final Report,
Report No. EUR-8996-EN, ISBN-92-825-42637
1983 150 pp. (PB85-100774/XAB)
  System integration studies, in Europe and
the USA, have shown that utility grid systems
can accept 10% to 20%, and possibly more, of
their total electricity needs from wind
energy systems without the need for energy
storage. Wind turbines are now becoming
available at prices in the range $1-2 per
watt. As a result the commercial applications of smaller wind turbines are
already starting to emerge, in locations
where the wind regime is good and government encouragement is provided. 10 multi-
megawatt machines have already been
constructed and several more will be completed
in the next three years.

1.3.4(31) WIND ENERGY IN PORTUGAL.
Da Fonseca, E. In Alternative Energy Sources
III, V. 4 - Indirect Solar/Geothermal Energy.
Proc. 3rd Miami Intl. Conf. on Alternative
Energy Sources, Miami Beach, USA 15-17 December
1980 T. Nejat Veziroglu (ed.) pp. 171-174.
Washington, USA, Hemisphere Publishing Corp.
1983.
  Interest in wind generated electricity is
revived and some prototypes of wind energy
converters of advanced design made. An unique
automatic variable pitch system has been
developed.

1.3.4(32) BELGIUM'S WIND ENERGY TEST SITE.
Windirections 4 (1) July 1984 pp. 6-7.
  The test site for Belgium's wind energy is
located near the northern part of Brussels.
The specifications for the three turbines are,
horizontal axis - Windmaster 50 kW 22, and two
vertical axis - STRO-V10 and STRO-V6. The
test site is 80 m above sea level and houses
a 24 m high mast with weather vane and
anemometers.

1.3.4(33) ITALIAN WIND ENERGY ACTIVITIES.
Gaudiosi, G. In Wind Energy Conversion, proc.
5th BWEA Wind Energy Conf., Reading, U.K.
23-25 March 1983. P. Musgrove (ed.) 1984
pp. 30-33. Cambridge, Cambridge University
Press 1984 375 pp. (Isbn. 0-521-26250-X)
  Lists the organisations involved in wind
power research and development in Italy, and
describes the range of their activities.
Outlines the work in progress on wind resources
and wind turbine siting evaluation, development of small wind turbines and organisation
of a National Test Station.

1.3.4(34) WIND ENERGY M.A.N. SYSTEMS AND
THEIR EFFICIENCY.
Siemer, J. Maschinenfabrik Augsburg-Nuernberg
A.G., Munich, F.R. Germany. Report No.
N84-34046/2/XAB 1984 12 pp. (In German)
Presented at Rahmen der Fachschau Energie Auf
der Hannover Messe 1984.
  Political and economic factors are analysed.
The Aeroman and WKA 100 (Growiar) systems are
presented. In Germany, 2% to 20% of annual
electricity consumption could be produced by
wind energy.

1.3.4(35) THE WIND ENERGY PROGRAMMES OF THE
FRENCH AGENCY FOR THE MANAGEMENT OF ENERGY (AFME).
Bremont, M. In Wind Energy Conversion, proc. 5th
BWEA Wind Energy Conf., Reading, U.K.
23-25 March 1983. P. Musgrove (ed.) pp. 19-22.
Cambridge, Cambridge University Press 1984
375 pp. (Isbn. 0-521-26250-X)
  Describes the aims of the wind power development programme in France. Manufacturers of small
wind turbines will be helped by support for
tests at the LANNION National Wind Test Site.
Medium power wind turbines are planned. Lists
the organisations involved in the programme.

1.3.4(36) WIND ENERGY PROJECTS IN AUSTRIA.
Szeless, A. In Wind Energy Conversion, proc.
5th BWEA Wind Energy Conf., Reading, U.K.
23-25 March 1983. P. Musgrove (ed.) pp. 23-29.
Cambridge, Cambridge University Press 1984
375 pp. (Isbn. 0-521-26250-X)

Present projects include:- 10 kW horizontal
axis wind turbine (Voest-Alpine Co.); 1 kW wind
power system for a buoy; 30 kW machine for an
alpine refuge. Describes some of the work of
the Leobersdorf experimental station for testing
alternative energy systems and suggests that
future applications of wind power in Austria may
focus on agricultural applications.

1.3.4(37) WINDPOWER IN CAP VERDE.
Windirections 4 (1) July 1984 pp. 22-24.

The projects and organisations involved in
using the wind resources are listed. Two
Vestas 55 kVA wind turbines from Denmark have
been bought and small scale mechanical wind
pumps for farms have been introduced. Other
projects have included wind powered ice
production, wind powered reverse osmosis
desalination and provision of instrumentation
to obtain reliable climatic data relating to
renewable energy resources.

1.4 Africa

1.4(1) ENERGY RESOURCES IN KENYA AND THEIR
ENVIRONMENTAL IMPACTS.
Marquand, C.J. and Githinji, P.M. (Univ. of
Nairobi, Kenya) Paper presented at U.N. Env.
Programme/Et Al Energy & Environment in East
Africa Conference, Nairobi 7-10 May 1979
pp. 159-178.

Kenya's energy resource base includes oil,
coal, electricity, wood, charcoal, geothermal
energy, small-scale hydropower, biofuels, solar
energy, wind energy. Environmental impacts
associated with the development and exploitation of these resources are identified.

1.4(2) REFLECTIONS ON ENERGY WITH SOME
REFERENCE TO TANZANIA.
Nkoma, J.S. and Asman, S.J. (Univ. of Tanzania,
Tanzania) Paper presented at U.N. Env.
Programme/Et Al Energy & Environment in East
Africa Conference, Nairobi 7-10 May 1979
pp. 267-281.

Tanzania's energy resources are analysed
to determine if future energy demand can be
satisfied with indigenous supplies of coal,
oil, natural gas, firewood, charcoal, hydropower, wind, solar radiation, biofuels, and
geothermal energy. Policy issues for energy
research and development and enery utilisation plans are identified. (12 refs.)

1.4(3) RURAL ENERGY DEMAND, PERSPECTIVE AND
PROSPECTS.
Olindo Perez, M. (East African Wildlife
Society, Nairobi) Paper presented at U.N.

Env. Programme/Et Al Energy & Environment in
East Africa Conference, Nairobi 7-10 May 1979
pp. 296-306.

In Africa's developing nations, such as
Kenya, a majority of the population lives in
rural areas and is almost entirely dependent
on firewood. To meet minimal future environmental requirements, a deliberate movement
away from this traditional source must be
achieved. Wind, geothermal, nuclear, and
other energy supply technologies can be
developed.

1.4(4) RURAL ENERGY NEEDS AND ALTERNATIVE
SOURCES.
Muchiri, G. (Univ. of Nairobi, Kenya) Paper
presented at U.N. Env. Programme/Et Al Energy
& Environment in East Africa Conference, Nairobi
7-10 May 1979 pp. 232-250.

Surveys energy requirements of rural areas in
developing nations. Currently, most are met by
animal and human energy, and by firewood.
Possible alternative sources of energy for
these areas are identified, including biomass,
biogas-slurry technology, alcohol fuels, wind
energy, solar power generation, and small-
scale hydroelectric power. (16 refs.)

1.4(5) WIND ENERGY IN SENEGAL: GEOGRAPHICAL
REPARTITION.
Gourieres, D. (Dakar Univ.) In proc. 3rd Inter-
national Symposium on Wind Energy Systems,
Lyngby, Denmark 26-29 August 1980. Cranfie
U.K. BHRA Fluid Engng. 1980 Paper X2
pp. 571-579.

Discusses wind energy in Senegal, concentrating on wind velocity and direction,
including maximum wind speeds, wind speed
frequencies, and the variations of speed with
altitude. The problem of calm spells is highlighted because the period must be covered by
storage when the smaller wind driven plants are
used autonomously

1.4(6) (NATIONAL REPORTS ON ENERGY RESOURCES
FROM AFRICAN COUNTRIES).
Papers presented at U.N. Env. Programme/Et Al
Energy & Environment in East Africa Conference,
Nairobi 10-21 August 1981 various pagination.

Reports from Cameroon, U.N. Economic
Commission on Africa (UNECA), Zambia, Sudan,
Nigeria, Sierra Leone and Liberia on various
forms of renewable energy resources, including
wind energy, and their application in the
various countries.

1.4(7) WIND ENERGY IN THE KINGDOM OF MOROCCO -
PAST, PRESENT AND FUTURE.
D'Aquanni, R.T. and Wung, H.T. (Chas. T. Main
Inc.) In proc. International Colloquium on Wind
Energy, Brighton, U.K. 27-28 August 1981.
Cranfield, U.K. BHRA Fluid Engng. 1981 Session 1
pp. 35-40.

Discusses wind energy conversion systems
(WECS) as an available technology to supply
small electric loads and to replace unreliable
and expensive diesel engines for water pumping.

Selected systems ranged from 10 to 40 kW's in peak capacity, used both AC and DC generators, and employed battery and water storage.

1.4(8) LEAVING NO STONE UNTURNED.
Okwanyo, J.H. (Kenya Minister for Energy)
Paper presented at OPEC Energy & Development Options for Global Strategies Symposium, Vienna November 1981 pp. 104-108.
Examines policy measures implemented in Kenya to reduce dependence on imported oil and to attain specified levels of energy self-sufficiency. Indigenous energy resources being exploited include hydropower, geothermal energy, wood fuel, biomass, solar energy, and wind energy. Sectoral energy conservation programmes and incentives are also discussed.

1.5 Middle East

1.5(1) DRAFT REPORT ON WIND ENERGY IN THE ARAB WORLD.
U.N. Economic Commission for Western Asia Report, August 1980 72 pp.
Examines the potential of wind energy for arab countries. Experts predict that harnessing wind energy will be of great importance to desert and rural communities distant from utilities, which have been using diesel engines and human and animal power for agriculture. Present demand and future needs of energy are charted and results indicate that wind energy holds a promising future in these communities. (32 refs.)

1.5(2) REPORT SUBMITTED BY THE ECONOMIC COMMISSION FOR WESTERN ASIA, LEBANON AND EGYPT.
Paper presented at U.N. New & Renewable Sources of Energy Conference, Nairobi 10-21 August 1981 various pagination.
Gives detailed breakdowns of energy aspects affecting various arab countries, focusing on availability of renewable energy including wind energy. Policies, activities, economics, constraints, and environmental and social implications of energy-related issues are discussed. National and regional recommendations are outlined.

1.5(3) WIND ENERGY UTILISATION POSSIBILITIES IN TURKEY.
Oney, S. and others. (Tech. Univ., Istanbul, Turkey) In Alternative Energy Sources 2, proc. of 2nd Miami International Conference V 4, Indirect Solar Energy, Miami Beach, Florida 10-31 December 1979. Hemisphere Publ. Corp., Washington, D.C. 1981. Distributed outside USA by McGraw-Hill pp. 1635-1656.
Presents an overview of the wind energy utilisation problems and possibilities of wind energy utilisation in Turkey are discussed. (5 refs,)

1.5(4) SOLAR AND OTHER ALTERNATIVE ENERGY IN THE MIDDLE EAST.
Perera, J. Economist Intelligence Unit Report 108, September 1981 92 pp.
Reviews the need for solar and other alternative forms of energy for Middle East nations. A country-by-country analysis of alternative energy potential is provided. Important factors affecting the adoption of many alternative energy forms are regional cooperation, technology transfer, and international aid. Includes wind energy research and development.

1.5(5) THE HASHEMITE KINGDOM OF JORDAN.
OAPEC B, 8 (5) May 1982 pp. 8-12.
Presents an overview of Jordan's energy requirements. The country depends entirely on crude oil imports from Saudi Arabia. Electricity demand has soared because of economic projects including housing, services, and rural electrification. Jordan's 5 year plan for 1981-85 includes development of the following domestic energy sources - hydropower, solar and wind energy, and geothermal energy.

1.6 Far East

1.6(1) (NATIONAL REPORTS SUBMITTED BY FAR EASTERN COUNTRIES).
Papers presented at U.N. New & Renewable Sources of Energy Conference, Nairobi 10-21 August 1981 various pagination.
Reports from Afghanistan, Bangladesh, Indonesia, Japan, Pakistan, Thailand, China and the Economic and Social Commission for Asia and the Pacific (U.N.). All cover various forms of renewable energy resources, including wind energy.

1.6(2) WIND ENERGY UTILISATION IN CHINA.
He, D. and Wang, H. (Chinese Aerodyn. Research & Development Centre (CARDC)) In proc. 5th Biennial Wind Energy Conference & Workshop, Washington, D.C., USA 5-7 October 1981 I.E. Vas (ed.) V. 2, Palo Alto, USA, Solar Energy Research Institute, 1981 Session 2 pp. 145-148. Report No. SERI/CP-635-1340 CONF-811043.
Brief review of the status and future of the wind power programme in China.

1.6(3) POTENTIAL OF WIND ENERGY IN TAMIL NADU STATE - A CASE STUDY.
Jagadeesh, A., Varshaneya, N.C. and Chand, I. (Cent. Building Research Institute, Roorkee, India) Energy Management (New Delhi), 5 (4) October-December 1981 pp. 293-299.
Report of a study of the available data at 13 meteorological stations in Tamil Nadu State showing that wind velocities are high in Kodaikanal, Nagapattinam, Pamban and

Tiruchirapalli, while moderate in Cuddalore, Palayamkottai and Madras. The wind energy that can be generated in the state is more than the electricity consumed annually. It is suggested that manufacture of windmills of type TOOL-ORP and "Anila-l" on a large scale will provide energy to lift water for irrigation and drinking in rural areas. (8 refs.)

1.6(4) THE WIND PROGRAMME IN A TYPHOON ENVIRONMENT.
Tsao, Y.S. (National Taiwan Univ.) In proc. 5th Biennial Wind Energy Conference & Workshop, Washington, D.C., USA 5-7 October 1981, I.E. Vas (ed.) V. 1, Palo Alto, USA, Solar Energy Research Institute, 1981 Session 1B pp. 297-303.
Gives details of Taiwan's wind energy programme: prospecting, especially in coastal and offshore areas; installation of experimental machines; evaluation of potential wind energy in Taiwan, and the effects of typhoons on the safety, cost and operation of machines.

1.6(5) PROGRESS AND PROJECTION OF INDONESIAN WIND ENERGY PROGRAMME.
Djojodihardjo, H. (Indonesian National Institute Aeronaut. & Space) In papers presented at 4th International Symposium on Wind Energy Systems, Stockholm, Sweden 21-24 September 1982. V. 1, Cranfield, U.K. BHRA Fluid Engng. 1982 Session A, Paper A5 pp. 47-66.
Review of research and development effort in wind energy in Indonesia started in 1979. Objectives set for the first 5 years were assessing wind energy potential, development of small to medium wind pumpers for applications in rural areas, development of small wind energy systems for electricity generation and assessment of possible application of larger scale wind energy systems. Existing data from meteorological stations indicates that the monthly average velocities vary between 3 to 5 m/sec. Several prototypes ranging from the Savonius rotor to cambered-blade multivane fan windmills have been developed, which are presently undergoing field tests. Darrieus type VAWT and propeller type HAWT prototypes have been built and are currently undergoing system tests, and a pilot plant utilising 10 kW wind generator is being installed and was expected to be operational in 1983.

1.6(6) PROSPECTS FOR WIND ENERGY UTILISATION IN MAHARASHTRA STATE, INDIA.
Jagadeesh, A., Varshaneya, N.C. and Chand, I. (Univ. of Roorkee, India) Natural Resources Forum, (1) January 1982 pp. 93-101.
Large-scale wind power use is possible at several sites in India, but the relatively high cost of locally designed windmills and lack of expertise in manufacturing windmills to meet local wind conditions and pumping requirements has inhibited wider use of windmills. Total wind power that could be generated in Maharashtra State equals 21.25 109 kW-H and using escarpments 74.75 109 kW-H which could be

employed in irrigation and for drinking water purposes in rural areas. In areas of the State where wind velocities are fairly good, efforts to exploit this energy should be made. (14 refs.)

1.6(7) SEASONAL CONSTRAINTS IN MANAGEMENT OF ALTERNATIVE SOURCES OF ENERGY IN RURAL AREAS.
Malhotra Kulbir, S. and Nahar, N.M. (Central Arid Zone Research Institute, India) Intl. J. Energy Research, 6 (3) July-September 1982 pp. 283-292.
Formulates an energy plan incorporating energy potential and requirements for a typical village of India's arid zone. A proposal for management of alternative energy sources - biogas, wind power, and solar energy - is included. (16 refs.)

1.6(8) WIND - A SOURCE OF ENERGY.
Sharma, M.P. and Gomkale, S.D. (Central Salt & Marine Chemicals Research Institute, Bhavnagar, India) Energy Manage. (New Delhi), 6 (4) October-December 1982 pp. 257-268.
Reviews the developments in the field of wind energy and attention is also paid to some theoretical aspects and discussions of various components used in wind energy are included. It is concluded that because of some constraints the use of wind energy even for pumping is still not attractive in developing countries like India, owing to lack of awareness of its potential, high capital investment and non-availability of experts for repair and maintenance. More attention should be given to the development of a simple wind machine (whether used for pumping or for power generation) which can be manufactured at low cost by the developing countries. (17 refs.)

1.6(9) POWER GENERATION IN 2000 A.D.
Sunavala, P.D. (Indian Institute Technology, Bombay) URJA, 12 (5) November 1982 pp. 293-296.
Gives electricity supply and demand projections to 2000 and identifies available options for satisfying future demands. Contributions of domestic and imported coal, oil, and natural gas to electric power production are estimated and the potential of renewable sources of energy for power generation is assessed, including geothermal energy, solar energy, wind energy, and biomass. (5 refs.)

1.6(10) PROSPECTS OF WIND ENERGY UTILISATION IN INDIA.
Singhal, O.P. (Central Institute of Agricultural Engineering, India) URJA, 13 (4) April 1983 pp. 219-224.
Examines the prospects of wind energy utilisation in India. Wind energy is desirable because it is inexhaustible, free, and clean. Drawbacks of wind energy include high initial costs, windmills generate only part of the power requirement for irrigation water pumping during peak period of the Rabi season, and windmills cannot operate as an

independent source of energy because of the
need for electrical storage systems for
constant supply. An estimation of power
availability from windmills in India, and
windmills for the generation of electricity
is discussed. (10 refs.)

1.6(11) A REVIEW ON THE WIND ENERGY DEVELOP-MENT IN CHINA.

Changhai, D. and Dexin, H. (Chinese Aerodynamics
Society, China) In American Wind Energy Assoc.
Wind Energy Expo. Natl. Conf., San Francisco
17-19 October 1983 pp. 418-446.

The wind resource potential of the nation is
about 10 billion kW. Horizontal axis wind
turbines fabricated from helicopter rotor blades
are currently used for pumping and electrical
applications. Further research will be
directed towards construction of small-scale
wind energy conversion systems for rural
electrification. (9 refs.)

1.6(12) EVALUATION OF THE WIND DATA TO HARNESS WIND ENERGY.

Jagadeesh, A. (Dept. of Phys., Roorkee Univ.,
India) Indian J. Power & River Val. Dev.,
34 (2) February 1984 pp. 46-49, 45.

Detailed wind analysis of the continuous wind
speed recordings at Veeraval is undertaken and
the results presented. The hours in a month,
year, during which winds are favourable are
indicated for possible wind energy utilisation.

1.6(13) WIND ENERGY POTENTIAL IN PAKISTAN.

Nagrial, M.H. (Dept. of Electrical Engng.,
University of Garyounis, Benghazi, Libya)
Intl. J. Ambient Energy (GB) 5 (2) April 1984
pp. 97-100.

Mean wind speed data for 17 locations has
been obtained from meteorological measurements
over a period of 30 years. Rayleigh distribution has been employed to predict the wind
power potential. It has been found that the
southern and south-western parts of Pakistan
have exploitable wind power potential.

1.7 Australia/New Zealand

1.7(1) WIND ENERGY RESEARCH AND DEVELOPMENT AT LINCOLN.

Chilcott, R.W. (Univ. College of Agriculture,
Canterbury, New Zealand) Wind Eng., 3 (3) 1979
pp. 187-196.

Describes wind energy research and development work at Lincoln College, Canterbury,
New Zealand, including wind energy surveying
and resource assessment, wind environment
amelioration and wind power utilisation.
The Wind Energy Resource Survey of New Zealand
is providing a clearer picture of complex
local wind flows and the survey results are

used in long-term energy planning for the
future. (14 refs.)

1.7(2) THE EFFECT OF CHANGING ENERGY PATTERNS ON EMPLOYMENT.

Roberts, M. (New Zealand Dept. Labour) In proc.
University of Auckland/Et Al 4th New Zealand
Energy Conference, Auckland May 1979 2 (4)
pp. 47-59.

Looks at future energy options open to
New Zealand and their impacts on employment.
Resrouces and manpower implications are assessed
for the development of oil, gas, coal,
geothermal electricity, solar energy, wind
energy and biomass. Concludes that skill
requirements for many of these resources are
high, and the different power industries must
continue to train and attract staff to remain
viable. (7 refs.)

1.7(3) ENERGY PROSPECTS FOR COUNTRY TOWNS.

Roby, K.R. (Murdoch Univ., Australia) In proc.
of CSIRO Energy & Agriculture Symposium,
Australia 14-18 October 1979 pp 185-200.
(Commonwealth Scientific and Industrial
Research Organisation)

Outlines energy conservation strategies for
rural towns in Australia and advocates a scheme
based on the careful use of fossil fuels,
rigorous energy conservation, and a planned
shift to renewable energy resources including
wind energy. Research required for its
implementation is considered - energy analysis,
assessment of conservation and renewable
resource potentials, and the development of
coherent integrated schemes. (25 refs.)

1.7(4) NATIONAL REPORT SUBMITTED BY AUSTRALIA.

Paper presented at U.N. New & Renewable Sources
of Energy Conference, Nairobi 10-21 August 1981
30 pp.

Assesses Australia's energy consumption and
needs by sector, and lists government organisations concerned with the development of
renewable energy sources. Solar energy, oil
shale, biomass, wind power, and hydropower are
discussed with constraints on the development
of these technologies determined and national
policy for development of renewable energy
detailed.

1.7(5) NATIONAL REPORT SUBMITTED BY NEW ZEALAND.

Paper presented at U.N. New & Renewable Sources
of Energy Conference, Nairobi 10-21 August 1981
8 pp.

Extensive research on biomass, hydropower,
geothermal energy, solar energy, wind energy,
and ocean energy is presented. Data on primary
energy consumption is provided.

1.7(6) WIND POWER BETTER THAN WE THOUGHT.

Bell, A. ECOS, (27) February 1981 pp. 21-28.

Although the concept of an array of wind
generators feeding renewable pollution-free
into electricity grids is attractive, wind is

a resource characterised by intermittency, making associated conversion systems unreliable. The wind power potential in Australia is surveyed and although the wind varies considerably in the span of a day or year in most regions, the hour-to-hour and year-to-year variations are relatively small. Wind energy is particularly suitable for development in isolated and rural communities.

1.7(7) WIND POWER PROJECTS IN WESTERN AUSTRALIA.
Crawford, T.S. (Western Australia State Energy Commission, Perth) In proc. International Colloquium on Wind Energy, Brighton, U.K. 27-28 August 1981. Cranfield, U.K. BHRA Fluid Engng. 1981 Session 1 pp. 29-34.
Gives brief descriptions of electricity supply systems in Western Australia and the State Energy Commission's Project RAPSI (Remote Area Power Supply Investigation). Three operating wind energy converters ranging in size from 5kW to 50kW are reported on and costs of these units and future prospects for wind energy discussed.

1.7(8) SOLAR AND WIND ENERGY MONITORING IN THE NORTHERN TERRITORY.
Edwards, P.R. (Northern Territory Dept. Mines & Energy, Australia) In proc. of Australian Inst. Engineers/Et Al Solar Energy for the Outback Conference, Australia 28-30 September 1981 pp. 25/1-25/12.
The Northern Territory Dept. of Mines & Energy initiated a programme to establish a solar and wind monitoring network. This will be supplemented by site-specific monitoring stations at the location of various solar and wind installations. Techniques for obtaining insolation and wind resource data are explained.

1.7(9) PROSPECTS FOR SMALL SCALE WIND ELECTRIC CONVERSION SYSTEMS IN REMOTE AREAS OF AUSTRALIA.
Langworthy, A.P., Inall, E.K. and Bandopadhayay, P.C. (Australia Dept. National Development & Energy) In proc. of Australian Inst. Engineers/Et Al Solar Energy for the Outback Conference, Australia 28-30 September 1981 pp. 11/1-11/7.
In remote parts of Australia wind energy conversion systems are often viable altern- atives to the use of diesel systems for electric power. Further research and development is required to assure reliable, economic evaluation of such systems. Recent research in computer simulation, wind energy analysis, and optimal system design is surveyed.

1.7(10) WIND ENERGY UTILISATION IN NEW ZEALAND.
Wood, J.R. and Chasteau, V.A.L. (New Zealand Energy Research and Development Committee, Auckland) Report No. NZERDC-67 November 1981 45 pp.
Covers aspects of utilisation of wind energy for supplying electricity to the national grid. However, large-scale utilisation of wind energy for electricity generation in New Zealand appeared to be remote, nevertheless, New Zealand should maintain an active interest in wind power. Duration curves for system load and wind power are presented and the degree of correlation between windy spells and power demand is dealt with. It is concluded that wind occurrence closely approximates random behaviour in an annual cycle, as far as its effect on the grid is concerned. Short-term hydro storage run in conjunction with wind plant is briefly discussed and aspects of wind-turbine technology are presented.

1.8 South America

1.8(1) EVALUATION AND USE OF THE WIND ENERGY IN NORTH EAST BRAZIL.
Albino de Souza, A. (Inst. de Atividades Espaciais, Sao Jose dos Campos, Brazil) In proc. 11th Intersociety Energy Conversion Engng. Conference, State Line, Nevada 12-17 September 1976. New York, N.Y. AIChE 1976 V 2 SAE Paper 769300 pp. 1741-1745.
Experience with a 20 to 30 kW unit for winds of 7 to 10 metres per second is reported. Planning of wind plant is to be competitive with hydropower. The variability of the wind in Brazil is assessed and the first utilisation of the wind energy resources prototype is shown and the economics of its performance analysed. (5 refs.)

1.8(2) WIND ENERGY IN ARGENTINA.
Bastianon, R.A. (Research & Development Naval Service) In proc. 3rd International Symposium on Wind Energy Systems, Lyngby, Denmark 26-29 August 1980. Cranfield, U.K. BHRA Fluid Engng. 1980 Paper X1 pp. 565-570.
Discusses the high levels of wind speed in Southern Argentina together with the poss- ibility of an advantageous use of wind energy. The Science and Technology Secretariat of State has included wind energy in the Non-Conventional Energy National Programme. The high powered turbines connected to the national grid can provide energy at a very low cost due to the existing high winds while numerous low powered wind generators, between 10 and 20 kW, can be installed to provide energy to small communities in isolated areas, who have no hope of being connected to the distribution grid.

1.8(3) NATIONAL REPORT SUBMITTED BY BRAZIL.
Paper presented at U.N. New & Renewable Sources of Energy Conference, Nairobi 10-21 August 1981 32 pp.
Brazil plans to obtain energy self-sufficiency by the end of the century through exploitation of untapped coal and oil reserves, and development of new and renewable energy sources, including solar energy, ocean energy, and wind energy.

1.8(4) NATIONAL REPORT SUBMITTED BY GUYANA.
Paper presented at U.N. New & Renewable Sources
of Energy Conference, Nairobi 10-21 August 1981
9 pp.
  Oil imports provide about 85% of Guyana's
energy needs, and bagasse, firewood, and
charcoal make up the remainder. Developments
in hydropower, wood waste, biogas, bagasse,
solar energy, wind energy and peat are
investigated, and energy conservation
programmes have been initiated.

1.8(5) REPORT SUBMITTED BY THE ECONOMIC
COMMISSION FOR LATIN AMERICA.
Paper presented at U.N. New & Renewable Sources
of Energy Conference, Nairobi 10-21 August 1981
26 pp.
  The U.N. Economic Commission for Latin
America (CEPAL) investigates energy planning
and policies, renewable energy sources, and
energy conservation. It also collects and
publishes energy statistics. CEPAL's
activities for the next 3 years are outlined
and brief summaries of activities in each
renewable energy area included.

1.8(6) SUMMARY OF THE NATIONAL REPORT
SUBMITTED BY CUBA.
Paper presented at U.N. New & Renewable Sources
of Energy Conference, Nairobi 10-21 August 1981
4 pp.
  Cuba relies heavily on imported oil to meet
its energy needs. Efforts to reduce oil
consumption, replace oil with renewable energy
sources, and expand the national electric
power system are currently underway. Solar
energy, biomass, wind power and hydropower are
the 4 main renewable energy sources under
development.

1.8(7) GERMAN-ARGENTINIAN R AND D - PROJECT
FOR THE UTILISATION OF WIND ENERGY.
Axenath, K.H. and others (Bundesministerium
fuer Forschung und Technologie, Bonn-Bad
Godesberg, F.R. Germany) Report No. BMFT-
FB-T-84-080 May 1984 90 pp. (In German)
(Dept. of Energy Report No. DE 84 751934)

2 Resource Assessment and Site Selection

2.1 Methodology

2.1(1) LOW COST INSTRUMENTATION FOR A WIND ENERGY SURVEY.
Cherry, N.J., Edwards, P.J. and Roxburgh, A.J. (Lincoln College, Canterbury, New Zealand) In proc. 22nd International Instrum. Symposium, San Diego, California 25-27 May 1976 pp. 109-115.
Describes a low cost system for remote sensing and automatic recording of mean wind speed over averaging periods of an hour or submultiples of an hour down to a minute or less recorded electronically on standard reel-to-reel or cassette tape decks or recorders. Averaging periods of 1, 2, 5, 15, 30 or 60 minutes may be selected. Density of data on the tape is increased by recording the data in blocks of 100 numbers out of a memory unit. (5 refs.)

2.1(2) WIND ENERGY RESOURCES AND METHODS FOR THEIR ESTIMATION.
Anapol'skaya, L.E. and Gandin, L.S. (Main Geophys. Obs., USSR) Soviet Meteorol. Hydrol., 7 1978 pp. 6-11.
Reports a proposed procedure for evaluation of climatic wind energy resources, potential and that used by wind engines of a given type, from data on the statistical frequency distribution of wind velocities and the parameters of the engine. It shows that the annual variation of air density should be neglected in estimation of annual average wind energy resources. Climatic annual average wind energy resources are computed for several localities with strong winds and it is demonstrated that, for regions of the USSR with the richest wind energy resources, engine-control specifications sharply limit opportunities for their utilisation.

2.1(3) WIND ENERGY ASSESSMENT.
Hardy, D.M. and Walton, J.J. (Solar Energy Research Institute, Golden, Colorado, USA) In International Conference on Alternative Energy Sources, International Compend., Miami Beach, Florida, USA 5-7 December 1977. Hemisphere Publ. Corp., Washington, D.C. 1978 V. 4 pp. 1835-1863.
Covers major aspects of wind energy assessment. Spatial and temporal scales of wind variations and meteorological data requirements are discussed and meteorological conditions in several coastal and mountainous areas with high wind energy densities summarised. Methods used to map wind energy variations over the island of Oahu, Hawaii, are used as an example of wind energy assessment in a complex terrain. Wind resources by means of coordinated field data collection and numerical modelling efforts are currently being documented and a numerical windfield model is being used to calculate three-dimensional velocities over the island. Field measurement and numerical model results obtained for Oahu, Hawaii, are given to illustrate how this general methodology might be applied to other mountainous or hilly regions. (17 refs.)

2.1(4) ESTIMATION OF WIND CHARACTERISTICS AT POTENTIAL WIND ENERGY CONVERSION SITES. VOLUME 1. TECHNICAL REPORT.
Howard, S.M. and Chen, P.C. (Battelle Pacific NorthWest Labs., Richland, WA) Report Nos: PNL-3199-Vol. 1/2 March 1978 166 pp. and 363 pp. (U.S. Dept. Energy DE 82005481; DE 82005487.)
To achieve the DOE goal of placing the most economically viable wind energy conversion system at a site, it becomes necessary to determine the accuracy of the climatology development technique, and to determine the extent of its applicability to sites of differing geographic settings and data station locations. To achieve this goal, the Climatology Development Methodology was applied to eight candidate wind energy sites of differing site geographic settings and data station locations. Performance was evaluated by examining the accuracy of the hub-height wind climatologies generated for each type of site and station geographic setting. The Climatology Development Methodology developed in this study is presented and the techniques of the methodology discussed. Data for the C.D.M. programmes are given in Volume 2.

2.1(5) METHOD FOR ESTIMATING WIND CHARACTERISTICS AT POTENTIAL WIND-ENERGY-CONVERSION SITES.
Koch, R.C. and Pickering K.E. (Battelle Pacific NorthWest Labs., Richland, WA) Geomet. Inc., Gaithersburg, MD. Washington, D.C. Department of Energy. Report no: PNL-3197 March 1978 388 pp. DE 82008675.

Presents and discusses a method of estimating wind characteristics at potential wind energy conversion sites for which no meteorological data are available. The method extrapolates standard National Weather Service wind observations to a defined system hub height using a power-law relationship. Several observations are averaged using an inverse distance squared weighing factor to obtain a site estimate. The site estimate is then adjusted for local topographical effects. Long histories of hourly observations (15 to 20 years) are used to generate statistical characteristics of the estimated site winds, including mean wind speed, frequency distributions of wind speed (including seasonal and diurnal variations), wind rises and run durations of selected classes of wind speeds.

2.1(6) WIND ENERGY CONVERSION SYSTEMS.
South Dakota Office of Energy Policy
Report no: NP-2902754 1978 41 pp. DE 82902754.

Describes the steps for determining the practicality of a wind system. Includes evaluating potential legal and environmental problems, evaluating energy requirements, evaluating the wind resource at the proposed location, evaluating the application, selecting the system and components, evaluating the cost of the system, and evaluating alternatives in buying, installing, and owning a wind system. Procedures for choosing the best available site for a wind machine and for estimating the pertinent wind characteristics and performance and price information on commercially available electrical wind machines with outputs less than 100 kW is presented. Bibliography.

2.1(7) WIND CHARACTERISTICS FOR FIELD TESTING OF WIND ENERGY CONVERSION SYSTEMS.
Akins, R.E. (Sandia Labs., Albuquerque, NM)
Department of Energy Report no: SAND-78-1563
November 1979 55 pp.

Techniques are presented to determine placement of instrumentation to be used in measurement of wind characteristics for field testing of Wind Energy Conversion Systems (WECS). Potential errors in the measurement of a reference wind velocity as a result of physical separation between an anemometer and a WECS and interference between the WECS and the reference anemometer are outlined. Methods of correcting errors caused by these sources are developed.

2.1(8) WIND SPEED MEASUREMENTS FOR WIND TURBINE TESTING.
Bridson, D.W. (Exeter Univ.) In proc. 1st BWEA Wind Energy Workshop, Cranfield, U.K. April 1979. London, U.K. Multi-Science Publ. Co. Ltd. 1979 pp. 208-217.

There is a requirement for wind speed and direction measurement in the vicinity of wind turbines during open air testing. Methods used should produce information appropriate to this task rather than for meteorological or wind tunnel purposes. System response rates should be related to real wind values and turbine

characteristics. Arrays of wind measuring positions both upstream of turbines and in their wakes may be required. The requirement for wind speed measuring equipment is discussed and the evaluation of possible choices recently undertaken at the University of Exeter reported. The performance of equipment constructed by the Wind Energy Unit is described. (Author abstract.)

2.1(9) METEOROLOGICAL AND TOPOGRAPHICAL INDICATORS OF WIND ENERGY FOR REGIONAL ASSESSMENTS.
Elliott, D.L. (Battelle Pacific NorthWest Labs., Richland. Washington) Pac. NorthWest Lab. Tech. Report PNL 3214, proc. of Conference and Workshop on Wind Energy Characteristics and Wind Energy Siting, Portland, Oregon 19-21 June 1979. Richland, Washington, Pac. NorthWest Lab. 1979 pp. 273-283.

Existing wind data provide the primary basis for assessing a region's wind energy potential, but this data must be evaluated carefully to determine the representativeness of the site and local area. Wind energy is very sensitive to variations in terrain, vegetation roughness, height above ground and instrument exposure. In analysing the wind energy in complex terrain and data-sparse areas, the author relies on the use of various indirect indicators of wind energy and an understanding of the physical processes and features that result in high winds in some areas but not in others.

2.1(10) SITE SELECTION FOR SMALL WIND ENERGY CONVERSION SYSTEMS FOR U.S. DEPARTMENT OF ENERGY FIELD EVALUATION PROGRAM.
Bailey, B.H. (State Univ. of New York at Albany, Research Foundation) New York State Energy Research and Development Authority, Albany. Report no: PB81-226862 October 1980 67 pp.

Details the site selection procedure followed to locate two qualified sites for the installation and monitoring of two commercial small wind energy conversion systems as part of the U.S. Department of Energy's Field Evaluation Programme. The aim of the evaluation programme is to gain operating experience with wind systems in actual locations and to identify the siting and operational issues involving wind energy conversion systems.

2.1(11) A PRACTICAL AND ECONOMIC METHOD FOR ESTIMATING WIND CHARACTERISTICS AT POTENTIAL WIND ENERGY CONVERSION SITES.
Bhumralkar, C.M. and others. (SRI Intl.)
Solar Energy, 25 (1) 1980 pp. 55-65.

To assess the economic viability of installing a wind energy conversion system (WECS) at a site it is necessary to know the wind characteristics at that site. It is usually impractical to measure wind at all potential sites over a suitably long period of time, and it is therefore necessary to develop a methodology that can provide accurate estimates of wind economically at potential sites from data that are already available. A three-dimensional model has been developed that incorporates the effect of underlying terrain and uses available, conventional wind information from selected nearby weather stations. This model, COMPLEX, is essentially an objective analysis computer programme that interpolates values of wind from observations at irregularly spaced stations.

2.1(12) WIND ENERGY RESOURCE SURVEY METHODOLOGY.
Cherry, N.J. (Lincoln College, Canterbury, New Zealand) J. Ind. Aerodyn. 5 (3-4) May 1980 pp. 247-280.

An essential element in a programme involving the extensive use of wind energy is the wind energy resource survey. Previous survey programmes serve as a guide for the procedure, starting with a review of the available information, survey methods and existing wind data. An observational programme is usually necessary if the resource becomes a serious option for development. A regional survey to obtain data from likely sites can be followed by detailed studies at the most promising locations. Initially the survey can be used to choose the best sites for supplying the cheapest energy, but a carefully set-out system might also provide capacity substitution, which is an important economic factor. The wind energy resource survey of New Zealand serves as the major illustration. (32 refs.)

2.1(13) ACCURACY OF MEAN WIND SPEED MEASUREMENTS IN THE WIND ENERGY CONTEXT.
Coppin, P.A.W. Statusber Windenerg. (Semin.), Hamburg, Germany 24-26 June 1980. Dusseldorf, Germany, VDI 1980 pp. 211-225.

Examines the theoretical background to anemometer errors and shows that the most commonly used type of cup anemometer for the gathering of routine wind speed data by the weather services is known to overestimate wind-speed by up to 10%. This gives a possible overestimate for the potential powe in the wind of over 30%.

2.1(14) WIND CHARACTERISTICS OVER COMPLEX TERRAIN RELATIVE TO WECS SITING.
Frost, W. and Shieh, C.F. (FWG Assocs. Inc.) In A Collection of Technical Papers from AIAA/SERI Wind Energy Conference, Boulder, USA 9-11 April 1980. New York, USA, American Inst. Aeronaut. & Astronaut. 1980 Paper no: 80-0645 pp. 185-193.

Presents a literature review of information on characteristic effects of terrain features on the wind motion and provides basic guidance in practical Wind Energy Conversion Systems siting.

2.1(15) WIND SYSTEMS SITING AND PERFORMANCE.
Justus, C.G. (Georgia Inst. of Technology) Paper presented at ISES/Et Al 1980 Annual Conference, Phoenix 2-6 June 1980 V3.2 pp. 1437-1442. (International Solar Energy Society)

The Author covers site selection, screening, and validation criteria for wind energy

conversion systems and general guidelines for high wind sites are summarised. Effects of diurnal, seasonal and interannual wind variations on system performance and siting are surveyed. Numerical modelling techniques can also be used for site selection. (36 refs.)

2.1(16) NETWORK FOR THE MEASUREMENT OF WIND ENERGY AT HIGH MASTS.
Kvick, T. and Zimmerman, T. Naemnden foer Energiproduktionsforskning, Stockholm (Sweden) Report no: NE-VIND-82-21 November 1980 25 pp. (In Swedish) DE 83750503.

Fourteen stations have been erected in order to measure wind energy at the height of 50 to 100 m. above ground level. The measurements from these stations should form the basis of the design of future wind power plants. The choice of the measurement sites is discussed and the geographic conditions around the stations are described. The recording systems cover the measurement of the temperature, the velocity and direction of winds.

2.1(17) WIND ENERGY PLANNING: DEVELOPMENT AND APPLICATION OF A SITE SELECTION METHOD FOR WIND ENERGY CONVERSION SYSTEMS (WECS).
Otawa, T. Intl. J. Energy Research, 4 (3) July/ September 1980 pp. 283-306.

The use of wind energy by means of its conversion to electricity involves a number of constraints such as economic, environmental, technical, legal, social and institutional requirements. The opportunities and constraints significant at the regional level were identified, and a systematic method was developed to select sites for large WECS by incorporating the identified factors: wind resource, proximity to load centres, proximity to tie-in-points, and exclusive land-use areas. The developed method was applied to the western Massachusetts region, and the first results of the study have been acquired.

2.1(18) METEOROLOGICAL ASPECTS AND WIND ENERGY: ASSESSING THE RESOURCE AND SELECTING THE SITES.
Pennell, W. and others. (Battelle Pacific NorthWest Lab., Richland, Washington, USA) J. Ind. Aerodyn. 5 (3-4) May 1980 pp. 223-246.

In this paper the authors look at work designed to improve understanding of the wind energy resource within the United States and the techniques and procedures that have been developed for finding wind machine sites as part of the United States wind energy programme. (15 refs.)

2.1(19) NUMERICAL WIND FIELD MODEL VALIDATION IN COMPLEX TERRAIN WITH APPLICATION TO POLLUTANT TRANSPORT.
Porch, W.M. (Calif. Univ. at Livermore) In 2nd Joint Conference on Applications of Air Pollution Meteorology, New Orleans, USA 24-27 March 1980 and 2nd Conference on Industrial Meteorology, New Orleans, USA 28 March 1980. Boston, USA American Meteorol. Society 1980 Session 9 Paper 9.R2 pp. 608-613.

Describes how remotely sensed spatially averaged winds from optical anemometers can be used to help the interaction of wind measurements made at points within a region and numerical regional wind field models. Emphasis is placed on model validation as wind field estimates for applications such as wind energy prospecting.

2.1(20) ACOUSTIC DOPPLER MEASUREMENTS FOR PROSPECTING OF WIND ENERGY.
Salomonsson, S. and Holmgren, B. (Uppsala Univ. (Sweden) Meteorologiska Institutionen) Report no: UUIM-64 1980 44 pp. DE 82900715.

Discusses results from Doppler SODAR (Sound Detecting and Ranging) measurements of horizontal and vertical winds carried out during a wind prospecting project aimed at finding the best location for a wind power station on the island of Gotland in Sweden. The Doppler system was rebuilt with two standard SODAR units, primarily designed for vertical monostatic soundings. The wind velocities derived from the Doppler shift are compared with simultaneous wind measurements from double theodolite pilot balloon trackings. The test shows quite good agreement within the lowest 100 m. of the atmosphere. A comparison is also made between local variation of the boundary-layer dynamics in thermally stable stratification obtained on a SODAR record and simultaneous Doppler measurements of horizontal wind speed and direction at the 60 m. level.

2.1(21) DOE/NASA WIND TURBINE DATA ACQUISITION SYSTEM (PART 1 - EQUIPMENT).
Strock, O.J. Washington, D.C., Department of Energy/NASA January 1980 56 pp. Report nos: EMR 827053 and NASA-CR-159779.

Large quantities of data must be collected, stored and analysed in research and development programmes on wind turbines. The hardware of the wind energy remote data acquisition system assembled by EMR data systems and used on the NASA/DOE wind energy programme is described.

2.1(22) WIND-POWER SITE-SCREENING METHODOLOGY FINAL REPORT.
Walton, J.J. and others (LLNL) NTIS Report UCRL-52938 October 1980 67 pp.

LLNL has developed and demonstrated a wind energy site-screening methodology suitable for defining the location, geographical extent, and strength of wind resources. Principal-components analytical techniques are used to classify types of regional flow fields. The authors describe the applications of the methodology in Hawaii.

2.1(23) FEDERAL WIND ENERGY PROGRAMME FOR WIND RESOURCE ASSESSMENT AND SITING: AN OVERVIEW.
Wendell, L.L. (Battelle Pac. NorthWest Lab., Richland, Washington, USA) In proc. Annual Meeting American Section International Solar Energy Society 1980, Solar Jubilee 25 Years of the Sun at Work, V3.2 Phoenix, Arizona, USA 2-6 June 1980. American Section of Inter-

national Solar Energy Society Inc., Newark, Del, USA 1980 pp. 1452-1458.

The technical progress in wind resource assessment and siting is highlighted from the work of the Pacific NorthWest Laboratory and its subcontractors. A major effort in the programme is the completion of the regional wind energy assessments covering the United States and its territories. Methods for selecting the specific locations of wind turbines have also been developed for both large and small machines.

2.1(24) METEOROLOGICAL ASPECTS OF THE UTILISATION OF WIND AS AN ENERGY SOURCE.
Geneva, Switzerland, World Meteorol. Organisation 1931 192 pp. (Tech. Note No. 175) (WMO-No. 575)

Includes chapters on: wind, economic and technical aspects of wind energy; wind energy meteorology; and sections on Wind Energy Conversion Systems (WECS) and the selection of sites.

2.1(25) REPORT SUBMITTED BY THE WORLD METEOR-OLOGICAL ORGANISATION.
Paper presented at U.N. New & Renewable Sources of Energy Conference, Nairobi 10-21 August 1981 2 pp.

The World Meteorological Organisation provides monthly and annual world maps on relative global radiation, indicating the distribution of solar energy potential throughout the world. The meteorological aspects of wind as an energy source are also discussed.

2.1(26) WORLD-WIDE WIND RESOURCE ASSESSMENT.
Cherry, N.J. and others. (Lincoln College; Pacific NorthWest Lab.) In proc. 5th Biennial Wind Energy Conference & Workshop (Washington, D.C., USA 5-7 October 1981) I.E. Vas (ed.) V. 2 Palo Alto, USA, Solar Energy Research Institute 1981 Session 3C pp. 637-648.

A world-wide wind energy resource assessment has been carried out by Pacific NorthWest Laboratory, using the methods recently developed to critically analyse all available wind data and previous assessments in order to estimate the broad-scale distribution of wind energy flux over the world.

2.1(27) FORECASTING WIND POWER OUTPUT.
Dub, W. (Regensburg Univ., Federal Republic Germany) In Implementing Agreement for Programme of Research & Development on Wind Energy Conversion Systems 1981, Meeting of Experts of Annex III and IIIa: Integration of Wind Power into National Electricity Supply Systems (Regensburg 29-30 January 1981) Julich, Fed. Rep. Germany, Kernforschungsanlage Julich GmbH. April 1981 Session B Paper B3 pp. 71-80.

Describes an attempt to forecast wind energy output, in order to incorporate wind power into supply systems. Time periods which allow a reoptimisation of base, intermediate and peak load plants are discussed.

2.1(28) ICING ON WIND ENERGY SYSTEMS.
Hoffer, T., Reale, T. and Elfigi, A. (Dept of
Energy, USA) Reno, USA, Nevada Univ. January
1981 74 pp. Report no: DOE/ET/23170-80/1.
   Describes the procedures for analysing the
icing data of meteorological stations within
the USA in order to determine the maximum
possible icing to be expected at specific
locations. Models using rainfall and cloud
water data are used.

2.1(29) PROSPECTING FOR WIND, WINDMILLS AND
WIND CHARACTERISTICS.
Meroney, R.N. (Colorado State Univ.)
Transportation Engineering J-ASCE PROC., 107
(4) July 1981 pp. 413-427.
   According to the Author, past experience
with power generation by windmills indicates
that the most important factor controlling
success or failure is site wind characteristics. Site selection procedures including
statistical climatology, numerical simulation,
and physical simulation in meteorological
wind tunnels are described. Laboratory
measurements of wind overspeed, streamline
patterns, and turbulence changes over
idealised topography are compared with frozen
velocity numerical models. (18 refs.)

2.1(30) METEOROLOGICAL PRECONDITIONS OF WIND
ENERGY CONVERSION.
Moeller, L. and Windheim, R. (Kernforschungsanlage Julich GmbH, Projektleitung Energieforschung) Report no: Juel-Spez-132 CONF-
8105189 November 1981 284 pp. (In German)
Seminar of research in progress on the
Energy Research and Technology Programme of
the Federal Minister for Research and Technology, Julich, Fed. Rep. Germany 18 May 1981.
   In the framework of the programme on energy
research and technologies, a seminar on wind
energy conversion projects in West Germany
was held on 18/19 May 1981. The 15 papers
dealt mostly with the problem of wind
conditions, especially in Northern Germany.

2.1(31) SITE CHARACTERISTICS FOR WIND ENERGY
CONVERSION DEVICES.
Norman, R.S. (Illinois Inst. Technol.) In
Lecture Series on Wind Energy Conversion
Devices, Rhode Saint Genese, Belgium 1-5 June
1981. Rhode Saint Genese, Belgium, Von Karman
Institute Fluid Dynamics 1981 3 pp. (Lecture
Series 1981-8)
   Includes comments on wind characteristics,
required parameters, comparison of alternative
sites, and short term measurements, utilising
anemometers prone to error.

2.1(32) A SELF-CONTAINED WEATHER STATION FOR
WIND AND SOLAR ENERGY PROSPECTING.
Anderson, R.S. (Meteorological Research Inc.)
In proc. IECEC 1982 17th Intersociety Energy
Conversion Engng. Conference, Los Angeles, USA
8-12 August 1982. V. 3, New York, USA, Inst.
Electr. & Electron. Engrs., 1982 Paper No.
829355 pp. 1443-1447.

Describes a microprocessor controlled weather
station suitable for wind energy prospecting
and site studies and for solar energy measurement incorporating suitable anemometers.

2.1(33) DESIGN AND STANDARDISATION OF METEOR-
OLOGICAL MEASUREMENTS FOR WIND ENERGY CONVERT-
ING SYSTEMS.
Coppin, P.A.W., Tetzlaff, G. and Roth, R.
(Hanover Univ., Fed. Rep. Germany, Inst. fuer
Meteorologie und Klimatologie) National Aero-
nautics and Space Administration, Washington,
D.C. Final Report November 1981 Report no:
BMFT-FB-T-82-168 September 1982 66 pp.
(In German)
   Investigates the systematic error for several
standard anemometers. Efficiency of wind
energy converters can only be measured if
several requirements for the topographic forms,
surface coverage, stationarity of the wind field
and turbulence intensity are met. Only then
are standard models describing wind field
properties applicable and errors do not exceed
3%. Wind speed sensors should not exhibit a
systematic error. Under simplified conditions
three anemometers are sufficient to obtain
reliable data on the overall efficiency of
almost every wind energy conversion system.

2.1(34) WIND TURBINE SITING: A SUMMARY OF THE
STATE OF THE ART.
Hiester, T.R. (Flow Ind. Inc.) In proc Workshop
on Large Horizontal-Axis Wind Turbines, Cleveland, USA 28-30 July 1981, R.W. Thresher (ed.)
Washington, D.C., USA, NASA 1982 pp. 195-213.
NASA Conference publication 2230;DOE
Report no: CONF-810752 SERI/CP-635-1273.
   Several techniques for assessing the wind
resource have been explored or developed in the
Federal Wind Energy Programme. Local topography and meteorology will determine which of
the techniques should be used in locating
potential sites. None of the techniques can
do the job alone, none are foolproof, and all
require considerable knowledge and experience
to apply correctly. Efficient siting requires
a strategy which is founded on the application
of several techniques without relying solely
on one specific field of experience, and
meteorological instrumentation errors must be
considered.

2.1(35) THE USE OF A TETHERED WIND MEASUREMENT
SYSTEM IN TVA'S WIND ENERGY DATA COLLECTION
STRATEGY.
Hunter, C.H. and Meyers, C.E. (Tennessee Valley
Authority) In proc. 1982 Wind and Solar Energy
Technology Conference, Kansas City, USA
5-7 April 1982. Columbia, Missouri-Columbia
Univ. 1982 pp. 347-351.
   Describes a tethered wind measurement system,
which provides at low cost useful information
on the vertical shear of horizontal wind
speeds which a turbine's rotor blades could
experience in a large wind energy conversion
utility.

2.1(36) DEVELOPMENT OF A SLIDE PROGRAM
DESCRIBING A SITE-SELECTION PROCESS FOR SMALL
WIND-ENERGY-CONVERSION SYSTEMS (SWECS).
FINAL TECHNICAL REPORT.
Otawa, T. (Ball State Univ.) Department of
Energy, Washington, D.C. Report no:
DOE/R5/1030-1-Final May 1982 79 pp.
DE 82017394. (See also DOE/R5/10301-1 for
1st half of project.)

This report covers the technical background of the programme development. The
objectives are to clarify a variety of
factors affecting the SWECS siting process;
to investigate in detail the effectiveness
of existing SWECS siting techniques; to
translate the factors into criteria for
site-selection; and to clearly present a
prototype site-screening procedure for
SWECS. The intent is to summarise and present
rules for SWECS siting as well as the major
findings of scientific and technical siting
studies conducted in the past. The major
issues involved in SWECS siting are: wind-
resource assessment techniques, land-use
constraints, potential hazards, potential
environmental impact, site accessibility
and proximity to load location. The method
utilises the technique of overlay mapping to
analyse different relationships among
patterns of land characteristics. Trans-
parent or semi-transparent maps depicting
the patterns are overlaid to show a pattern
of light and dark tones revealing areas of
site potentials and limitations for SWECS
placement. A case-study is conducted to
demonstrate the site-selection method developed
in this study. (A summary of the report
appears in proc. 1982 Wind and Solar Energy
Technology Conference, Kansas City 5-7 April
1982. Publ. Missouri/Columbia Univ. 1982
pp. 102-108.)

2.1(37) METEOROLOGICAL FIELD MEASUREMENTS AT
POTENTIAL AND ACTUAL WIND TURBINE SITES.
Renne, D.S. and others. (Pacific NorthWest Lab.,
U.S. Dept. Energy) In proc. 4th International
Symposium on Wind Energy Systems, Stockholm,
Sweden 21-24 September 1982. V. 1 Cranfield,
U.K. BHRA Fluid Engng. 1982 Session C Paper Cl
pp. 141-154.

An overview of experiences gained in a meteorological measurement programme conducted at a
number of locations around the United States
for the purpose of site evaluation for wind
energy utilisation is given. The paper
discusses the evolution of the measurement
programme from its inception in 1976 to the
present day, to outline some of the major
accomplishments and areas for improvement, and
to present some conclusions of research utilising data from this programme that may be useful
in future site evaluation exercises.

2.1(38) THE WIND RUSH SYNDROME: A PERSPECTIVE
ON SITING.
Sass, W.L. (Second Wind Inc.) In proc. 1982
Wind and Solar Energy Technology Conference,
Kansas City, USA 5-7 April 1982. Columbia, USA,
Missouri-Columbia Univ. 1982 pp. 215-220.

The author proposes a definition to assist
in the proper siting for wind energy conversion
systems to take account of environmental
factors, suitable wind availability and
consistency.

2.1(39) LEARNING TO FORECAST WIND AT REMOTE
SITES FOR WIND-ENERGY APPLICATIONS. FINAL
REPORT.
Notis, C. and others. (Battelle Pacific North-
West Labs., Richland, Washington) Washington,
D.C., U.S. Dept. Energy Report no: PNL-4318
January 1983 239 pp. DE 83008756.

Observed wind patterns at six selected sites
are correlated. Objectives of the analysis
are: to identify synoptic and/or mesoscale
weather patterns that are associated with
recognisable wind events at the sites; to
define a set of criteria that uniquely describes
such forecasting rules derived from the association of weather patterns and site winds; and
to attempt to separate any mesoscale effects of
local topography from the synoptic-scale
effects. One-to-one mapping of wind regimes
onto synoptic types was not found and it was
concluded that four factors should be examined
when stratifying wind regimes: synoptic situation, descriptive climatology, pressure
gradient vector, and winds above site.

2.1(40) TEST APPLICATION OF A SEMI-OBJECTIVE
APPROACH TO WIND FORECASTING FOR WIND-ENERGY
APPLICATIONS.
Wegley, H.L. and Formica, W.J. (Battelle
Pacific NorthWest Labs., Richland, WA.)
Washington, D.C. U.S. Department of Energy
Report no: PNL-4403 July 1983 99 pp.
DE 83016641.

Describes the test application of the semi-
objective wind forecasting technique at three
locations. The forecasting sites are described
as well as site-specific forecasting procedures. Verification of the wind forecasts
is presented, and the observed verification
results are interpreted. Comparisons are
made between S-O wind forecasting accuracy
and that of two previous forecasting models
which used subjective wind forecasts and
model output statistics. Rhode Island,
North Dakota and California are considered in
the text.

2.1(41) MAKING USE OF WIND ENERGY: A COURSE
IN EVALUATING AND CHOOSING SITES FOR SMALL
WIND MACHINES.
Zengerle, R. (Battelle Pacific NorthWest Labs.,
Richland WA.) Washington, D.C. U.S. Dept.
Energy Report No. PNL-SA-9566 July 1983 177 pp.
DE 83014945.

A course has been developed to present an
integrated, not overly technical approach to
siting a wind machine. In the context of this
course, the term siting is used in a broad
sense to include estimating the wind resource
at a site, selecting the best wind machine
site, estimating energy production and economic
value of a wind machine. The workbook treats
various topics in greater detail, illustrating

various problem-solving techniques with sample problems. In addition, a number of practice problems are given. Complete answers are provided for each problem. Suitable as both student text and practical handbook.

2.1.1 MODELS (MATHEMATICAL AND PHYSICAL)

2.1.1(1) WIND FABRIC DIAGRAMS AND THEIR APPLICATION TO WIND ENERGY ANALYSIS.
Davis, B.L. and Ekern, M.W. (S.D. School of Mines & Technology, Rapid City) J. Appl. Meteorol. 16 (5) May 1977 pp. 522-531.

Describes how by means of the Lambert projection, wind vector data may be plotted onto a frequency map. The resulting diagram portrays the wind fabric for the data sample of a single station or for groups of stations. The true area distribution of wind vectors is thus given in great detail and allows several standard tests for homogeneity and anomaly significance. Using a swinging plate instrument the wind energy density and wind power can be calculated for any velocity-compass heading sector of the diagram desired and allows for a universal scaling of the velocity variable of the diagram. (5 refs.)

2.1.1(2) FLOW FIELD ANALYSIS.
Cliff, W.C. and Verholek, M.G. (Battelle Pac. NorthWest Lab., Richland, Washington) NASA Conference Publ. CP 2034, for a Workshop: Wind Turbine Struct. Dyn. held at NASA Lewis Research Centre, Cleveland, Ohio 15-17 November 1977 pp. 71-76. NASA, Washington, D.C. March 1978.

The average mean wind speed integrated over a disc is shown to be extremely close to the mean value of wind speed which would be measured at the centre of a disc for most geometries in which a WECS would operate. Field test results are presented which compare instantaneous records of wind speed integrated over a disc with the wind speed measured at the centre of the disc.

2.1.1(3) POTENTIAL IMPACT OF AUTOMATED WIND GUIDANCE ON WIND ENERGY CONVERSION OPERATIONS.
Carter, G.M. and Gilhousen, D.B. (NOAA, Camp Springs, Md.) In proc. of Conference and Workshop on Wind Energy Characteristics and Wind Energy Siting, Portland, Oregon 19-21 June 1979. Pacific NorthWest Lab., Richland, Washington 1979 pp. 191-205. (Pac. NorthWest Lab. Technical Report PNL-3214.)

The National Weather Service has been producing numerical-statistical forecasts of surface winds for approximately 250 locations throughout the United States. Made by the Model Output Statistics (MOS) technique, they serve primarily as guidance for aviation and public weather forecasters. This paper describes how the various MOS wind forecasts are produced and show verification results to indicate how the quality of the guidance varies with forecast projection and geographic location. The advantages and disadvantages of using the MOS

approach for future energy-related applications are discussed. (15 refs.)

2.1.1(4) NEW APPROACH FOR WIND SPEED CHARACTERISATION FOR WIND ENERGY STUDIES.
Goh, T.N. and Nathan, G.K. (Univ. of Singapore) In Sun 2, proc. of International Solar Energy Society Silver Jubilee Congress, Atlanta, Ga. May 1979 V. 3. Pergamon Press, Elmsford, N.Y. and Oxford, England 1979 pp. 2267-2271.

Describes an investigation of the characteristics of wind speeds recorded at six meteorological stations in Singapore. The study illustrates the extraction of information from wind data through the use of appropriate methods of modelling and time series analysis. (3 refs.)

2.1.1(5) MOMENTUM THEORY ANALYSIS OF UNCONVENTIONAL WIND EXTRACTION SCHEMES, PART 10.
Landahl, M.T. (Cambridge, USA, Mass. Inst. Technol.) Report nos: ASRL-TR-194-2-Pt-10 and FFA-AU-1499-Pt-10, 12 October 1979 23 pp.

A momentum theory analysis was carried out for idealised wind energy extraction devices under the assumption of uniform wake velocity. The wind energy extraction problem was analysed on the basis of some simple idealised flow models which demonstrate that the 'Betz limit' can be exceeded with the aid of some unconventional extraction schemes. (From Author's abstract.)

2.1.1(6) WIND TUNNEL MODELLING AS A PROSPECTING TOOL FOR WIND ENERGY SITE SELECTION: A FIELD ASSESSMENT.
Lindley, D. and others. (Taylor Woodrow Constr. Ltd., Colorado State Univ., Canterbury Univ.) In proc. 1st BWEA Wind Energy Workshop, Cranfield, U.K. April 1979. London, U.K. Multi-Science Publ. Co. Ltd. 1979 pp. 218-229.

Describes an attempt to evaluate the accuracy of a wind tunnel investigation of a complex terrain model. Both terraced and contoured models of the Rakaia River Gorge region of New Zealand were prepared to an undistorted geometric scale of 1:5000. On two spring days, selected for strong adiabatic down valley wind flow, three teams of investigators surveyed up to 27 sites on either side and within the river gorge. Measurements consisted of wind speed and direction at a 10 m. height. The laboratory simulation results were compared with the available field data by means of statistical correlation and scatter diagrams.

2.1.1(7) APPLICATIONS OF A NUMERICAL MODEL TO WECS SITING RELATIVE TO TWO-DIMENSIONAL TERRAIN FEATURES.
Shieh, C.F. and Frost, W. (Tennessee Univ.) In proc. 5th International Conference on Wind Engineering, Fort Collins, USA 8-14 July 1979. J.E. Cermak (ed.) V. 2 New York, USA Pergamon Press 1979 Session IX Paper 5 10 pp.

A numerical model of wind characteristics about bluff-shaped geometries is developed, which solves the two-dimensional Navier-Stokes equations of fluid motion coupled with two transport equations, for turbulence kinetic energy and for turbulence length scales, respectively. Verification of the model is demonstrated by comparison with experimental data from both wind tunnel and full-scale field measurements. It is shown to give realistic predictions of velocity distributions, turbulence properties and separation flow zones associated with flow fields over bluff objects. Finally, the influence of bluff terrain features on the siting of wind energy conversion systems, WECS, is simulated with the model, and parametric studies are reported.

2.1.1(8) RUN DURATION ANALYSIS OF SURFACE WIND SPEEDS FOR WIND ENERGY APPLICATION.
Sigl, A.B., Corotis, R.B. and Won, D.J. (South Dakota State Univ., Brookings) J. Appl. Meteorol., 18 (2) February 1979 pp. 156-166.
Hourly wind speed records are used to develop a model for the probability distribution of wind speed persistent above and below fixed reference speeds. Examination of duration histograms from 19 sites for records varying from 5 - 24 years leads to the development of a simple composite distribution. Enforcement of smooth behaviour and a parameter sensitivity analysis allow the model to be interpreted in terms of a single free parameter, which is then shown to be highly correlated to the seasonal mean wind speed at a site. (18 refs.)

2.1.1(9) APPLICATION OF POWER LAWS FOR WIND ENERGY ASSESSMENT.
Sisterson, D.L. and Hicks, B.B. (Argonne National Lab., Ill.) In proc. of Conference and Workshop on Wind Energy Characteristics and Wind Energy Siting, Portland, Oregon, 19-21 June 1979. Pacific NorthWest Lab., Richland, Washington 1979 pp. 353-361. (Pacific NorthWest Lab. Technical Report PNL 3214.)
This investigation makes use of wind data from the ANL meteorology tower in northern Illinois. For convenience, power laws have been fitted to actual wind profiles, and the frequency distribution of the power-law exponent has then been determined hourly, seasonally, and annually.

2.1.1(10) ESTIMATION OF THE PARAMETERS OF THE WEIBULL WIND SPEED DISTRIBUTION FOR WIND ENERGY UTILISATION PURPOSES.
Stevens, M.J.M. and Smulders, P.T. (Univ. of Technol., Eindhoven, Netherlands) Wind Eng., 3 (2) 1979 pp. 132-145.
Considers methods for estimating the parameters of the Weibull wind speed distribution from a given set of wind speed data. Of the five methods presented, two are selected for wind energy evaluation studies: one uses Weibull probability, the other so-called percentiles. For a comparison both methods have been applied to the data from six meteorological stations. The simple graphical method using Weibull probability is preferred to that employing percentile estimators. (16 refs.)

2.1.1(11) UTILITY AND VERIFICATION OF MATHE-MATICAL WINDFIELD MODELS FOR WIND ENERGY REGIONAL SCREENING AND SITE SELECTION.
Traci, R.M., Phillips, G.T. and Rock, K.C. (Sci. Appl. Inc., La Jolla, Calif.) In proc. of Conference and Workshop on Wind Energy Characteristics and Wind Energy Siting, Portland, Oregon 19-21 June 1979. Pacific NorthWest Lab., Richland, Washington 1979 pp. 381-391. (Pacific NorthWest Lab. Report PNL 3214) (See also similar paper by authors DOE/ET/20280-79.)
Describes a WECS siting methodology that has been developed in recent years which makes use of numerical windfield models to objectively and accurately extrapolate historical or field test data from sites within a mesoscale region of interest to other potentially windier sites throughout the region. Two windfield models have been developed for complementary use in the methodology: SIGMET, a primitive equation, terrain conformal, mesoscale meteorology model and NOABL, a simplified physics, terrain con-formal windfield model. (5 refs.)

2.1.1(12) CHANGES IN THE POTENTIAL FOR WIND ENERGY GENERATION DUE TO TERRAIN MODIFICATION OF THE BOUNDARY-LAYER FLOW.
Arnold, J.E. College Station, USA, Texas A & M Univ. 1980 158 pp. (Ph.D. Thesis)

2.1.1(13) NOTE ON THE USE OF THE INVERSE GAUSSIAN DISTRIBUTION FOR WIND ENERGY APPLIC-ATIONS.
Bardsley, W.E. (Waikato Univ.) J. Appl. Meteorol., 19 (9) September 1980 pp. 1126-1130.
The inverse Gaussian distribution is suggested as an alternative to the three-parameter Weibull distribution for the description of wind speed data with low frequencies at low speeds. A comparison of the two distributions indicates a region of strong similarity, corresponding reasonably well to three-parameter Weibull distributions which have been fitted to wind data. Maximum likelihood estimation of the inverse Gaussian parameters is much simpler than the iterative technique required for the three-parameter Weibull distribution. In addition, the inverse Gaussian distribution features the mean wind speed as a parameter, a desirable property for wind energy investigations. (From Author's abstract.)

2.1.1(14) PRACTICAL AND ECONOMIC METHOD FOR ESTIMATING WIND CHARACTERISTICS AT POTENTIAL WIND ENERGY CONVERSION SITES.
Bhumralkar, C.M. and others. (Stanford Research Inst. Int., Menlo Park, Calif.) Solar Energy, 25 (1) 1980 pp. 55-65. (For original Report see PNL-3074 October 1979 149 pp.)
Describes the development of a three-dimensional model that incorporates the effect of underlying terrain and uses available, conventional wind information from selected nearby weather stations. Called COMPLEX - it is essentially an objective analysis computer programme that interpolates values of wind from observations at irregularly spaced

stations. Required statistical wind characteristics are estimated from the synthesised
hourly winds, obtained using the COMPLEX
model; the model is used in conjunction with a
method for reducing the number of variables
while still retaining most of the information
of the original data set. This involves eigen-
vectors of the covariance matrix of the
original data set. (8 refs.)

2.1.1(15) APPLICATION OF STATISTICAL TECHNIQUES
TO WIND CHARACTERISTICS AT POTENTIAL WIND
ENERGY-CONVERSION SITES. FINAL REPORT,
1ST OCTOBER 1978-30TH SEPTEMBER 1979.
Corotis, R.B. (Northwestern Univ., Evanston,
IL. Dept. of Civil Engineering) May 1980
172 pp. DOE/ET/20283-2. (For shorter handbook
on same topic see DOE/ET/20283-3.)

The distribution for the magnitude of the
vector sum of two orthogonal horizontal wind
velocity components is often modelled by the
Rayleigh distribution, which is derived
assuming that the components are independent,
identically distributed, zero-mean, Gaussian
random variables. The probability density
function for a more realistic case where the
two components are correlated and not equal in
variance is derived and it is found that the
derived distribution is adequately modelled by
the Rayleigh distribution. A 24-hour record of
20-second average wind speed was collected to
assess the effect of sampling rate and averaging time on computed wind speed means and
variances, autocorrelation, and run duration.
An approximate procedure is developed to
simulate the time sequence of wind speed at a
single site; the procedure uses a Weibull
distribution with conditional parameters updated each hour as a function of the previously
simulated value and the autocorrelation.

2.1.1(16) WIND ENERGY: DATA ACQUISITION AND
REAL-TIME ANALYSIS.
Dastmalchi, B. (Colorado State Univ., Fort
Collins, Dept. of Mechanical Engineering)
Washington, D.C. Department of Energy. Report
no: DOE/ET/23164-1 December 1980 134 pp.

Describes a minimcomputer-based data
acquisition system that supplies a curve fit
of data in real time and is small enough to be
easily carried and installed as a portable
unit. The algorithm for data analysis was
developed and the software provided for a
Hewlett Packard HP9845T minicomputer. It was
then tested on a Darrieus wind turbine
generator located at the Colorado State University dairy farm. Measurements include: wind
speed and direction, air temperature, and ac
power for each phase of the three-phase
induction generator. A linear model is
described for the power vs. wind speed curve,
and the effect of wind variation (gustiness)
is considered. Results are then presented
and compared.

2.1.1(17) PRACTICAL METHOD FOR ESTIMATING
WIND CHARACTERISTICS AT POTENTIAL WIND-ENERGY-
CONVERSION SITES.
Endlich, R.M. and others. (Battelle Pacific

NorthWest Lab., Richland, Washington) NTIS
Report no: PNL-3808 August 1980 165 pp.

A method is developed to compute local wind
characteristics for estimating the wind energy
available at any potential site for a wind
turbine. The method uses the terrain heights
for an area surrounding the site and a series
of wind and pressure reports from the nearest
weather service stations.

2.1.1(18) TECHNIQUES FOR SPATIAL EXTRAPOLATION
OF WIND DATA.
Meeker, L.D. and others. (New Hampshire Univ.)
ASCE proc. J. Engng. Mech. Div. 106 (EM2)
April 1980 pp. 201-212.

This paper is concerned with the estimation
of wind system parameters at a possible wind
generation site, by means of a short observed
record at that site and extrapolation of a long
term record at a regional weather station.
The estimation is accomplished by assuming that
the wind system at each site is generated by a
common non-stationary process modified by site-
specific scale factors and stationary noise
processes. The procedure is exemplified using
wind records from Boston Airport and Boston
light vessel.

2.1.1(19) PROSPECTING FOR WIND ENERGY: A FIELD
ASSESSMENT OF PHYSICAL MODELLING.
Neal, D. and Stevenson, D.C. (Canterbury Univ.)
In proc. 7th Australian Conference on Hydraulics
& Fluid Mechanics, Brisbane, Australia
18-22 August 1980. Barton, Australia. Inst.
Engrs. Aust. 1980 pp. 27-30. (National Conf.
Publication no: 80/4.)

For prospecting and evaluating potential
wind energy sites without extensive and long
term meteorological measurements, physical
simulation of wind regimes found over complex
terrain offer significant advantages in terms
of time, expense and control of independent
variables. Terraced and contoured models of
the Gebbies Pass on Banks Peninsula in the
South Island of New Zealand were prepared to
an undistorted geometric scale of 1:4000 and
surveyed in an atmospheric boundary layer
wind tunnel. The laboratory simulation
results are compared with field measurements
in the region to examine the viability of the
method. Preliminary results show promising
agreement between field and wind tunnel tests.

2.1.1(20) WIND ENERGY SITING METHODOLOGY WIND-
FIELD MODEL VERIFICATION PROGRAM. II NEVADA
TEST SITE DATA SET. INTERIM REPORT 15 JUNE 1979-
15 FEBRUARY 1980.
Traci, R.M. and others. La Jolla, USA. Sci.
Appl. Inc. February 1980 141 pp. DOE/ET/20280-
80/2. (See also DOE/ET/20280-79 and PNL 3214.)

Presents results from the second part of a two
part verification programme for a mathematical-
model based Wind Energy Conversion System Siting
Methodology. The objective of the present programme is to expand the model verification by
assessing the quantitative accuracy of the
models relative to observed data for a midcont-
inental region of complex topography and

meteorology. In particular, to determine the ability of the detailed-physics SIGMET model to describe dynamic atmospheric and boundary layer effects on the windfield in a mesoscale region with complex terrain and to determine the ability of the simplified-physics NOABL model to predict windfields utilising a small amount of available data. (See 2.1.1(11))

2.1.1(21) STABILITY ANALYSIS OF ANNUAL WIND SPEED CHARACTERISTICS.
Goh, T.N. and Nathan, G.K. (Singapore National University) In proc. International Colloquium on Wind Energy, Brighton, U.K. 27-28 August 1981. Cranfield, U.K. BHRA Fluid Engng. 1981 Session 1 pp. 17-22.
Considers data requirements in site selection of wind energy conversion systems, and some methods for evaluation of stability of wind characteristics from year to year are suggested. The approach is based on stochastic analyses quantifying in the time domain the dynamic nature of wind speed data.

2.1.1(22) A PARALLEL-ANEMOMETRIC APPROACH TO WINDMILL SITING.
Halperin, D.A. and Beckman, R.A. (Aeolian Kinetics, Providence) In proc. ISES-AS/Et Al 1981 Annual Conference, Philadelphia 26-30 May 1981 V. 2 pp. 1553-1557. (International Solar Energy Society - American Section)
The authors discuss the parallel-anemometric approach to the siting of wind energy conversion systems. Measurements of speed and direction for a three-month period at both a prospective wind energy site and a site with known wind characteristics are taken and data divided into sets by direction. Correlations between the two sites are computed for each set. The Weibull distribution is applied to the correlations to extrapolate the short-term data to a long-term prediction of available power at the prospective site. (5 refs.)

2.1.1(23) THE EFFECT OF TERRAIN AND CONSTRUCTION METHOD ON THE FLOW OVER COMPLEX TERRAIN MODELS IN A SIMULATED ATMOSPHERIC BOUNDARY LAYER.
Lindley, D. and others. (Taylor Woodrow Constr. Ltd., Canterbury Univ.) In proc. Third BWEA Wind Energy Conference, Cranfield, U.K. 9-10 April 1981. Cranfield, U.K. BHRA Fluid Engng. 1981 pp. 195-211. (See also earlier paper by same authors ref. 6 in this section.)
This paper gives details of measurements over a series of model two dimensional hills placed in a simulated 1:300 rural atmospheric boundary layer to determine the effect of slope, shape and surface roughness on mean wind velocity, RMS velocity fluctuations and energy spectra for the streamwise velocity component. Results of these tests are compared where possible with existing wind tunnel data. The authors go on to describe an extension of the work on two dimensional hills in which the flow over a 1:4000 undistorted scale model of Gebbies Pass in the South Island of New Zealand is investigated in an atmospheric boundary layer wind tunnel. Three approaches to modelling complex

terrain were employed. Velocity and turbulence profiles, Reynolds stresses and spectra were measured over terraced, contoured and 'roughness added' models and compared with field measurements taken in the modelled region.

2.1.1(24) WIND SPEED SIMULATION FOR ECONOMIC EVALUATION OF WIND ENERGY CONVERSION SYSTEMS.
Ramsdell, J.V., Athey, G.F. and Ballinger, M.Y. (Battelle Pacific NorthWest Lab., Richland, Washington) Washington, D.C. U.S. Department of Energy Report no: PNL-SA-9149 CONF-810742-3 July 1981 14 pp. In U.S. National Conference on Wind Engineering Research, Seattle, WA., USA 26 July 1981. U.S. Dept. Energy Report no: DE 81030077.
A time series model has been developed for the simulation of wind speeds. It provides for the incorporation of systematic seasonal variation of the mean speed, standard deviation, and correlation of speeds and also provides for incorporation of the systematic diurnal variation of the mean speed and the standard deviation. A number of simulations have been made using model parameters derived from data collected at the Hanford Meteorology Station. Results of analyses of both sets of data, the simulated set and the real data, have been compared. Generally, the major features found in the analyses of the real data are identifiable in the corresponding analyses of the simulated data. The primary difference is in the frequency of high wind speeds.

2.1.1(25) WIND MEASUREMENT SYSTEM AND WIND TUNNEL EVALUATION OF SELECTED INSTRUMENTS.
Ramsdell, J.V. and Wetzel, J.S. Richland, USA, Battelle Pacific NorthWest Labs. May 1981 79 pp. PNL-3435.
Discusses wind measurement systems and presents the results of wind tunnel tests of seven systems carried out as part of a programme to evaluate the accuracy and reliability of instruments for use with small wind energy conversion system siting studies.

2.1.1(26) SIMULATION OF WIND-SPEED TIME SERIES FOR WIND-ENERGY CONVERSION ANALYSIS. FINAL REPORT.
Corotis, R.B. (Battelle Pacific NorthWest Labs., Richland, WA.) Washington, D.C. U.S. Department of Energy Report no: PNL-4349 June 1982 78 pp. DE 83000043.
A simple wind speed simulation model, WEISIM, is developed based on the Weibull probability distribution for wind speeds with a correction based on the lag-one auto-correlation value. The model can simulate at rates from one a second to one an hour, and wind speeds can represent short-term averages (e.g., 1-sec. averages) or longer-term averages (e.g., 1-min. or 1 hr. averages). The validity of the model is verified with PNL data for both histogram characteristics and persistence characteristics. (See 2.1.1(15))

2.1.1(27) GENERALISED CHARACTERISTICS AND
APPLICABILITY OF VARIOUS PROBABILITY DISTRIB-
UTIONS FOR WIND ENERGY APPLICATIONS.
Eskinazi, S. and Cramer, D.E. (Syracuse Univ.,
NY., USA) J. Energy, 6 (6) November-December
1982 pp. 384-392.

The authors attempt to show the presence of
certain important general behaviour characteristics of hourly wind speed variations in the
atmospheric surface layer, in spite of differences that exist in site roughness, seasons,
and thermal stability at each of the 28
different sites considered. Four different
types of probability time data of each of the
nearly 100 site-months are processed and a
new definition of "best fit" is proposed; on
this basis, comparisons and recommendations are
made. (12 refs.)

2.1.1(28) DISCRETE GUST MODEL FOR USE IN THE
DESIGN OF WIND ENERGY CONVERSION SYSTEMS.
Frost, W. and Turner, R.E. (Univ. of Tenn.,
Tullahoma, USA) J. Appl. Meteorol. 21 (6)
June 1982 pp. 770-776.

Describes the discrete gust model which
incorporates the 'number-of-crossings' theory,
i.e. an expression for the number of times per
unit time the wind exceeds a specific value is
derived as a function of mean wind speed and
the standard deviation of turbulence. A
technique for determining the cut-off frequency
which represents the upper limit of integration
used to estimate the standard deviation of wind
fluctuation accelerations is described. Utilising this definition, comparison of prediction
and experiment is carried out. The number of
times the wind speed exceeds a certain value
over a yearly period is then estimated by integrating the product of the number-of-crossings
joint probability times the Weibull distribution. A mathematical filter to isolate those
disturbances in the atmosphere of a characteristic size is developed based on vertical and
lateral coherence functions. (14 refs.)

2.1.1(29) MASS-CONSISTENT WIND FIELD MODELINGS
FOR SITING WIND ENERGY CONVERSION SYSTEMS.
Huang, C.H. and Ardis, C.V. (Toledo Univ.)
In proc. 1982 Wind and Solar Energy Technology
Conference, Kansas City, USA. 5-7 April 1982.
Columbia, USA, Missouri-Columbia Univ. 1982
pp. 413-419.

Two mass-consistent wind field models
developed to compute the wind field over complex
terrain are described: a direct method for
adjusting the wind field to compute air flow
over complex terrain in a conformal space, and
a mass-consistent, interpolated wind field model.
Both are less expensive and easier to use than
the widely-applied MATHEW-type model.

2.1.1(30) THE ESTIMATION OF WIND ENERGY
RESOURCE USING MESOSCALE MODELLING TECHNIQUES.
Johnson, R.F., Burch, S. and Newton, K.
(ERA Technol. Ltd., AERE Harwell) In proc.
Fourth BWEA Wind Energy Conference, Cranfield,
U.K. 24-26 March 1982. P.J. Musgrove (ed.)
Cranfield, U.K. BHRA Fluid Engng. 1982
pp. 189-196.

Describes the evaluation of a computer model
to predict wind flow over complex terrains
for assessment of the onshore wind energy
resource, and to identify the necessary
assumptions and methodology. The model was
applied to a 98 x 84 km region of S.W. Scotland,
and used to predict the distribution of annual
mean wind speeds, and wind speed statistics.
An algorithm was developed to provide calculations of technically extractable wind
energy resources taking into account wind
turbine output characteristics and first-
order siting restrictions.

2.1.1(31) DIRUNAL WIND SPEED PROFILES IN THE
SURFACE LAYER AND THEIR INFLUENCE ON AVAILABLE
WIND ENERGY.
Kuffel, L. In proc. 1982 Wind and Solar Energy
Technology Conference, Kansas City, USA
5-7 April 1982. Columbia, USA, Missouri-Columbia
Univ. 1982 pp. 352-358.

Suggests a simplified diurnal model of
horizontal mean wind speeds for varying
vertical heights to provide an improved
description of available wind resources in the
lower atmosphere.

2.1.1(32) MATCHING WECS SURVIVABILITY WITH
SEVERE WIND CHARACTERISTICS - A RATIONAL
APPROACH.
Liu, H. (Missouri-Columbia University) In proc.
1982 Wind and Solar Energy Technology Conf-
erence, Kansas City, USA 5-7 April 1982.
Columbia, USA, Missouri-Columbia Univ. 1982
pp. 109-113.

Proposes an approach to match the survival
wind speed of wind energy conversion systems
with the probability of the occurrence of high
winds, based on the probability concept of
severe wind.

2.1.1(33) TIME SERIES MODELS FOR HORIZONTAL
WIND.
McWilliams, B. and Sprevak, D. (Queen's Univ.
of Belfast, Dept. of Engineering Mathematics,
Northern Ireland) Wind Energy, 6 (4) 1982
pp. 219-228.

Describes a modified version of an existing
time series modeling procedure, which models
orthogonal statistically independent components
of wind velocity and subsequently provides
information on wind speed and wind direction.
An alternative model which is based on the
Weibull distribution and describes wind speed
is also discussed. The two modeling procedures
are used to generate data which is compared
with observations to determine each model's
relative strengths and weaknesses.

2.1.1(34) WIND STUDY: FIELD WORK AND RESULTS.
Peterson, J.N. (Idaho Univ.) In proc. 1982
Wind and Solar Energy Technology Conference,
Kansas City, USA 5-7 April 1982. Columbia,
USA, Missouri-Columbia Univ. 1982 pp. 368-370.

Describes a study conducted to identify
sites for anemometer installations, to operate
and maintain anemometers, and to assess the

wind energy potential from the collected data, as part of an evaluation of wind energy potential in Southern Idaho.

2.1.1(35) BAYESIAN PERSISTENCE ANALYSIS FOR WIND ENERGY.
Rao, H.G. and Corotis, R.B. (Old Dom. Univ., Norfolk, Va., USA) ASCE Proc., J. Energy Div., 108 (EY2) June 1982 pp. 116-127.

Presents a practical Bayesian formulation whereby regional and site-specific mean wind speed may be used to calibrate a probability model of persistence. Prior estimates for the principal model parameter are obtained from the ratio of wind speed level considered for persistence to site mean wind speeds using weighed regression. The Bayesian approach is in terms of specific persistence events, such as a wind speed run above a given wind speed exceeding a particular length of time.
(11 refs.)

2.1.1(36) BAYESIAN ANALYSIS OF REGIONAL WIND ENERGY POTENTIAL.
Rao, H.G. and Corotis, R.B. (Old Dom. Univ., Johns Hopkins Univ.) ASCE Proc. J. Engng. Mech. Div. 108 (EM6) December 1982 pp. 1198-1214.

Describes the application of the Bayesian statistical analysis technique to the problem of preliminary site assessment for wind energy potential, assuming a Rayleigh wind speed distribution. (See also previous reference by sane authors.)

2.1.1(37) DEVELOPMENT AND EVALUATION OF WIND FORECASTS FOR WIND-ENERGY APPLICATIONS.
Wegley, H.L. (Battelle Pacific NorthWest Labs., Richland, WA.) Washington, D.C. U.S. Dept. of Energy. Pacific NorthWest Lab. Report no: PNL-SA-10351 CONF-820683-1 March 1982 9 pp. (Conference on Weather Forecasting and Analysis, Seattle, WA., USA 28 June 1982.)

Describes the development of 24-hour wind speed forecasts for wind energy applications using both MOS and a semi-objective approach. Results of detailed analysis of MOS forecasts and preliminary verification of the semi-objective forecasts are given. The forecasts are compared and their value to electrical utilities for power generation estimated.

2.1.1(38) WEIBULL DISTRIBUTION FUNCTION AND WIND POWER STATISTICS.
Bowden, G.J. and others. (Univ. of New South Wales, Sch. of Physics, Kensington, NSW, Australia) Wind Eng., 7 (2) 1983 pp. 85-98.

Uses the properties of the Weibull distribution function to show that there are some simple relationships between the Weibull parameters and the wind speed moments. These results are then used to examine the relationship between seasonal, yearly and multi-yearly Weibull parameters for a given site. Comments are also made concerning the choice of distribution function to model wind speed statistics.

2.1.1(39) VECTOR STATISTICS OF HOURLY WIND ADVECTION DATA FOR ENERGY TRANSPORT APPLICATIONS.
Eskinazi, S. and others. (Syracuse Univ., Mechanical & Aerospace Engineering, Syracuse, NY, USA) J. Energy 7 (3) May-June 1983 pp. 264-271.

Use is made of the joint-probability density function in representing vector properties of wind advection. It is shown that these functions, when plotted, reveal qualitative and quantitative wind transport characteristics dependent on surface site topography as well as diurnal heating and cooling cycles. The vector "wind print" developed at the site for a given season has many permanent features and, therefore, is amenable to reliable engineering and scientific assessment of wind transported physical quantities.

2.1.1(40) WIND-LOAD CORRELATION AND ESTIMATES OF THE CAPACITY CREDIT OF WIND POWER: AN EMPIRICAL INVESTIGATION.
Martin, B. and Carlin, J. (Australian National Univ., Dept. of Mathematics, Canberra, ACT, Australia) Wind Eng., 7 (2) 1983 pp. 79-84.

Calculations made with several years of data from Western Australia indicate that empirical estimates of capacity credit may vary greatly with small changes in the joint distribution of wind and load. It is shown that capacity credit estimates are sensitive to the availability of wind power at a few periods of high load, and a simple summary measure is described which gives an indication of the strength of wind-load association in relation to capacity credit.

2.1.1(41) INFLUENCE OF MODEL SCALE ON A WIND-TUNNEL SIMULATION OF COMPLEX TERRAIN.
Neal, D. (Vickers Dawson, Crayford, Kent, U.K.) J. Wind Eng. Ind. Aerodyn. 12 (2) July 1983 pp. 125-143.

The results of wind-tunnel simulations involving scale models of Gebbies Pass (New Zealand) are presented and compared with full scale measurements. The results from the two models are compared to investigate the possible effects of wind-tunnel blockage and to test the validity of segmenting a model for analysis. Full-scale wind-structure measurements were made to height of 20 m. at selected sites in the region. The full-scale and model results are compared for velocity and turbulence-intensity profiles, energy spectra and length scales. In addition, velocity-profile data collected by means of Tala kites are presented and compared with the results obtained in the wind-tunnel simulations. In all cases there is a high degree of compatibility, which suggests that wind-tunnel modeling as for Gebbies Pass complex terrain is a viable tool in the evaluation of potential wind-energy sites. (See also earlier reference by Neal - 2.1.1(19))

2.1.1(42) DIFFICULTIES IN USING POWER LAWS FOR WIND ENERGY ASSESSMENT.
Sisterson, D.L. and others. (Argonne National Lab., Atmospheric Physics Section, Argonne, Ill., USA) Solar Energy, 31 (2) 1983 pp. 201-204.

A 1/7 power law is often used to estimate atmospheric wind profiles. The use of this relationship frequently results in serious underestimates of wind speeds aloft at night. Several years of wind speed data from a 45 m. tower and from use of a Doppler acoustic sounder indicate that low-level wind speed maxima seem to form on 50 percent of summer nights and 15 percent of winter nights in northeastern Illinois. Even with 24 hr. averages, the extrapolation of 6 m. wind speeds to those at 45 m. calculated with the 1/7 power law expression are found to be approximately 15 percent too small, corresponding to a 40 percent underestimate of wind power potential.

2.1.1(43) THE INFLUENCE OF TURBULENCE ON THE DYNAMIC BEHAVIOUR OF LARGE WTG'S.
Swansborough, R.H. (ERA Technol. Ltd.) In Wind Energy Conversion, proc. 5th BWEA Wind Energy Conference, Reading, U.K. 23-25 March 1983. P. Musgrove (ed.) pp. 129-136. Cambridge, Cambridge University Press 1984 375 pp. (Isbn. 0-521-26250-X)
Describes some of the work for NMI being undertaken by ERA on the dependence of wind turbine generator design of meteorological parameters particularly wind turbulence characteristics. The synthetic wind model used for time domain analysis is a time-stepping model based on the power spectrum of wind fluctuations. Outlines the modules used for the turbine generation and transmission systems.

2.1.1(44) SUBHOURLY WIND FORECASTING TECHNIQUES FOR WIND TURBINE OPERATIONS.
Wegley, H.L. Battelle Pacific NorthWest Labs., Richland, WA. Report No. PNL-4894 August 1984 41 pp. (Dept. of Energy Report No. 84016205/XAB) (See also similar paper on hourly forecasts by same author PNL-4900 1984 38 pp.)
Three models for making automated forecasts of subhourly wind and wind power fluctuations were examined to determine the models' appropriateness, accuracy, and reliability in wind forecasting for wind turbine operation. Such automated forecasts appear to have value not only in wind turbine control and operating strategies, but also in improving individual wind turbine operating strategies.

2.2 Feasibility Assessments and Studies

2.2.1 U.S.A.

2.2.1(1) COASTAL-ZONE WIND ENERGY. PART II. FREQUENCY DISTRIBUTION OF WINDS BY DIRECTION FOR EAST AND GULF COAST STATIONS.
Garstang, M. and others. (Virginia Univ., Charlottesville, Dept. of Environmental Sciences) Dept. of Energy Report no:
DOE/ET/20274-77/78/79-8 May 1979 53 pp. (DE 82000336)
Frequency distributions of observed wind speeds by direction for 42 East and Gulf coast stations are given over ten years, 1955 to 1964. Observations for all hours and for all months are combined; the only stratification is the direction from which the wind blows. No adjustment has been made for any change in anemometer height. Distribution for each of the 16 cardinal wind directions is displayed on a speed scale (metres per second, mps) by vertical line segments and the percent of the total number of observations occurring in each of the 16 directional categories and the percent of calm winds are shown separately, numerically and by horizontal line-plot.

2.2.1(2) WIND ENERGY RESOURCE DEVELOPMENT IN CALIFORNIA.
Ginosar, M., Cook, C. and Waco, D. (California Energy Comm., Sacramento, California) In proc. of Conference and Workshop on Wind Energy Characteristics and Wind Energy Siting, Portland, Oregon, USA 19-21 June 1979. Richland, Washington, Pac. NorthWest Lab. PNL Report no: PNL-3214 1979 pp. 229-241.
Reports the work of the California Energy Commission (CEC) engaged in wind resource development throughout the State. Economic feasibility of wind energy conversion systems depends heavily on the total amount of power available in a specific area, so the goal is to characterise the nature of the wind resource for the many diverse wind regimes in California.

2.2.1(3) WIND ENERGY ASSESSMENT OF THE SAN GORGONIO PASS REGION.
Walker, S.N. and Zambrano, T.G. (Aerovironment Inc., Pasadena, California) In proc. of Conference and Workshop on Wind Energy Characteristics and Wind Energy Siting, Portland, Oregon, 19-21 June 1979. Richland, Washington, Pac. NorthWest Lab. PNL Report no: PNL-3214 1979 pp. 405-415.
Report on the progress made in the Southern California Edison- and California Energy Commission-sponsored field investigation of the wind energy potential of the San Gorgonio Pass Region. The field programme is organised to obtain wind information, which can be directly interpreted in terms of WECS design requirements. (5 refs.)

2.2.1(4) WIND RESOURCE ASSESSMENT IN CALIFORNIA.
Berry, E.K. (Atmospheric Research & Technology, Inc., Sacramento) NTIS Report no: PB80-195167 May 1980 77 pp.
Gives an analysis for wind energy data, a design for a wind prospecting instrument, and a strategy for wind energy prospecting. An improved basis for the calculation of wind machine performance is also included.

2.2.1(5) SOUTHERN CALIFORNIA DESERT WIND-ENERGY
ASSESSMENT.
Berry, E.K. (Atmospheric Research & Technology,
Inc., Sacramento) In 3rd International Symposium
on Wind Energy Systems, Lyngby, Copenhagen,
Denmark 26-29 August 1980. Cranfield, U.K.
BHRA Fluid Engng. 1980 Paper B5 12 pp.

Project Windesert is a large-area wind-energy
prospecting project encompassing the southern
California desert, characterised by a desert
climate, complex mountain ranges, and gradual
slopes that channel air flow. Small changes in
terrain and air density sometimes cause abrupt
wind changes in less than a few kilometres.
Wind energy is generally uncorrelated with site
elevation. Wind power measurements show significant deviations from the Rayleigh distribution
prediction using monthly mean wind speed.
These findings address the way wind power should
be measured in large area projects and provide
atmospheric science insights for the inter-
polation of station data to the topographically
complex area between stations. (See also paper
on AIAA/SERI Wind Energy Conference, Boulder,
La. 9-11 April 1980 paper no: 80-0647.)

2.2.1(6) WIND RESOURCE ASSESSMENT AND SITING.
Bortz, S.A., Fieldhouse, I and Budenholzer, R.A.
(Illinois Inst. of Technology) In proc. ISES-AS/Et
Al 1980 Annual Conference, Phoenix 2-6 June 1980
V. 3.2 pp. 1459-1464. (International Solar Energy
Society - American Section)

Reports a study designed to investigate the
feasibility of employing wind power as a possible
energy source for the New Hampshire power grid.
Wind data obtained from various sources were
used as a data base for indicating potential
wind energy conversion sites and sixty potential
sites were identified in the area surrounding
Mount Washington. Economic analyses show that
such wind energy generation stations would
result in significant savings in fuel oil
consumption.

2.2.1(7) COASTAL ZONE WIND ENERGY. PART I.
POTENTIAL WIND POWER DENSITY FIELDS BASED ON
3-D MODEL SIMULATIONS OF THE DOMINANT WIND
REGIMES FOR THREE EAST AND GULF COAST AREAS.
Garstang, M., Pielke, R.A. and Snow, J.W.
(Battelle Pacific NorthWest Labs., Richland,
Washington. Virginia Univ., Charlottesville,
Dept. of Environmental Sciences) Pacific
NorthWest Lab. Report no: PNL-3905 April 1980
35 pp.

Presents the results of applying a numerical
model of the atmosphere to the problem of
locating areas of maximum wind power. Three
U.S. coastal regions are investigated. For
each region the spatial distribution of daily
average power density for the lowest 100 m. of
the atmosphere is given for the three most
prevalent weather regimes. These are then
combined to form an estimate of the annual
average power density for each region.

2.2.1(8) WIND ENERGY ASSESSMENT OF THE PALM
SPRINGS-WHITEWATER REGION, CALIFORNIA, USA.
Lissaman, P.B.S., Zambrano, T.G. and Walker, S.N.
(Aerovironment Inc., Pasadena, California, USA)

In 3rd International Symposium on Wind Energy
Systems, Lyngby, Copenhagen, Denmark
26-29 August 1980. Cranfield, U.K. BHRA Fluid
Engng. 1980 Paper B2 pp. 91-106.

Report of a major survey carried out in 1978
and 1979, sponsored by the California State
Energy Commission and the Southern Californian
Edison Company by Aerovironment Inc. Data was
collected from historical and existing meteorological records, from vegetation flagging
studies, and from field stations erected for
this programme providing continuous records at
the 10 m. level. The methodology is described
as a reliable and cost-effective method of
assessing the wind potential of a region. New
techniques for rapid initial screening using
readily available short-term data were
developed as a result. (6 refs.)

2.2.1(9) WIND CHARACTERISTICS IN SOUTHERN
WYOMING.
Martner, B.E. and others. (Univ. of Wyoming,
USA) In proc. ISES-AS/Et Al 1980 Annual
Conference, Phoenix 2-6 June 1980 V. 3.2
pp. 1468-1473. (International Solar Energy
Society - American Section)

A large array of giant wind turbine
generators is planned for the Medicine Bow,
Wyoming area. Wind measurements have been
recorded from a network of near-surface
anemometers and a meteorological tower.
Annual mean wind speeds range from 4 mps in
summer months to 10 mps in winter. (5 refs.)

2.2.1(10) HIGH YIELD ENERGY RESOURCES IN
NEW YORK STATE.
Sforza, P.M. (Flowpower Inc., NY.) NTIS Report
PB81-142754 June 1980 177 pp.

Gives an inventory of wind energy resources
in New York State based on data collected at
various meteorological stations and anemometers
throughout the State.

2.2.1(11) WIND ENERGY IN MARYLAND.
Tompkins, D. (Maryland Dept. of Natural
Resources, Annapolis. Power Plant Siting Pro-
gramme) Report no: PB81-199937 July 1980 37 pp.

Gives a state-of-the-art review of applications, economic, and environmental issues.
Results of a preliminary wind energy resource
assessment completed in April of 1979 are
discussed, and recommendations are made for
further resource definition.

2.2.1(12) A CASE STUDY OF WIND TURBINE GENERATOR
SITING IN COMPLEX TERRAIN.
Vachon, W.A., Downey, W.T. and Madio, F.R.
(Arthur D. Little Inc., Massa. USA) In proc. of
ISES-AS/Et Al 1980 Annual Conference, Phoenix
2-6 June 1980 V. 3.2 pp. 1481-1485. (Inter-
national Solar Energy Society - American Section)

A cost-effective wind turbine generator siting
method is described, and the results of a preliminary application of the technique in the
complex terrain of the New Hampshire mountains
are examined. Stresses the use of advanced
planning, map interpretation, potential site

aerial examination, and onsite studies for recording.

2.2.1(13) THE SOUTHERN CALIFORNIA EDISON COMPANY WIND RESOURCE ASSESSMENT PROGRAM.
Yinger, R.J. (Southern California Edison Co., California, USA) In proc. of ISES-AS/Et Al 1980 Annual Conference, Phoenix 2-6 June 1980 V. 3.2 pp. 1464-1470. (Internat. Solar Energy Society - American Section)

Wind turbine generator systems are expected to make cost-effective electricity contributions to the Southern California Edison Co. utility system. The wind resource in this utility's service area is being investigated to determine what type of wind energy systems to develop and where. Studies using computer modeling, field measurements, and analyses of existing meteorological data are reviewed.

2.2.1(14) EXECUTIVE SUMMARY. WIND-ENERGY ASSESSMENT STUDIES IN THE GOODNOE HILLS AND CAPE BLANCO AREAS. PROGRESS REPORT, OCTOBER 1980-SEPTEMBER 1981.
Baker, R.W. and others. (Oregon State Univ., Corvallis, Dept. of Atmospheric Sciences) Department of Energy Report no: DOE/BP-107, BPA-81-7 December 1981 26 pp. (DE 82015670) (See also DOE/BP-106 December 1981 110 pp.)

Spatial wind surveys in the Goodnoe Hills and Cape Blanco area, wind turbine generator wake measurements at the Goodnoe Hills site, and developing a methodology for sampling the wind flow using a kite anemometer are described.

2.2.1(15) WIND ENERGY DATA BASE.
Barchet, W.R. (Pacific NorthWest Lab., Richland, Washington) In proc. 5th Biennial Wind Energy Conference & Workshop, Washington, D.C., USA 5-7 October 1981. I.E. Vas (ed.) V. 2. Palo Alto, USA, Solar Energy Research Inst. 1981 Session 3C pp. 661-671. Report nos: SERI/CP-635-1340, CONF-811043.

Describes six data files giving information on many properties of the wind as a resource for power generation.

2.2.1(16) NATIONAL WIND RESOURCE ASSESSMENT.
Elliott, D.L. and Barchet, W.R. (Pacific NorthWest Lab.) In proc. 5th Biennial Wind Energy Conference & Workshop, Washington, D.C., USA 5-7 October 1981. I.E. Vas (ed.) V. 2. Palo Alto, USA, Solar Energy Research Inst. 1981 Session 3C pp. 649-660. Report nos: SERI/CP-635-1340, CONF-811043.

Describes the results of the national wind energy resource assessment and how to interpret it with a brief summary of the data sources and methodology employed.

2.2.1(17) AN ASSESSMENT OF THE SOLAR AND WIND RESOURCE IN NORTHEAST UTILITIES SERVICE TERRITORY.
Goodrich, R.W. and Law, S.H. (NorthEast Utilities) Paper presented at ANS Alternative Energy Sources for Electrical Power Conference, MA. 4-7 October 1981 14 pp.

Presents a solar and wind energy resource assessment in the NorthEast Utilities Service territory - concentrated in Connecticut and Massachusetts. Anemometers have been used to survey viable wind energy sites; wind sheer or wind profile as a function of height has been examined; and modeling of wind patterns has been carried out. The company has also embarked on a solar resource measurement programme to evaluate options for solar-powered electricity. (4 refs.)

2.2.1(18) FEDERAL APPLICATIONS FOR WIND ENERGY SYSTEMS.
Sklar, H. (Solar Energy Research Institute) In proc. 5th Biennial Wind Energy Conference & Workshop, Washington, D.C., USA 5-7 October 1981 I.E. Vas (ed.) V. 2. Palo Alto, USA, Solar Energy Research Inst. 1981 Session 3A pp. 525-534. Report nos: SERI/CP-635-1340, CONF-811043.

Reports on a study to discover commercially attractive sites for wind energy systems to be used by Federal agencies in the United States. Potential sites were identified and brief case studies were performed for two of them.

2.2.1(19) WIND-ENERGY ASSESSMENT STUDIES FOR SOUTHERN CALIFORNIA. DATA SUPPLEMENT FOR THE WIND MONITORING STATIONS IN THE PALM SPRINGS-WHITEWATER REGION, AUGUST 1978-MARCH 1981.
Zambrano, T.G. and Arcemont, G.J. (AeroVironment Inc., Pasadena, CA.) Report no: P-500-81-034-V.4-App.E, August 1981 462 pp. (Dept. Energy Report no: DE 82903686)

Data are summarised from April 1977 for one station that records wind speeds at 75 feet above ground and other stations that record speeds at 33 feet above ground. The vicinity of each site is mapped and its location given. Wind speeds are given hourly for each site, and each daily average and peak is given.

2.2.1(20) WIND-ENERGY-ASSESSMENT STUDIES FOR SOUTHERN CALIFORNIA. VOLUME 2 OF 4. DATA SUPPLEMENT FOR THE WIND-RESOURCE LOCATIONS IN SOUTHERN CALIFORNIA. APPENDIX C.
Zambrano, T.G. and Arcemont, G.J. (AeroVironment Inc., Pasadena, CA.) Report no: P-500-81-032-App.C, August 1981 378 pp. (Dept. Energy Report no: DE 82903692)

The western sector of southern California was divided into nine geographical regions. For each, a map is provided to identify the location of each wind monitoring station, along with a list of stations within the region. Station longitude and latitude, instrument height, and period of data collection is given on each wind data summary. The wind data was recorded by many different organisations for a variety of applications (e.g. air quality, fire weather, and airways reports). Since the use of wind data for wind energy applications is relatively new, a standardised data format was developed specifically for wind energy resource evaluations. The summary displays three important factors of wind flow at the site: diurnal variations of wind speed, seasonal variations of wind speed, and the dominant strong wind direction (wind speeds from that direction average greater than 10 mph.).

2.2.1(21) NEW YORK STATE'S WIND ENERGY POTENTIAL.
Bailey, B.H. (New York State Univ. at Albany) In proc. 1982 Wind and Solar Energy Technology Conference, Kansas City, USA 5-7 April 1982. Columbia, USA., Missouri-Columbia Univ. 1982 pp. 242-248.

Summarises the activities and findings related to several years of investigation into the State's wind energy resources. The current status of, and long range outlook for, wind energy development is also considered by examining the State's energy profile and the existing incentives and constraints to WECS development.

2.2.1(22) REGIONAL WIND-ENERGY-ASSESSMENT PROGRAM. PROGRESS REPORT, OCTOBER 1981-SEPTEMBER 1982.
Baker, R.W. and others. (Oregon State Univ., Cornvallis, Dept. of Atmospheric Sciences) Department of Energy Report no: DOE/BP-167, BPA-82-8 December 1982 115 pp. (See also previous annual reports nos. DOE/BP-89 & 90, BPA-81-6 + Appendix) (DE 83018259)

The research activities done in 1982 on the Regional Wind Energy Assessment Program are described. Includes work completed in large area wind power prospecting and data collection and analysis. The wind data network, climatological wind speed and energy analysis, and wind statistical analysis are discussed as is a wind power prospecting helicopter survey over the entire state of Washington.

2.2.1(23) REGIONAL WIND-ENERGY-ASSESSMENT PROGRAM. PROGRESS REPORT, OCTOBER 1981-SEPTEMBER 1982. APPENDIX. WIND STATISTIC SUMMARIES.
Baker, R.W., Wade, J.E. and Persson, P.O.G. (Oregon State Univ., Corvallis, Dept of Atmospheric Sciences) Department of Energy Report no: DOE/BP-168, BPA-82-8-App. December 1982 68 pp. (DE 83018258)

Tabulates mean monthly wind speeds at the wind power data sites in the states of Washington, Oregon, Montana and Nevada for the period of June 1981 to May 1982 and presents wind speed frequency distribution summaries.

2.2.1(24) COASTAL ZONE WIND ENERGY. PART III: A PROCEDURE TO DETERMINE THE WIND POWER POTENTIAL OF THE COASTAL ZONE.
Garstang, M., Pielke, R. and Snow, J.W. (Battelle Pacific NorthWest Labs., Richland, WA.) Virginia Univ., Charlottesville, Dept. of Environmental Sciences. Department of Energy, Washington, D.C. Pacific NorthWest Labs. Report no: PNL-3903 March 1982 49 pp. (Dept. Energy Report no: DE 82014334)

Presents a stepwise procedure for determining the seasonal and/or annual mean potential wind power density for any location on the East and Gulf coasts of the United States. Includes

reference to the dominant wind regimes and mean power densities already obtained to estimate the wind power potential of the location under consideration; methods to calculate the potential wind power distributions and steps to be taken to locate the best site in the area of interest. Can be best applied where the atmospheric systems which produce most of the wind energy at the surface are relatively persistent.

2.2.1(25) A WIND ENERGY RESOURCE ASSESSMENT FOR THE SOUTH CENTRAL UNITED STATES.
Graves, L.F. and Schnebelt, J.J. (Inst. Storm Research, Houston) In proc. 1982 Wind and Solar Energy Technology Conference, Kansas City, USA 5-7 April 1982. Columbia, USA, Missouri-Columbia Univ. 1982 pp. 196-204.

2.2.1(26) WINDPOWER POTENTIAL IN NEW HAMPSHIRE.
Kraft, L.G. and Hodgkins, P. (New Hampshire Univ.) In proc. 1982 Wind and Solar Energy Technology Conference, Kansas City, USA 5-7 April 1982. Columbia, USA, Missouri-Columbia Univ. 1982 pp. 340-346.

A survey of wind power potential using a network of wind energy data collection systems has shown that New Hampshire has an abundant wind energy resource at intermediate elevations above sea level.

2.2.1(27) APPLICATION OF U.S. UPPER WIND DATA IN ONE DESIGN OF TETHERED WIND ENERGY SYSTEMS.
O'Doherty, R.J. and Roberts, B.W. (Solar Energy Research Inst., Golden, CO.) Solar Energy Res. Inst. Report no: SERI/TR-211-1400 February 1982 133 pp. Department Energy Report no: 82042880.

The upper atmospheric wind resource for the continental United States, Hawaii and Alaska is addressed. The raw data were obtained from the National Centre for Atmospheric Research, Boulder, Colorado. Probability distributions of velocity are presented for 54 sites, and detailed calm wind analyses have been undertaken for five of these locations.

2.2.1(28) WIND STUDY: FIELD WORK AND RESULTS.
Peterson, C.N. (Idaho Univ.) In proc. 1982 Wind and Solar Energy Technology Conference, Kansas City, USA 5-7 April 1982. Columbia, USA, Missouri-Columbia Univ. 1982 pp. 368-370.

Describes a study conducted to identify sites for anemometer installations, to operate and maintain anemometers, and to assess the wind energy potential from the collected data, as part of an evaluation of wind energy potential in Southern Idaho.

2.2.1(29) ASSESSMENT OF POTENTIAL WIND ENERGY SITES ON BUREAU OF LAND MANAGEMENT LAND IN THE CALIFORNIA DESERT.
Zalay, A.D., Arcemont, G. and Waco, D.E. (Aero-Vironment Inc., Calif. Energy Comm.) In proc. 1982 Wind and Solar Energy Technology Conference, Kansas City, USA 5-7 April 1982. Columbia, USA, Missouri-Columbia Univ. 1982 pp. 249-257.

Reports a wind energy survey consisting of an evaluation of climatological data, field observations and cost analysis for siting an array of wind machines in one of the potentially high wind resource areas of the desert.

2.2.1(30) THE FEASIBILITY OF ELECTRIC POWER GENERATION BY THE WIND ON THE UNIVERSITY OF NEW ORLEANS CAMPUS.
Hilbert, L.B. and Janna, W.S. (Exxon Corp. and New Orleans Univ.) In Intl. Wind Energy Symposium 5th Annual Energy-Sources Technol. Conf., New Orleans, USA 7-10 March 1982. New Yor, USA, Am. Soc. Mech. Engrs. 1982 pp. 285-292.

Discusses some general considerations for modern wind power systems and examines analysis of wind speed and direction data for evaluation of the University of Orleans site. Outlines energy analysis of the future MOD 2 (Boeing) and General Electric MOD 1 wind plants. Discusses energy calculations and economic evaluation of four small wind turbine systems. Tabulates the windmill design specification, available energy (kWh/year and month) and costing results.

2.2.1(31) WINDFARMS WIND POWER PROJECTS IN HAWAII AND CALIFORNIA.
Laessig, R.R. (Windfarms Ltd.) In Intl. Wind Energy Symposium 5th Annual Energy-Sources Technol. Conf., New Orleans, USA 7-10 March 1982. New York, USA, Am. Soc. Mech. Engrs. 1982 pp. 113-128.

The first phase of the Kahuku Pt (Hawaii) project includes five MOD-2 wind turbines rated at 2.5 mW. These have an upwind rotor of 300 ft. diameter. The second phase plan calls for MOD-5 machines, approximately 10 units to meet the wind farm rating of 80 mW. Outlines procedures for site acquisition, meteorological data acquisition and environmental permits. The California project comprises a 350 mW wind farm in Solano County. 60 machines of various design should be included. Outlines some economics, regulatory and environmental constraints on such wind farms in the USA.

2.2.1(32) WISCONSIN POWER AND LIGHT COMPANY'S WIND ENERGY RESEARCH AND DEMONSTRATION PROGRAM.
DeWinkel, C.C. (Wisconsin Power & Light Co.) In Intl. Wind Energy Symposium 5th Annual Energy-Sources Technol. Conf., New Orleans, USA 7-10 March 1982. New York, USA, Am. Soc. Mech. Engrs. 1982 pp. 57-59.

Outlines the test programme, which involves installation of six windmills on customers premises in rural areas. Describes field experience with the 5 operating windmills, noting problems which occurred and certain design improvements. Preliminary results on cumulative energy production and performance are presented.

2.2.1(33) LEARNING TO FORECAST WIND AT REMOTE SITES FOR WIND-ENERGY APPLICATIONS. FINAL REPORT.
Notis, C. and others. Battelle Pacific North-West Labs., Richland, WA. Report No. PNL-4318 January 1983 239 pp.

Observed wind patterns were correlated with synoptic or mesoscale weather systems. Six test sites in the USA are described.

2.2.1(34) MONITORING AND APPRAISAL EVALUATION OF WIND ENERGY POTENTIAL FOR ELECTRIC POWER GENERATION IN THE BRISTOL BAY AREA. FINAL REPORT.
Zambrano, T.G. and Arcemont, G.J. (Aero-Vironment Inc., Pasadena, USA) Report No. AV/FR-82-594 February 1983 252 pp.

An analysis was made of the technical feasibility of developing wind power in the Dillingham/Naknek-King Salmon area of Bristol Bay, Alaska. The analysis involved a one-year wind monitoring programme using data from existing government weather stations as well as anemometers installed by AeroVironment Inc. A technical analysis was made of expected energy outputs from selected, promising sites. Since Naknek was the most promising site, an energy production calculation for three representative turbines of 25 kW, 65 kW and 125 kW rating was made. Conclusions are that a 350 kW turbine wind farm with a capacity factor of 0.34 is feasible at Naknek.

2.2.1(35) THE STATUS OF WIND TURBINE DEVELOPMENT IN SOUTHERN CALIFORNIA EDISON'S SERVICE TERRITORY.
Yinger, R.J. (Southern California Edison, CA.) In American Wind Energy Assn. Wind Energy Expo Natl. Conference, San Francisco 17-19 October 1983 pp. 276-282.

Describes assessment studies for the wind energy potential in the Southern California Edison Service area. Wind energy data are being collected at over 30 sites. Demonstration projects are also underway, testing 4 turbines in the 50 - 1300 kW capacity range. Commercialisation of this technology is being encouraged as 28.4 mW of wind-generated electricity from private developers is now connected to the utility system.

2.2.1(36) THE WEIBULL DISTRIBUTION APPLIED TO WIND SITES IN CALIFORNIA.
Waco, D.D. and Hennessey, J.R. (California Energy Commission) In American Wind Energy Assn. Wind Energy Expo Natl. Conf., San Francisco 17-19 October 1983 pp. 153-170.

The expected generating output is calculated for selected wind turbines and sites. Although the Weibull-calculated mean annual available power density compares favourably with measured mean power density data, the distribution fails to approximate the annual wind turbine output at many sites to within 10% of measured values. (12 refs.)

2.2.1(37) ANALYSIS OF TIME AND SPACE VARIATIONS IN LONG-TERM MONTHLY AVERAGED WIND SPEEDS IN THE UNITED STATES.
Balling, R.C. and Cerveny, R.S. (Dept. of Geography, Univ. of Nebraska, Lincoln, USA) Wind Eng. 8 (1) 1984 pp. 1-8.

Results from harmonic and principal components analyses of long-term monthly averaged wind speeds for a large number of locations in the United States are presented. These reveal the order that underlies the time and space variations in the wind speed data. The identification of the basic variance structures is crucial in developing physical, dynamical, or synoptic explanations of the climatological wind patterns in the United States. The results are equally critical to engineers and planners who are designing the equipment systems to make maximum use of the wind energy resources. (39 refs.)

2.2.1.1 UNITED STATES WIND ENERGY RESOURCE ATLAS PROJECT.

2.2.1.1(1) PUTTING WIND RESOURCE ATLASES TO USE.
Elliott, D.L. (Pacific NorthWest Labs.) In proc. Workshop on Large Horizontal-Axis Wind Turbines, Cleveland, USA 28-30 June 1981. R.W. Thresher (ed.) Washington, D.C., USA, NASA 1982 pp. 141-145. (NASA Conf. Publication 2230; DOE Publication CONF-810752 & SERI/CP-635-1273)

Describes how the twelve wind energy resource atlases for the United States and its territories can be used to evaluate various aspects of an area's wind resource. Interpretation of information in the atlas on various geographic scales and annual, seasonal and diurnal time scales is discussed. In addition to techniques for extracting the magnitude of the wind resource, methods are presented for estimating the seasonal and diurnal variations of the wind resource for an area, the certainty with which the resource has been estimated and the fraction of land area with a given wind resource.

2.2.1.1(2) TECHNIQUES FOR ASSESSING THE WIND ENERGY RESOURCE IN ALASKA.
Wentink, T.Jr. and Wise, J.L. (Battelle Pacific NorthWest Labs., Richland, WA.) Alaska Univ., Fairbanks, Geophysical Inst. Alaska Univ., Anchorage. Arctic Environmental Information and Data Centre. Report no: PNL-3519 April 1981 90 pp. (Dept. of Energy Report no: DE 82017182)

Technical procedures specifically used to develop Wind Energy Resource Atlas: The Alasaka Region are described. Includes a discussion of the methods used to identify, screen, evaluate, and analyse the various types of data, and to produce the wind energy resource graphical presentations. The analyses and assumptions used in preparing the atlas are described and characteristics of the data base are outlined; data values are tabulated in an appendix. Summary descriptions of major wind resource areas follow the discussion of the data base.

2.2.1.1(3) TECHNIQUES FOR ASSESSING THE WIND ENERGY RESOURCE IN THE EAST CENTRAL REGION.
Brode, R. and Stoner, R. (Battelle Pacific NorthWest Labs., Richland, WA.) Pacific North-West Lab. Report no: PNL-3451 Sept. 1980 128 pp. (Dept. of Energy Report no: DE 81025105)

Describes the methods specifically used to produce the wind energy resource atlas of the East Central region. Screening procedures were developed to identify stations with the most useful data and to eliminate stations that would not significantly contribute information on the distribution of the wind resource. A total of 488 stations were screened and 228 stations retained for analysis. Unsummarised data were used only when no summarised or digitised data were available. Seasonal and annual values of wind power density were calculated from the screened wind data set. In a single number, wind power density incorporates the combined effect of the distribution of wind speeds and the dependence of the power density on air density and on the cube of the wind speed.

2.2.1.1(4) TECHNIQUES FOR ASSESSING THE WIND ENERGY RESOURCE IN THE GREAT LAKES REGION.
Paton, D.L., Bass, A. and Smith, D.G. (Environmental Research and Technology Inc., Concord, MA.) Pacific NorthWest Labs. Report no: PNL-3668 February 1981 132 pp. (Dept. of Energy Report no: DE 81025103)

The atlas of the wind resource in the Great Lakes region is one of 12 regional assessments sponsored by the Department of Energy. As a supplement to the wind energy resource atlas for the Great Lakes region this report includes a detailed discussion of the methodologies and data employed to complete the atlas. A summary of the regional annual average wind power, highlighting the major wind resource areas is included.

2.2.1.1(5) TECHNIQUES FOR ASSESSING THE WIND ENERGY RESOURCE IN HAWAII AND PACIFIC ISLANDS REGION.
Schroeder, T.A. and Hori, A.M. (Battelle Pacific NorthWest Labs., Richland, WA.) Pacific North-West Lab. Report no: PNL-3673 August 1980 42 pp. (Dept. of Energy Report no: DE 81025101)

Explains the procedures utilised in preparing the Wind Energy Resource Atlas: Hawaii and Pacific Islands Region and contrasts these methods with those used in the other regional assessments. Techniques generally paralleled those of the northwest wind resource assessment. Quality of data bases differed drastically between Hawaii and the Pacific Islands.

2.2.1.1(6) TECHNIQUES FOR ASSESSING THE WIND ENERGY RESOURCE IN THE NORTH CENTRAL REGION.
Freeman, D.L. (Battelle Pacific NorthWest Labs., Richland, WA.) Pacific NorthWest Lab. Report no: PNL-3667 March 1981 84 pp. (Dept. of Energy Report no: DE 81025102)

The US Department of Energy has sponsored the development of regional wind energy resource atlases for twelve regions of the United

States. The North Central Region consists of North Dakota, South Dakota, Nebraska, Minnesota and Iowa. This report describes the observational and analytical techniques used in the development of the North Central Region's wind energy resource atlas.

2.2.1.1(7) TECHNIQUES FOR ASSESSING THE WIND ENERGY RESOURCE IN THE NORTHEAST REGION. Pickering, K.E. and others. Rockville, USA, Geomet. Technol. Inc. Pacific NorthWest Lab. Report no: PNL-3452 June 1980 191 pp.

Describes techniques developed by the Pacific NorthWest Laboratory and Geomet for assessing wind energy resources in the NorthEast United States, and preparing the Wind Energy Resource Atlas.

2.2.1.1(8) TECHNIQUES FOR ASSESSING THE WIND ENERGY RESOURCE IN THE SOUTHEAST REGION. Zabransky, J. and others. (Battelle Pacific NorthWest Labs., Richland, WA.) Pacific North-West Lab. Report no: PNL-3669 October 1980 138 pp. (Dept. of Energy Report no: DE 81025100)

Describes techniques developed by Pacific NorthWest Laboratory and Geomet to assess the wind energy resource in the SouthEast. Wind data locations were identified and the data obtained were screened to determine their usefulness. Summarised data were processed and wind power values were estimated for all data locations. Upper air data were used to estimate wind power on mountaintops and ridge crests. Maps of annual and seasonal wind power classes were constructed from the computed wind power values along with the aid of landform and relief maps.

2.2.1.1(9) TECHNIQUES FOR ASSESSING THE WIND ENERGY RESOURCE IN THE SOUTHERN ROCKY MOUNTAIN REGION. Freeman, D.L. and Anderson, S.R. (Battelle Pacific NorthWest Labs., Richland, WA) Pacific NorthWest Lab. Report no: PNL-3671 April 1981 97 pp. (Dept. of Energy Report no: DE 81025099)

The US Department of Energy has sponsored the development of regional wind energy resource atlases for twelve regions of the United States. The Southern Rocky Mountain Region consists of Arizona, Colorado, New Mexico and Utah. The report describes the observational and analytical techniques used in the development of the Southern Rocky Mountain region's wind energy resource atlas.

2.2.1.1(10) TECHNIQUES FOR ASSESSING THE WIND ENERGY RESOURCE IN THE SOUTHWEST REGION. Simon, R.L. and Norman, G.T. (Battelle Pacific NorthWest Labs., Richland, WA) Pacific North-West Lab. Report no: PNL-3672 December 1980 43 pp. (Dept. of Energy Report no: DE 81025098)

Report prepared as a technical supplement to the SouthWest Regional Wind Energy Resource Atlas (Simon and Norman 1980), discusses the methodology used to generate the resource assessment and the results obtained in a manner useful and instructive in the wind resource assessment field. Designed to be used in conjunction with the Atlas and the maps and tables contained within it.

2.2.1.1(11) WIND-ENERGY ASSESSMENT FOR THE WESTERN PACIFIC BASED ON SHIP REPORTS. Schroeder, T.A. and Hori, A.M. (Hawaii Univ., Honolulu, Dept. of Meteorology) Report no: UHMET-82-05 November 1982 28 pp.

From over 468,000 wind reports from ships traversing the Pacific Islands maps were prepared of annual and seasonal average wind speed and wind energy density and wind rose summaries for 100 2 exp 0 by 5 exp 0 (latitude by longitude) boxes. The Northern Marshall Islands possess the best wind energy resource in the region, the Northern Marianas the next best. Tropical storms exert a limited influence on the wind statistics.

2.2.1.1(12) WIND ENERGY RESOURCES ATLAS. VOLUME 1. NORTHWEST REGION. Elliott, D.L. and Barchet, W.R. (Battelle NorthWest Labs., Richland, WA) US Department of Energy. Pacific NorthWest Lab. Report no: PNL-3195-WERA-1 April 1980 192 pp.

Presents wind resource data for the North-West Region and individual state assessments for Idaho, Montana, Oregon, Washington and Wyoming.

2.2.1.1(13) WIND ENERGY RESOURCE ATLAS. VOLUME 2. THE NORTH CENTRAL REGION. Freeman, D.L. and others. (ERT/Western Scientific Services, Inc., Fort Collins, CO) US Department of Energy. Pacific NorthWest Lab. Report no: PNL-3195-WERA-2 February 1981 191 pp.

The North Central atlas assimilates six collections of wind resource data: one for the region and one for each of the five states that compose the North Central region (Iowa, Minnesota, Nebraska, North Dakota and South Dakota).

2.2.1.1(14) WIND ENERGY RESOURCE ATLAS. VOLUME 3. GREAT LAKES REGION. Patón, D.L. and others. (Environmental Research and Technology, Inc., Concord, MA.) US Department of Energy. Pacific NorthWest Lab. Report no: PNL-3195-WERA-3 February 1981 187 pp.

The Great Lakes Region atlas assimilates six collections of wind resource data, one for the region and one for each of the five states that compose the region: Illinois, Indiana, Michigan, Ohio, Wisconsin. At the state level, features of the climate, topography, and wind resource are discussed in greater detail and the data locations on which the assessment is based are mapped. Variations over several time scales in the wind resource at selected stations in each state are shown on graphs of monthly average and interannual wind speed and power, and of hourly average wind speed for each season. Other graphs present speed, direction, and duration frequencies of the wind.

2.2.1.1(15) WIND ENERGY RESOURCE ATLAS. VOLUME 4. THE NORTHEAST REGION. Pickering, K.E. and others. (Geomet, Inc., Gaithersburg, MD.) US Department of Energy. Pacific NorthWest Lab. Report no: PNL-3195-WERA-4 September 1980 230 pp.

This atlas of wind energy resource is composed of introductory and background information, a regional summary of the wind resource, and assessments of the wind resource in each state of the region. Background is presented on how the wind resource is assessed and on how the results of the assessment should be interpreted. A description of the wind resource on a regional scale is then given. The results of the wind energy assessments for each state are presented as an overview and summary of the various features of the regional wind energy resource. Assessments for individual states are presented. The state wind energy resources are described in greater detail than is the regional wind energy resource, and features of selected stations are discussed.

2.2.1.1(16) WIND ENERGY RESOURCE ATLAS. VOLUME 5. THE EAST CENTRAL REGION. Brode, R. and others. (NUS Corp., Rockville, MD.) US Department of Energy. Pacific NorthWest Lab. Report no: PNL-3195-WERA-5 1980 214 pp.

Assessments for individual states are presented as separate chapters. The state wind energy resources are described in greater detail than is the regional wind energy resource, and features of selected stations are discussed for states including Delaware, Maryland, Kentucky, North Carolina, Tennessee, Virginia and West Virginia.

2.2.1.1(17) WIND ENERGY RESOURCE ATLAS. VOLUME 6. THE SOUTHEAST REGION. Zabransky, J. and others. (Geomet Technologies, Inc., Rockville, MD.) US Department of Energy. Pacific NorthWest Lab. Report no: PNL-3195-WERA-6 January 1981 175 pp.

The SouthEast atlas assimilates six collections of wind resource data: one for the region and one for each of the five states: Alabama, Florida, Georgia, Mississippi and South Carolina. At the state level, features of the climate, topography and wind resource are discussed in great detail. Variations, over several time scales, in the wind resource at selected stations in each state are shown on graphs of monthly average and interannual wind speed and power, and hourly average wind speed for each season. Other graphs present speed, direction and duration frequencies of the wind.

2.2.1.1(18) WIND ENERGY RESOURCE ATLAS. VOLUME 7. THE SOUTH CENTRAL REGION. Edwards, R.L. and others. (Battelle Pacific NorthWest Labs., Richland, WA.) US Department

of Energy. Pacific NorthWest Lab. Report no: PNL-3195-WERA-7 March 1981 238 pp.

This atlas of the South Central region combines seven collections of wind resource data: one for the region, and one for each of the six states: Arkansas, Kansas, Louisiana, Missouri, Oklahoma and Texas. At the state level features of the climate, topography, and wind resource are discussed in great detail. Variations, over several time scales, in the wind resource at selected stations in each state are shown on graphs of monthly average and interannual wind speed and power, and hourly average wind speed for each season. Other graphs present speed, direction, and duration frequencies of the wind at these locations.

2.2.1.1(19) WIND ENERGY RESOURCE ATLAS. VOLUME 8. THE SOUTHERN ROCKY MOUNTAIN REGION. Andersen, S.R. and others. (Battelle Pacific NorthWest Labs., Richland, WA) US Department of Energy. Pacific NorthWest Lab. Report no: PNL-3195-WERA-8 March 1981 178 pp.

The Southern Rocky Mountain atlas assimilates five collections of wind resource data: one for the region and one for each of the four states: Arizona, Colorado, New Mexico and Utah. At the state level, features of the climate, topography and wind resource are discussed in great detail. Variations, over several time scales, in the wind resource at selected stations in each state are shown on graphs of monthly average and interannual wind speed and power, and hourly average wind speed for each season. Other graphs present speed, direction and duration frequencies of the wind at these locations.

2.2.1.1(20) WIND ENERGY RESOURCE ATLAS. VOLUME 9. THE SOUTHWEST REGION. Simon, R.L. and others. (Global Weather Consultants, Inc., San Jose, CA. Battelle Pacific NorthWest Labs., Richland, WA.) US Department of Energy. Pacific NorthWest Lab. Report no: PNL-3195-WERA-9 November 1980 128 pp.

This atlas of the wind energy resource is composed of introductory and background information, a regional summary of the wind resource, and assessments of the wind resource in Nevada and California. Background on how the wind resource is assessed and on how the results of the assessment should be interpreted is presented. A description of the wind resource on a regional scale is then given. The results of the wind energy assessments for each state are assembled into an overview and summary of the various features of the regional wind energy resource.

2.2.1.1(21) WIND ENERGY RESOURCE ATLAS. VOLUME 10. ALASKA REGION. Wise, J.L. and others. (Alaska Univ., Anchorage, Arctic Environmental Information and Data Centre) US Department of Energy. Pacific NorthWest Lab. Report no: PNL-3195-WERA-10 December 1980 181 pp.

This atlas of the wind energy resource is composed of introductory and background information, a regional summary of the wind resource, and assessments of the wind resource in each subregion of Alaska. Background is presented on how the wind resource is assessed and on how the results of the assessment should be interpreted. The results of the wind energy assessments for each subregion are assembled into an overview and summary of the various features of the Alaska wind energy resource. Assessments for individual subregions are presented as separate chapters.

2.2.1.1(22)WIND ENERGY RESOURCE ATLAS. VOLUME 11. HAWAII AND PACIFIC ISLANDS REGION. Schroeder, T.A. and others. (Hawaii Univ., Honolulu, Dept. of Meteorology) US Department of Energy. Pacific NorthWest Lab. Report no: PNL-3195-WERA-11 February 1981 127 pp.

Background on how the wind resource is assessed and on how the results of the assessment should be interpreted is presented. Assessments for individual divisions are presented as separate chapters. Much of the information in the division chapters is given in graphic or tabular form. The sequences for each chapter are similar, but some presentations used for Hawaii are inappropriate or impractical for presentation with the Pacific Islands.

2.2.1.1(23) WIND ENERGY RESOURCE ATLAS. VOLUME 12. PUERTO RICO AND US VIRGIN ISLANDS. Wegley, H.L. and others. (Battelle Pacific NorthWest Labs., Richland, WA) US Department of Energy. Pacific NorthWest Lab. Report no: PNL-3195-WERA-12 January 1981 93 pp.

The Puerto Rico/US Virgin Island atlas assimilates three collections of wind resource data: one for the region as a whole and one for each of the Commonwealth of Puerto Rico and the US Virgin Islands. Variations, over several time scales, in the wind resource at selected stations in both subregions are shown on graphs of monthly average and interannual wind speed and power, and hourly average wind speed for each season. Other graphs present speed, direction and duration frequencies of the wind at these locations.

2.2.2. OTHER COUNTRIES.

2.2.2.(1) WIND ENERGY IN THE U.K. Rayment, R. (Build. Res. Establishment, U.K.) Build. Serv. Engng., 44 (3) June 1976 pp. 63-69.

Analyses of the frequency distribution and geographical variation of wind speeds for the U.K. are used to produce a new wind-energy map for the U.K. Based on the method for predicting the availability of wind-energy between different wind speeds for any location in the U.K. The relevance of this tc the operation of wind generators in determining their mean annual output and operating time is presented. (22 refs.)

2.2.2(2) WIND ENERGY RESOURCE ASSESSMENT OF NEW ZEALAND. Cherry, N.J. (Lincoln College, Canterbury, New Zealand) Pacific NorthWest Lab. Tech. Report PNL-3214, proc. of Conference and Workshop on Wind Energy Characteristics and Wind Energy Siting, Portland, Oregon 19-21 June 1979. Richland, Washington, Pacific NorthWest Lab. 1979 pp. 261-271.

Sufficient wind data exists in the archives of the New Zealand Meteorological Service to broadly characterise most areas of the country and to indicate a number of possible pilot plant sites. (8 refs.)

2.2.2(3) OTAGO WIND-ENERGY RESOURCE SURVEY, FIRST REPORT. WIND-ENERGY RESOURCE IN OTAGO. SECOND REPORT. DATA ACQUISITION AND PROCESSING. Edwards, P.J. and others. New Zealand Energy Research and Development Committee, Auckland. April 1979 60 pp. (US Dept. of Energy Report nos: DE 82904574 & DE 82904573)

The methodology of the Survey is described, the selection of anemometer sites is discussed and details of the individual sites given. Annual mean wind speed and wind energy flux at each site are tabulated. From these results, an inventory of the wind energy resource is estimated. (The second report NZERDC-P-13 covers data acquisition and processing, operation of anemometers and wind speed instrumentation.)

2.2.2(4) EGYPTIAN WIND ENERGY RESOURCES STUDY. PHASE II. FINAL REPORT. Hughes, W.L. (Oklahoma State Univ., Stillwater. Engineering Energy Lab. Washington, DC. US Department of Energy) Report no: DOE/ET/ 20607-T1 November 1979 139 pp. (DE 81026222)

The data gathered in Egypt in Phase I of the programme indicated favourable wind energy possibilities along the Mediterranean Coast west of Alexandria and along the Red Sea south of Suez. In Phase II, several continuous wind recording instruments were established on the North Coast and Red Sea Coasts. Locations selected for the North Coast were distributed from Mersa Matruh to Borg El Arab, a coastal community about seventy kilometers west of Alexandria. Recorded data from the monitoring stations are presented.

2.2.2(5) POTENTIAL OF WIND ENERGY IN ANTARTICA. Bowden, G.J. and others. (Univ. of NSW, Kensington, Australia) Wind Eng., 4 (3) 1980 pp. 163-176.

Mawson, Australian Antarctic Territory, possesses a high average windspeed of 47.6 km/hr a characteristic of coastal bases which experience strong Katabatic winds. Six years of windspeeds, recorded every 3 hours, have been analysed and used to compute wind, storm and lull frequency histograms. These are used to compute the expected annual power outputs and availabilities from the 300 kW and 500 kW ALCOA ALVAWT Darrieus windgenerators. (9 refs.)

2.2.2(6) WIND-ENERGY RESOURCE SURVEY OF CANTER-
BURY: PROJECT DESCRIPTION AND INITIAL RESOURCE
ASSESSMENT.
Cherry, N.J. and Smyth, V.G. (New Zealand Energy
Research and Development Committee, Auckland)
NZERDC-P-41 June 1980 66 pp. (US Dept. of Energy
Report no: DE 82905066)

The background of the wind energy survey of
Canterbury; the sites chosen, the instrumentation used and the methods of data analysis are
described and the initial results of the survey,
which spanned the period 1975 to 1978, are
presented in the form of mean wind speeds and
wind energy fluxes. While it is very difficult
to specify the total resource potential, an
indication of the magnitude of Canterbury's
wind energy resource can be gauged from this
study which estimated that at least 26% of the
land area could be utilised.

2.2.2(7) WIND MEASUREMENTS IN AN EQUATORIAL
REGION (SINGAPORE)
Nathan, G.K. and Goh, T.N. (Singapore National
University) Sol. Energy, 26 (3) 1981
pp. 275-278.

Presents a summary of wind data for a few of
the islands around Singapore to provide a basis
for evaluation of wind energy potential.
Available power was found to be affected by the
monsoon weather pattern, and off-shore islands
experience a more uniform wind speed throughout the day and night reaching a peak around
noon each day.

2.2.2(8) WIND ENERGY IN SOUTH ARGENTINA 1:
RESOURCE STUDY.
Barros, V. and Erramuspe, H.J. In proc. of
Fourth International Symposium on Wind Energy
Systems, Stockholm, Sweden 21-24 September 1982
V. 1 Cranfield, U.K., BHRA Fluid Engng 1982
Session A Paper A2 pp. 11-25.

Reports on the first part of a resource assessment programme in Patagonia. Measurements began
in March 1980 at four sites and digitised information covering wind speed at three levels (up to
90 metres) and wind direction at two levels were
recorded on magnetic tapes averaging values over
10 minutes. One year's results show that
monthly mean wind values range from 7 m/s to
12 m/s. Velocity probability density functions
have a maximum at around 9 m/s. Autocorrelations
are significantly higher than those of the same
latitude in North America. Implications of
the different sampling techniques on wind
energy evaluation are discussed.

2.2.2(9) WIND ENERGY IN THE LOWER ST. LAWRENCE
RIVER VALLEY AND SITE EVALUATION FOR PROJECT
AEOLUS.
Berry, R.L. and Taylor, P.A. (Atmos. Environ.
Serv.) In proc. of Fourth International
Symposium on Wind Energy Systems, Stockholm,
Sweden 21-24 September 1982 V. 1. Cranfield,
U.K., BHRA Fluid Engng. 1982 Session D
Paper D3 pp. 229-244.

A preliminary site selection programme has
resulted in the identification of several
potential sites in the Gulf of St. Lawrence,

Canada at which to erect a megawatt-scale wind
power system. A general overview is presented
discussing several climatological factors that
will impact on both the site selection and
turbine design processes. A programme of data
collection from 60 m towers at three sites is
described and some preliminary results are
presented. Indications are that there are
many good locations in the Gulf area for
exploiting wind energy, especially if natural
features are utilised beneficially.

2.2.2(10) THE WIND RESOURCE ASSESSMENT PROGRAM
IN QUEBEC, CANADA,
Kahawita, R. and others. (Montreal Ecole
Polytech.; Hydro-Quebec) In proc. of Fourth
International Symposium on Wind Energy Systems,
Stockholm, Sweden 21-24 September 1982 V. 1.
Cranfield, U.K., BHRA Fluid Engng. 1982
Session C Paper C2 pp. 155-166.

Provides an overview of the wind resource
assessment programme undertaken by the
provincial power utility Hydro-Quebec, Canada.
The methodology used is explained and the
results discussed. Supplementary studies of
airflow over complex terrain using numerical
modelling are described and the results
evaluated. Conclusive statements cannot yet
be made about the viability of the wind energy
resource, but tentative conclusions are that
wind energy as an alternate source of energy
is likely to be commercially viable since a
good wind regime and the availability of suitable land are present.

2.2.2(11) HAWAII'S WIND DATA BANK.
Neill, D.R. (Hawaii Univ. at Manoa) In proc.
of Fourth International Symposium on Wind
Energy Systems, Stockholm, Sweden
21-24 September 1982 V. 1. Cranfield, U.K.,
BHRA Fluid Engng. 1982 Session C Paper C4
pp. 175-194.

The long-term wind data measurement programme
is described in detail, including the current
programme which has 18 long-term wind data
stations. These, coupled with a short-term
measurement assistance programme and wind data
collection, reduction, processing, reporting,
storing, and using, make up the Hawaii Wind
Data Bank Program. It is designed to assess
Hawaii's wind energy resource and to provide
assistance for planning of wind energy conversion systems.

2.2.2(12) MEASUREMENTS FOR WIND ENERGY PROSPECT-
ING IN SWEDEN.
Olsson, L.E. and Kvick, T. In proc. of Fourth
International Symposium on Wind Energy Systems,
Stockholm, Sweden 21-24 September 1982 V. 1.
Cranfield, U.K., BHRA Fluid Engng. 1982
Session C Paper C1 pp. 167-174.

Wind and temperature profile measurements are
performed on 14 tall television masts throughout Sweden and data collected and analysed for
use in different wind energy projects in Sweden.
Studies include: improvement of wind energy
assessment, development of siting techniques,
turbulence characteristics and integration with

grid and meteorological measurement at prototype sites. Data are also collected by automatic weather station systems utilising the
public telephone network. Some results from the
two years of measurements are also presented.

2.2.2(13) WIND ENERGY IN CHINA.
Pengfei, S. (Xining Res. Inst. Mech. & Electr.
Engrs. Alt.) In proc. of Fourth International
Symposium on Wind Energy Systems, Stockholm,
Sweden 21-24 September 1982 V. 1. Cranfield,
U.K., BHRA Fluid Engng. 1982 Session A
Paper A3 pp. 27-34.
   According to the data collected by the
Meteorological Research Institute, the distribution of wind potential in China may be
classified into four different wind energy
regions. The best region includes the northern
part of China and the southeastern coast where
the wind energy density at 10 m is high and
available wind speed of 3-20 m/s occurs more
than 5,000 hours per year. A map of the
distribution of wind potential shows the
preliminary assessment of available wind energy
in China. The history of wind energy utilisation, research activities, vital characteristics of some wind machines and guiding
principles of developing wind energy in China
are presented.

2.2.2(14) NUMERICAL STUDY OF WIND ENERGY
CHARACTERISTICS OVER HETEROGENEOUS TERRAIN -
CENTRAL ISRAEL CASE STUDY.
Segal, M. and others. (Virginia Univ.; Jerusalem
Hebrew Univ.) Boundary Layer Meteorol., 22 (3)
March 1982 pp. 373-392.
   A numerical mesoscale meteorological model has
been applied to study wind energy character-
istics of three typical synoptic situations.
The supportive nature of this method for
observationally oriented wind energy studies
has been emphasised.

2.2.2(15) ANALYSIS OF SOME WIND SPEED DATA
FROM SOUTHERN AUSTRALIA WITH REFERENCE TO
POTENTIAL WIND ENERGY CONVERSION.
Carlin, J. and Diesendorf, M. (C.S.I.R.O.)
Wind Engng., 7 (3) 1983 pp. 147-160.
   Presents an analysis of special
measurements of hourly wind speeds and wind
directions relevant to the assessment of wind
energy potential in Western Australia, South
Australia and Tasmania. The summary statistics
derived from the wind speed data include annual
mean, standard deviation, frequency distribution, diurnal and seasonal variations, auto-
correlation function, mean and maximum lull
periods, and wind energy availability. The
results show Australia, including the northwest coast of Tasmania, has high wind energy
potential.

2.2.2(16) WIND-ENERGY INVENTORY FOR EUROPEAN
NORTH OF USSR.
Minin, V.A. and Stepanov, I.P. Power Eng.
(New York), 21 (1) 1983 pp. 98-105.
   Presents method for calculating the aero-
dynamic and energy characteristics of wind.

Ten-year series of eight standard observations
of wind speed at 164 meteorological stations
have been processed and the results used to
give the basic elements of the wind-energy
inventory for the European North of the USSR.
The areas most promising from the viewpoint
of utilisation of wind energy are indicated
and the wind-energy resources estimated.
(122 refs.)

2.2.2(17) A STUDY OF WIND SPEED STATISTICS
OF 14 DISPERSED U.K. METEOROLOGICAL STATIONS
WITH SPECIAL REGARD TO WIND ENERGY.
Halliday, J.A. Science & Engng. Research
Council, Rutherford, Appleton Lab. Report no:
RL-83-124 December 1983 103 pp.

2.2.2(18) STUDY OF WIND SPEED STATISTICS OF
14 DISPERSED U.K. METEOROLOGICAL STATIONS
WITH SPECIAL REGARD TO WIND ENERGY.
Halliday, J.A. Chilton, U.K., Rutherford
Appleton Lab. 1983 107 pp. Report no:
RL-83-124 (PB84-160415)
   Reviews the U.K. Meteorological Office techniques for measuring and recording wind speed
data, and the methodology for extrapolating
wind speed data from one height to another.
A statistical analysis of the wind speed was
carried out and Weibull parameters calculated.
A summary of the extreme hourly mean speeds and
gust speeds recorded is presented, as is a
limited analysis of the frequency and duration
of periods of calm and high winds.

2.2.2(19) METEOROLOGICAL MEASUREMENTS IN THE
NORTHERN GERMAN COASTAL AREA FOR WIND ENERGY
PROSPECTION.
Tetzlaff, G. and others. Bundesministerium
fuer Forschung und Technologie, Bonn-Bad
Godesberg, F.R. Germany. Report No. BMFT-FB-T-
84-017 January 1984 221 pp. (In German)
(Dept. of Energy Report no: DE 84751422)
   Wind speed and direction were measured for
the period from April 1979 to February 1982.
The wind sensors were mounted up to a maximum
height of 46 m at six different sites in the
area. The results for the measuring period
were compared to long term geostrophic wind
data and then matched. The wind data then
served to compute the wind power potential and
the energy yield of a wind energy converter
(GROWIAN).

2.2.2(20) TETHERED WIND SYSTEMS FOR THE
GENERATION OF ELECTRICITY.
Riegler, G. and Riedler, W. Inst. for Applied
Systems Technology, Austria. J. Solar Energy
Engineering 106 (2) May 1984 pp. 177-181.
   Power density increases considerably with
altitude, enhancing the feasibility of proposed
tethered wind energy conversion systems. Such
systems comprise a balloon-borne generator platform, a cable for tethering and conducting the
current, and a ground station for control and
energy distribution. The turbine unit,
generator unit, and other system components
are described. Energy availability and economic
considerations are also surveyed. (10 refs.)

3 Design of Wind Energy Conversion Systems

3.1 Systems Design and Siting

3.1(1) WIND ENERGY: A SUPPLEMENT TO HYDRO-
ELECTRIC ENERGY USING THE COLUMBIA RIVER
VALLEY AS AN EXAMPLE.
Chen, P.I. and Garg, V.K. (Portland State
Univ., Oregon, USA) In proc. Annual Meeting
of the International Solar Energy Society
American Section V. 1. Orlando, Fla.
6-19 June 1977. Cape Canaveral, USA.
ISES/AS 1977 Sect. 19 Session B. 2, 6 pp.

Explores a conceptual wind energy conversion system which consists of a wind power
conversion unit that pumps water from the tail
water level to the reservoir level in order to
store it there in a form of potential energy
that can be converted into electrical energy
through existing hydro-electric power plant.
Site selections are constrained not only by
the availability of wind power but also the
existence of hydro-power facilities. Economic
assessments of the wind power systems are
presented. (13 refs.)

3.1(2) DETERMINATION OF OPTIMUM ARRAYS OF
WIND ENERGY CONVERSION DEVICES.
Bragg, G.M. and Schmidt, W.L. (Univ. of
Waterloo, Ont., Canada) J. Energy, 2 (3)
June 1978 pp. 155-159.

Discusses the use of large-scale wind
energy conversion systems consisting of
arrays of individual wind machines. The
arrays have been analysed in some detail with
the aid of a rough boundary-layer velocity
profile model. The analysis indicates inter-
machine spacings that will provide for maximum
output from either the total array or individual
machines within the array. Using the results
obtained, detailed optimisation and economic
analyses can be made for large-scale wind
systems. (9 refs.)

3.1(3) FEATURE REVIEW OF SOME ADVANCED AND
INNOVATIVE DESIGN CONCEPTS IN WIND ENERGY
CONVERSION SYSTEMS.
Weisbrich, A.L. (Kaman Aerosp. Corp., Bloom-
field, Conn., USA) In Alternative Energy
Sources: An International Compendium V. 4.
Indirect Solar Energy edited by T.N. Veziroglu.
Washington, USA Hemisphere Publ. Corp. 1978
pp. 1663-1679.

Numerous alternative energy generation
schemes are being proposed and pursued. In
the area of wind energy much effort is being

expended to develop a superior energy conversion
system. Some advanced wind energy conversion
system (WECS) concepts are presented and their
features highlighted. (21 refs.)

3.1(4) OPTIMISATION OF WIND ENERGY CONVERSION
SYSTEMS.
Smith, M.C. (Mich. State Univ., East Lansing,
USA) Policy Anal. Inf. Syst., 2 (1) 15 July
1978 pp. 149-171.

Describes technical optimisation of wind
energy conversion systems, including site
selection. From the example given it is
shown that only system efficiency, wind
characteristics, energy desired, and the system
cost structure are required. (6 refs.)

3.1(5) RECENT DEVELOPMENTS IN WIND ENERGY.
Divone, L.V. (US Department of Energy)
In 2nd International Symposium on Wind Energy
Systems, Amsterdam, Netherlands 3-6 October
1978. Cranfield, U.K., BHRA Fluid Engng. 1978
pp. A3. 17-A3. 28.

Sums up the changing trend over the past few
years from studies and estimates to actual
experiments. Testing programmes on both small
wind turbines for dispersed, private use,
systems and large utility class machines are
presented and non-technical issues such as
demand charges for the small systems and the
potential for TV interference with large
systems are discussed. Present experimental
and prototype systems produce energy at 10
cents to 20 cents per kilowatt hour. The
potential for research and development and
production maturity to reach the cost requirements is discussed.

3.1(6) WIND ENERGY CONVERSION IN THE MW RANGE.
Lois, L. (Univ of Maryland, USA) In Alternative
Energy Sources: An International Compendium
V. 4. Indirect Solar Energy edited by
T.N. Veziroglu. Washington, USA. Hemisphere
Publ. Corp. 1978 pp. 1659-1667 (McGraw-Hill
outside USA)

The paper aims to show that certain wind
patterns above the continental United States
are particularly suited for wind energy conversion utilising wind powered stations in the
MWe range; to describe a system specifically
designed for such stations, and to present

calculations which show that such a system is within the range of existing technology. (14 refs.)

3.1(7) WIND ENERGY GENERATION WITH HYPERBOLIC COOLING TOWERS.
Rogers, P. (Calif. State Univ., Los Angeles, USA) In Los Angeles Council Eng. Sci. Proc. Ser. V. 4: Greater Los Angeles Area Energy Symposium, California 23 May 1978. North Hollywood, Calif., USA. West Period Co. 1978 pp. 68-72.

Describes the secondary utilisation of the exterior of hyperbolic cooling tower shells for supporting wind-rotors, and for producing electric energy directly and at a very reduced cost.

3.1(8) WIND POWER GENERATION.
In Alternative Energy Sources: An International Compendium V. 4. Indirect Solar Energy edited by T.N. Veziroglu. Washington, USA. Hemisphere Publ. Corp. 1978 pp. 1811-1961.

Includes a review of wind electric conversion technology; wind energy assessment; clean energy from humid air; wind energy conversion systems (WECS) for central station and dispersed power applications; energy analysis of a wind energy conversion system for fuel displacement; and aeroelastic pumps.

3.1(9) WIND TURBINES AND SITING.
In Alternative Energy Sources: An International Compendium V. 4. Indirect Solar Energy edited by T.N. Veziroglu. Washington, USA. Hemisphere Publ. Corp. 1978 pp. 1615-1808.

Includes chapters on an axial flow wind turbine with delta wing blades, windmills with increased power output due to tipvanes; feature review of some advanced and innovative design concepts in wind energy conversion systems; flap-cone control of windmill speed; theoretical studies of a hybrid wind turbine; analysis of the effect of turbulence on wind turbine generator rotational fluctuations; toroidal accelerator rotor platforms for wind energy conversion; the Campbell Chinese type windmill; location of windmills; and site selection for optimum wind power systems.

3.1(10) ANALYSIS OF THE POTENTIAL OF WIND ENERGY CONVERSION SYSTEMS.
Reed, J.W. (Sandia Lab., Albuquerque, NM, USA) In Renewable Energy Prospects: proc. of Conf. on Non-Fossil Fuel and Non-Nuclear Fuel Energy Strategies, Honolulu, Hawaii, USA 9-12 January 1979. Oxford and New York, Pergamon Press 1980. (Energy, 4 (5) 1979 pp. 811-822)

Wind energy conversion systems (WECS) are classed as solar energy systems because the sun drives atmospheric wind calculations. This paper reviews the flow of solar energy and wind, describes the time and space distribution of useful wind power, and considers some of the modern machinery that has been conceived to capture wind energy. Some limitations to practical wind energy extraction and use are

pointed out and available and projected wind power hardware systems summarised. (31 refs.)

3.1(11) SITING SMALL WIND MACHINES.
Wegley, H.L. and Pennell, W.T. (Pacific North-West Lab., Washington, USA) Paper presented at DOE Small Wind Turbine Systems R & D Requirements Conference, Boulder, USA. 27 February-1 March 1979 V. 1 pp. 198-212 (US Dept. of Energy).

Discusses siting strategy for small wind machines. Feasibility must be determined by estimating power outputs and power needs and economic analyses performed to formulate working budgets and consider initial system costs. Wind resource assessments should be conducted to select candidate sites and to choose suitable wind energy conversion systems. (8 refs.)

3.1(12) TOWER DYNAMICS ANALYSES AND TESTING.
Butterfield, S. and Sexton, J. (Rockwell Intl. Rocky Flats Plant) Paper presented at DOE Small Wind Turbine Systems R & D Requirements Conference, Boulder, USA. 27 February-1 March 1979 V. 1 pp. 171-178 (US Dept. of Energy).

Rockwell International has initiated a wind energy tower dynamics supporting research and technology study designed to evaluate structural computer codes capable of static and dynamic analyses of towers. Various codes applicable to the wind energy conversion industry in terms of simplicity and low cost are identified. (6 refs.)

3.1(13) WIND ENERGY CONVERSION SYSTEM TRANSIENT PERFORMANCE ANALYSIS.
Price, W.W. and Macklis, S.L. (Gen. Electr. Co.) In proc. Workshop on Economic and Operational Requirements and Status of Large Scale Wind Systems, Monterey, USA. 28-30 March 1979. Palo Alto, USA. Electr. Power Res. Inst. 1979 pp. 366-377. (Special Report no: ER-1110-SR) (DOE Conf. 790352)

Various parameters such as voltage fluctuations, drive train efficiency and the effects of gusts of wind are considered.

3.1(14) FEASIBILITY STUDY FOR A HIGH-ALTITUDE WIND POWER PLANT.
Reigler, G., Riedler, W. and Horvath, E. (Institut fur Angewandt Systemtechnik, Austria) Paper presented at DOE/SERI 2nd Conference on Wind Energy Innovative Systems, Colorado Springs, USA 3-5 December 1980 pp. 93-111.

A high-altitude wind power plant consists of a balloon-borne platform for electrical power generators and a cable connection for tethering and energy transfer. A ground station is also envisaged as part of the system for meteorological and technical control, and energy distribution. The aim for such generating stations is a continuous power output of 80-100 MW for a platform with capital and operating costs competitive with conventional thermal power stations.

3.1(15) NEW APPROACHES IN SMALL WIND SYSTEM
DESIGN.
Rogers, E. (Gale Co., Wisconsin, USA) Paper
presented at International Solar Energy Society
American Section 1980 Annual Conference,
Phoenix 2-6 June 1980 V 3.2 pp. 1478-1480.

A wind system that is optimised for the
most frequently occurring moderate wind site
is proposed. This system is considered
suitable for sites with average annual wind
speeds of 10-12 mph. Return on investment
is used as a measure of system cost effectiveness.

3.1(16) PRELIMINARY REPORT ON AN ASSESSMENT
OF A TETHERED WIND ENERGY SYSTEM.
Furuya, O. and Maekawa, S. (Tetra Tech Inc.)
In SERI Second Wind Energy Innovative Systems
Conference, Colorado Springs, USA 3-5 December
1980 V. II. Golden, USA Solar Energy Res.
Inst., 1980 Session 3 pp. 37-54. (Solar Energy
Res. Inst. Report no: SERI/CP-635-1061)

Describes a study begun in September 1980
for assessing a tethered wind energy system as
a possible energy source under contract from
the Solar Energy Research Institute. Key
items for study include the relationship
between the rated power and altitude as a
function of the seasonal wind speed; required
lifting force for carrying the platform and
cable weight; and generated power and power
cable diameter (i.e. power transmission
problem). The survivability of the TWES in
a wind gust or other accident, and the
problem of maintenance and repair are also
carefully studied.

3.1(17) SCREENING METHOD FOR WIND ENERGY
CONVERSION SYSTEMS.
McConnell, R.D. (Solar Energy Res. Inst.,
Golden, Colorado, USA) In proc. Annual
Meeting International Solar Energy Society
American Section 1980, Solar Jubilee 25 Years
of the Sun at Work V. 3. 2, Phoenix, Arizona,
USA 2-6 June 1980. Del., USA ISES/AS 1980
pp. 1507-1511.

Describes a screening method developed for
evaluating wind energy conversion systems
logically and consistently. The method uses
both value indicators and simplified cost
estimating procedures. Value indicators are
selected ratios of engineering parameters
involving energy, mass, area and power.

3.1(18) TECHNICAL AND MANAGEMENT SUPPORT FOR
THE DEVELOPMENT OF SMALL WIND SYSTEMS, 1980.
PROGRAM SUMMARY.
Golden, USA, Rockwell Intl. 1 March 1980 41 pp.
(RFP-3121/3533/80/8)

An overview of the Rocky Flats Small Wind
Systems Programme operated by the Rockwell
International Energy Systems Group for the US
Department of Energy. It provides technical
and management support for the development of
small wind systems. Overall objective is to
stimulate the manufacture of small wind energy
conversion systems by the private sector and
utilisation of these systems by the public.

The current programme in terms of its objectives
and role, interprogramme relationships, administration activity highlights and plans for
1980 with projections through 1982 is
presented.

3.1(19) WIND ENERGY - HOW RELIABLE.
Sherman, D.J. (Aeronautical Research Labs.,
Melbourne, Australia) ARL-Structures Report
no: ARL/STRUC-380 January 1980 36 pp.
(AD-A094988/3)

Reliability of a wind energy system depends
on the size of the propeller and the size of
the back-up energy storage. Design of the
optimum system for a given reliability level
can be performed if a time series of wind speed
data is available. However, a design based on
conventional meteorological records, which
sample the wind speed with a ten minute averaging time at three-hourly intervals, will overestimate the storage by a factor of 2, and if
the wind speed is only available on a daily
basis the storage will be over-estimated by a
factor of 2.5 to 4.0. This is because a
propeller can respond to wind speed changes in
much less than ten minutes and also because
three-hourly sampling does not often pick up
the brief high-speed incidents which generate
a significant part of the wind energy. A nomo-
gram is presented, based on some continuous wind
speed measurement, which enables storages
calculated from three-hourly or daily data to
be appropriately reduced because of these two
effects. (Author's abstract)

3.1(20) ASPECTS OF A WIND ENERGY CONVERSION
SYSTEM.
Sexon, B. and others. (Reading Univ.) In proc.
of Conference on Energy for Rural and Island
Communities, Inverness, U.K. 22-24 September
1980. J. Twidell (ed.) Oxford, U.K. Pergamon
Press Ltd. 1981 Topic D Paper D4 pp. 153-158.

Describes a wind turbine backed up by a short
term battery storage facility and a diesel
generator. The batteries store any excess
energy from the turbine, delivering it to the
load if the turbine is unable to produce
sufficient power at any instant. The diesel
generator ensures the continuation of supply
during any lengthy period of no wind condition
which may occur.

3.1(21) DESIGN CONSIDERATIONS FOR SMALL WIND
ENERGY CONVERSION AND STORAGE SYSTEMS.
Chang, G.C. (Cleveland State Univ., Ohio, USA)
In proc. 16th Intersoc. Energy Conversion
Engng. Conference V. 2. Atlanta, Ga., USA
9-14 August 1981. New York, NY., USA, ASME
1981 pp. 2070-2074.

Describes a small wind turbine generator
system capable of providing electricity to
an all electric residence. Several major
design considerations are examined including
the characteristics of a wind turbine
generator, the nature of the electric load
demand, the available wind resources, siting
considerations, and other related socioeconomic factors. (6 refs.)

3.1(22) ENERGY REQUIREMENTS OF LARGE WIND TURBINE SYSTEMS.
Dixon, J.C. and Lowe, R.J. (Open University, U.K.) In proc. 3rd BWEA Wind Energy Conference, Cranfield, U.K. 9-10 April 1981. Cranfield, U.K. BHRA Fluid Engng. 1981 pp. 111-119.

Gives results of a full process analysis of large land and offshore wind energy systems. The land system is found to have a payback of 0.47 years and an energy gain of 50. The offshore system in shallow water (20 m) has a payback period of 0.37 years and an energy gain of 55. These figures appear to be superior to those exhibited by nuclear systems, and therefore as a fuel-saving technology, from the energy analysis aspect, wind power is very competitive according to the authors.

3.1(23) GENERAL INTRODUCTION TO WIND ENERGY CONVERSION.
De Vries, O. In Lecture Series on Wind Energy Conversion Devices, Rhode Saint Genese, Belgium 1-5 June 1981. Rhode Saint Genese, Belgium, Von Karman Institute of Fluid Dyn., 1981 126 pp. (Lecture Series 1981-8)

A short survey of different wind turbine and wind concentrator concepts is given and the aerodynamic characteristics of wind turbines discussed. Includes a survey of the choice of the various parameters, determining a WECS, such as wind data, turbine control, conversion system, structural, dynamic and cost aspects and environmental impediments. The review concludes with a short survey of the design criteria and the desirability of formulating building codes for WECS.

3.1(24) HYBRID SOLAR-WIND ENERGY CONVERSION SYSTEMS METEOROLOGICAL ASPECTS.
Aspliden, C.I. (Battelle Pacific NorthWest Labs., Richland, Washington, US Department of Energy) Report no: PNL-SA-10063 CONF-811175-1 December 1981 16 pp. (WMO Technical Conference on Meteorology and Energy, Mexico City, Mexico 3 November 1981) US Dept. of Energy Report no: DE 82005798.

Solar-wind hybrid systems may be more attractive and suitable to meet specific power demands and/or to avoid expensive storage systems (or expansion of conventional systems) than wind or solar energy extracting systems alone. To investigate the feasibility of operation of hybrid wind-solar systems at a particular site, requires detailed and simultaneously acquired information on solar and wind energy availability on time scales from minutes to hours over a period of at least a year.

3.1(25) OVERVIEW OF WIND ENERGY SYSTEMS: ISSUES IN DEVELOPMENT AND APPLICATION.
Moretti, P.M. and Thresher, R.W. (Okla. State Univ., Stillwater) ASME Trans., J. Solar Energy Engng., 103 (1) February 1981 pp. 3-10.

Presents an overview of the current status of wind energy technology and system development with major emphasis placed on the key issues which face the commercialisation of wind technology. Basic fundamentals of the tech-

nology are reviewed and the direction of current development is outlined. Economic considerations are discussed both from a machine development point of view and the electricity utility-industry cost-of-service approach. (32 refs.)

3.1(26) POSSIBILITIES AND LIMITATIONS OF WIND ENERGY UTILISATION.
Feustel, J.E. (Maschinenfabr. Augsburg-Nuremberg, Munich, Germany) Intl. J. Ambient Energy, 2 (4) October 1981 pp. 197-205.

With higher fuel prices and reduced energy resources, wind energy again becomes attractive for many applications. In almost all coastal regions where mean annual wind speed exceeds 5 m/s the utilisation of wind energy is promising. The applications range from small size water pumping stations at low rotational speed up to large size electricity generating wind energy converters at high rotational speed. Well designed systems should be able to achieve 25-30% annual electrical efficiency. Available wind energy sites, possible applications of wind energy and design and construction of wind energy converters are reviewed. (10 refs.)

3.1(27) SMALL WIND ENERGY CONVERSION SYSTEMS: THE OPTIONS.
Cuddy, J. (Minnesota Energy Agency, USA) Paper presented at International Solar Energy Society American Section 2nd Conference on Wind Power Energy Alternatives for MidWest, Minnesota 3-4 April 1981 pp. 5-7.

Describes various applications of small wind energy conversion systems. Overall efficiency of a particular system will depend on the choice of system components and the end use of the power. SWEC's with battery systems for storing DC electrical energy can serve remote telecommunications equipment or residences located far from existing powerlines; SWEC's connected directly into a utility grid can feed electricity in excess of the onsite demand into the utility lines. A thermal system may replace solid fuel for space and water heating.

3.1(28) SMALL WIND SYSTEMS TECHNOLOGY ASSESSMENT.
Shepherd, D.C. (Rockwell Intl., USA) In proc. 5th Biennial Wind Energy Conference & Workshop, Washington, DC., USA 5-7 October 1981 I.E. Vas (ed.) V. 1. Palo Alto, USA Solar Energy Research Institute 1981 Session 1B pp. 215-226. (Report nos: SERI/CP-635-1340 & CONF-811043)

An assessment of small wind systems technology is conducted periodically at Rocky Flats to determine the technical advancements and needs of the SWECS industry. Performance, reliability, costs of currently available systems, adequacy of design analysis methods are examined. Results of a recently completed assessment are presented in this paper. Hardware performance has been quantified using a number of 'Figures of Merit' (FOM's). FOM's are presented for:

system annual energy yield, system availability, average power train efficiency, and the ratio of actual to expected annual energy production. The adequacy of rotor aerodynamic performance and tower dynamics prediction methods is assessed by comparing analytical and test results.

3.1(29) SOLAR TECHNOLOGY ASSESSMENT PROJECT. VOLUME VIII. WIND ENERGY.
Hughes, W.L., Ramakumar, R.G. and Lingelbach, D.D. (Oklahoma State Univ., Stillwater, Engineering Energy Lab.; University of Central Florida, Orlando, Florida Solar Energy Centre) Washington USA Dept. of Energy Report no: DOE/CS/30278-T10 April 1981 143 pp. (DE 81029009)
Gives a brief historical perspective of wind energy utilisation followed by a discussion of the potential uses, economic costs and technical difficulties. A discussion of the statistical characteristics of the wind follows for a moderate to high wind area in the United States (Oklahoma City).
Information on average available energy on an annual basis is presented, along with approximately monthly variations. A number of the various types of windmills in existence are discussed briefly and data on efficiencies and power coefficients for a variety of turbines is presented. Small and large WEC systems are then discussed with some detail given on small system applications and economics.

3.1(30) SOME PRACTICAL ASPECTS OF SMALL WIND ENERGY CONVERSION SYSTEMS.
Watson, G.R. (Energy Centre Ltd.- Northumbrian Energy Workshop Ltd.) In proc. Third BWEA Wind Energy Conference, Cranfield, U.K. 9-10 April 1981. Cranfield, U.K., BHRA Fluid Engng. 1981 pp. 145-152. (British Wind Energy Association)
The paper considers the quest for efficiency in small wind energy conversion systems which may take different directions from those considered important in high megawatt systems. The system must be considered as a whole and the importance of rotor performance is less than factors that will influence the actual usable energy output.

3.1(31) UTILITY-SIZED MADARAS WIND PLANTS.
Whitford, D.H. and Minardi, J.E. (Univ. of Dayton Research Inst., USA) Intl. J. Ambient Energy, 2 (1) January 1981 pp. 3-22.
The Madaras Rotor Power Plant Concept utilises rotating cylinders, mounted on flat cars, to achieve an interaction with the wind and propel a train of cars around a track at constant speed. The Concept has been updated to determine whether it can compete with horizontal axis wind-turbine generators. The study consisted of a wind-tunnel test series, a performance analysis, and a cost analysis. The findings indicated that the most efficient Madaras plants should be equipped with race-track platforms and that utility-sized plants could be built. (5 refs.)

3.1(32) THE WIND POWER BOOK.
Park, J. Palo Alto, USA, Cheshire Books 1981 255 pp. (ISBN 0-917352-05-X)
Covers wind power systems, wind energy resources, wind machine fundamentals and turbine design, building and siting a wind power system, and wind power economics, legal and social issues.

3.1(33) WIND SYSTEM EXPERIENCES OVERCOMING THE TECHNOLOGY LAG.
Schmidt, J. Alternative Sources of Energy, July-August 1981 (50) pp. 10-12.
Briefly describes operating experiences of two wind energy conversion systems sited in Minnesota to illustrate the technology lag associated with wind energy. This derives from the marketing of new designs with little or no testing to ensure their ability and reliability. Poor assembly and inadequate operating instructions and inefficient components were the primary causes of failure for the two machines discussed.

3.1(34) WINDPOWER: A HANDBOOK ON WIND ENERGY CONVERSION SYSTEMS.
Hunt, V.D. New York, USA., Van Nostrand Reinhold Co., 1981 626 pp.
Topics covered include wind characteristics and operation of wind energy conversion systems; system design characteristics; towers; electric power; conversion and storage systems; WECS applications; the US Federal Wind Energy Programme; commercialisation; environmental, institutional and legal barriers; and international developments.

3.1(35) AEROELASTIC STABILITY AND DYNAMIC RESPONSE CALCULATIONS FOR WIND ENERGY CONVERTERS.
Vollan, A. In papers presented at Fourth International Symposium on Wind Energy Systems, Stockholm, Sweden 22-24 September 1982 V. 1. Cranfield, U.K. BHRA Fluid Engng. 1982 Session G Paper G2 pp. 427-444.
More or less severe vibration problems may occur for almost all types of wind energy converters. Describes a rational and automatic method for the aeroelastic stability and dynamic response calculations which are based on a combination of the finite element method with a component mode coupling technique of the rotor and tower substructures. The resulting equations with periodic coefficients are solved by using the Floquet method for the stability problem. In the time history calculations, dynamic forces and stress values may be assessed.

3.1(36) DESIGN LESSONS FROM THE U.S. DOE SMALL WIND SYSTEMS DEVELOPMENT PROJECTS.
Healy, T.J. and Dodge, D.M. (Rockwell Intl. Corp., USA) In Fourth International Symposium on Wind Energy Systems, Stockholm, Sweden 21-24 September 1982 V. 1. Cranfield, U.K. BHRA Fluid Engng. 1982 Session F Paper F1 pp. 351-363.

A series of wind system prototype development projects were initiated at Rocky Flats by the Rockwell International Energy Systems Group under the U.S. DOE Federal Wind Energy Programme. The first eight prototypes in the 1=2 kW, 8 kW and 40 kW size ranges have been tested and modified to correct early design problems which have involved control systems and startup, vibration, fatigue, fasteners, and noise propagation. Four prototypes in the 4 kW and 15 kW size ranges were shipped to Rocky Flats in early 1982.

3.1(37) EFFECT OF WIND LOADING ON THE DESIGN OF A KITE TETHER.
Varma, S.K. and Goela, J.S. (Indian Inst. Technol., Kanpur) J. Energy, 6 (5) September-October 1982 pp. 342-343.
In a tethered wind energy conversion system, the tether is usually designed with the assumption that the tether static profile is a catenary. However, a catenary cannot accurately represent the tether profile in situations where the wind drag is of the same order of magnitude as the weight of the tether. The tether drag varies as the tether diameter, whilst the tether weight is proportional to the square of the diameter. Consequently, for small-diameter tethers or when the tethers are made of materials with high strength-to-weight ratios, the deviation from a catenary behaviour may become appreciable. Considers the effect of wind drag on the static profile and force transmission efficiency of a tether connected to a kite-like system.

3.1(38) REVIEW OF LARGE WIND TURBINE SYSTEMS.
Lerner, J.I. and Selzer, H. ASES Advances in Solar Energy, 1982 pp. 175-188.
Reviews the technical and economic status of large wind turbine systems nearing commercial readiness. Emphasis is placed on preliminary operating results to date and on the design tradeoffs that illustrate why large turbines are designed on the basis of compromise. Energy capture, production costs, and operating-maintenance costs are considered. Rotor blades, drive trains and other system components and materials are surveyed.
(22 refs.)

3.1(39) A TOTAL ENERGY WIND CONVERSION SYSTEM.
Bandopadhayay, P.C. (Commonwealth Scientific & Industrial Research Organisation) Wind Engng. 6 (2) 1982 pp. 85-94.
A Total Energy Wind Conversion System is proposed which supplies airconditioning and hot water, as well as electrical load. It is shown that such a system can improve the economic justifications for the use of wind energy to supply electricity to an isolated consumer.

3.1(40) THE UNIVERSITY OF DAYTON ALTERNATE ENERGY HYBRID RESEARCH FACILITY.
Boehman, L.I. and others. (Dayton Univ., Ohio, USA) In proc. 1982 Wind and Solar Energy Technology Conference, Kansas City, USA 5-7 April 1982. Columbia, USA, Missouri-Columbia Univ., 1982 pp. 121-131.
Describes a hybrid solar and wind energy system including a wind energy converter, solar photovoltaic cell array, and storage battery. A computer for system monitoring is being used to identify cost-effective combinations of solar and wind energy components which satisfy a specific load in a given location.

3.1(41) WIND ENERGY AND DESIGN PARAMETERS OF AUTONOMOUS WIND DIESEL POWER UNITS.
Fritzche, A. (Dornier Syst. GmbH) In Fourth International Symposium on Wind Energy Systems, Stockholm, Sweden 21-24 September 1982 V. 2. Cranfield, U.K., BHRA Fluid Engng. 1982 Session M Paper M3 pp. 297-304.
Presents some design criteria for autonomously operating power supply systems consisting of a wind energy converter and a diesel generator. Specific speeds and pressure coefficients of wind energy converters have been calculated and will be compared with those of other turbomachinery. The ratio of kinetic energy of a rotor and rated power is a time-parameter connected with non-stationary operation.

3.1(42) WIND POWER PLANTS: THEORY AND DESIGN.
Le Gourieres, D. Oxford, U.K., Pergamon Press 1982 285 pp.
Chapters are included on the wind, the notions of fluid mechanics necessary to the understanding of wind energy problems, horizontal axis and vertical axis wind installations, the use of wind energy for pumping water and producing electricity, the problem of adapting the wind rotor to electrical generators or to pumps and the design of various types of turbine and overall power plant.

3.1(43) PERFORMANCE AND SIZE ESTIMATING FOR WIND SYSTEMS.
Moment, R.L. (Rockwell International, Golden, Colorado, Rocky Flats Plant) Rocky Flats Plant Report no: RFP-3586 February 1983 72 pp.
(US Dept. of Energy Report no: DE 84006980)
A method is described for estimating wind energy conversion system performance, given wind speed, and annual energy production in common wind regimes for different rotor diameters ranging from 10 to 140 ft. Results obtained using this method were verified for one WECS using a rotor performance code coupled with drive train efficiency tables. Curves show that a minimum rotor diameter exists for a specified level of energy production. Plots are also presented which illustrate the effect of wind regime on annual energy generated and the relationships between rated wind speed and average wind speed.

3.1(44) STRATEGY FOR OPTIMISATION OF WIND ENERGY SYSTEMS.
Westberg, S. (Univ. of Linkoping, Dept. of

Mechanical Engineering, Linkoping, Sweden)
Wind Engineering, 7 (2) 1983 pp. 104-114.

Presents a strategy to deal with wind energy systems as an optimisation problem. An attempt is made to create a general procedure to describe a complete system, divided into three independent parts: the application, the wind turbine and the state of wind. A wind rotor directly connected to the compressor of a heat pump is used as an example.

3.1(45) TRANSFORMATION OF WIND ENERGY BY A HIGH-ALTITUDE POWER PLANT.
Riegler, G., Riedler, W. and Horvath, E. (Research Centre Graz, Austria) J. Energy, 7 (1) January-February 1983 pp. 92-94.

Describes attempts to optimise a high-altitude power plant in respect of capacity and energy yield and to estimate the development and operation costs to be expected. The costs given are only guiding principles which may be corrected by further investigations, including tests on the pilot plant.

3.1(46) WIND ENERGY.
Taylor, R.H. In Alternative Energy Sources for the Centralised Generation of Electricity, Bristol, U.K., Adam Hilger Ltd. 1983 Chapter 2 pp. 9-67.

One chapter of a book on alternative energy sources. The author gives a historical introduction to wind energy and includes sections on wind characteristics, simple aerodynamics of windmills, machine design options, materials for wind turbines, environmental effects and siting (including offshore) of wind turbines, economics and world research and development in wind energy.

3.1(47) WIND ENERGY CONVERSION SYSTEMS.
Mikhail, A. (Solar Energy Research Inst., USA) Electr. Conserv., 3 (4) April 1983 pp. 6-9

The latest development in wind system technology is the development of wind energy conversion systems that can produce direct current or alternating current power synchronously or in a stand alone mode. Technical characteristics and cost data are given for large-scale centralised WECS, and commercially available stand-alone small WECS.

3.1(48) WIND TURBINE PARKS. ANALYSIS OF SITES.
(Jysk-Fynske Elsamarbejde, Fredericia, Denmark) Report no: NP-3751372 April 1983 170 pp. (In Danish) (DE 833751372)

Sites for 46 wind turbine parks are selected in the area of the Jutland-Funen electric utilities - ELSAM, where it is possible to place about 600 wind turbines of 2.5 MW each. A wind turbine park can have about 15 wind turbines. The report briefly describes the wind conditions in Denmark, the landscape, and the problems connected with siting of wind turbines. The two main parts of the report contain calculations of the expected wind power production, the connection costs, and detailed descriptions of the selected wind turbine parks.

3.1(49) FIVE POWER GENERATING WINDMILLS. FINAL REPORT.
DeBever, O.J. (O.J. DeBever Assoc., Bay Saint Louis, MS.) US Dept. of Energy Report no: DOE/R4/10163-T1 1984 2 pp. (DE 84000329)

The windmills are being installed on two 80 ft. self supporting steel towers and three on 90 ft. guyed steel towers. The generators are rated 1800 watts at a wind velocity of 24 mph and have a maximum capability of 2100 watts at 28 mph. The prime power is a 3-bladed prop having a peripheral diameter of 13 ft. The windmill will start turning when the wind velocity reaches 9 mph at which speed it will generate about 200 watts. Power increases approximately 130 watts with each 2 mph increase in wind velocity. (Author)

3.1(50) THE ORKNEY PROJECT: WIND TURBINE PROJECT.
Armstrong, J.R.C. and others. In Fourth International Conference on Energy Options. The Role of Alternatives in the World Energy Scene, London, England 3-6 April 1984 pp. 105-111. IEE. London, England IEE 1984 421 pp. (SBN O-85296-290-8)

Reviews the first findings of the wind measurement programme and describes in outline the commissioning and initial operation of both the 250 kW machine and the data acquisition system. An overview is given of the design analysis of the monitoring programme and the configuration of the 3 MW machine is described. (8 refs.)

3.1(51) UNTAPPED MIDWESTERN WIND ENERGY RESOURCE: NOCTURNAL LOW-LEVEL WIND MAXIMA.
Frenzen, P. and Sisterson, D.L. (Argonne National Lab., IL.) US Dept. of Energy Report no: CONF-840469-1 1984 14 pp. (Also in Recent Developments in Electric Utility Research Conference, Chicago, IL., USA 3 April 1984.) (DE 84006424)

Demonstrates that the use of the seventh-root power-law profile for wind energy resource assessment underestimates wind power density aloft in the U.S. MidWest. Much of this error is caused by the failure of this relationship to simulate the contributions of nocturnal wind maxima, a regional circulation which frequently forms above 100 m. over extensive, flat terrains. This phenomenon is shown to contribute significant additional energy at the 50 m. level during the Summer and early Autumn periods when midwestern wind energy assessments indicate a deficiency. It is also shown that increasing a wind system's tower height could gain significant additional energy.

3.1(52) WIND ENERGY.
Jufer, M. (EPF, Lausanne, Switzerland) Bull. Assoc. Suisse Electr. (Switzerland) 3 March 1984 75 (5) pp. 245-249. (In French, German)

Author surveys the possibilities of wind energy and the factors decisive for its exploitation, then describes the latest types of windmills and their main characteristics.

Performance is compared in a wind installation developed by the Ecole Polytechnique Federale of Lausanne (EPFL). Formulae are given for computing recoverable wind energy. Doubts are also expressed about the future of large aerogenerators.

3.1(53) NEW ELEMENTS IN WIND ENERGY CONVERSION SITING.

Lois, L. (Maryland University at College Park) In Alternative Energy Sources III, V. 4 - Indirect Solar/Geothermal Energy, proc. 3rd Miami Intl. Conference on Alternative Energy Sources, Miami Beach, USA 15-17 December 1980. T. Nejat (ed.) pp. 21-30. Washington, USA, Hemisphere Publishing Corp. 1983 (ISBN 0-89116-226-0)

Quantifies a number of site characteristics for site selection for wind powered generators. It is assumed that the wind powered generators will be integrated in a utility grid which is in addition supplied from conventional power sources and does not have energy storage capabilities. The objective is to enable the selection of the available sites in such a manner as to maximise the desirable benefits from the installation of the wind powered generators.

3.1(54) WIND SHEAR MEASUREMENTS AND SYNOPTIC WEATHER CATEGORIES FOR SITING LARGE WIND TURBINES.

Kirchhoff, R.H. and Kaminsky, F.C. (Univ. of Massachusetts) J. Wind Engineering & Industrial Aerodynamics 15 1983 pp. 287-297.

A histogram of 173 observed shear coefficients is shown together with a plot of the normal density function that has hypothesised to be the appropriate model. Random behaviour of the coefficient within three of five possible weather categories is examined, which have important applications in modelling the energy produced from specific wind energy conversion systems. (13 refs.)

3.1(55) A UNIFIED SITE EVALUATION SYSTEM FOR WIND ENERGY CONVERSION.

Biro, G.G. (Gibbs Hill Inc.) In Alternative Energy Sources III, V. 4 - Indirect Solar/ Geothermal Energy, proc. 3rd Miami Intl. Conference on Alternative Energy Sources, Miami Beach, USA 15-17 December 1980. T. Nejat (ed.) pp. 31-67. Washington, USA, Hemisphere Publishing Corp. 1983 (ISBN 0-89116-226-0)

The described evaluation system includes all field and office engineering work needed for proper site selections and for writing the environmental impact statement.

3.1.1 OFFSHORE SYSTEMS

3.1.1(1) EVALUATION OF OFFSHORE SITE FOR WIND ENERGY GENERATION.

Kirschbaum, H.S., Sulzberger, V.T. and Somers, E.V. (Westinghouse Electr. Corp.,

Pittsburgh, Pennsylvania, USA) In IEEE Power Eng. Soc., Text of Paper from the Summer Meeting, Portland, Oregon 18-23 July 1976. Publ. by IEEE, New York, NY 1976 Pap A-76-398-8 7 pp.

An analysis of the potential for wind generation at an offshore site indicates a potential in excess of 5700 kwh/kw for a 1Mw windmill rated at 20 mi/h and a hub height of 235 feet. As a limited supplement to base loaded nuclear and other forms of generation, it appears to offer enough promise for a further study to be justified to determine the overall economic, technical and environmental feasibility of such an application.

3.1.1(2) OFF-SHORE MULTI-MW SIZE WIND TURBINE SYSTEM DEVELOPMENT IS THE KEY TO COST-EFFECTIVE WIND ENERGY FOR SWEDEN.

Ljungstrom, O. (FFA, Sweden) In Energy and Aerospace proc. of Anglo American Conference organised by R. Aeronaut. Soc. and the AIAA, London 5-7 December 1978. London, U.K. Royal Aeronautical Society, 1978 13 pp.

The author stresses the advantage of off-shore wind turbines over land-based wind power systems for Sweden. Based on current system studies of the vertical axis type wind turbine in multi-MW sizes, it should be possible to develop very large off-shore units, up to 15 MW or even 25 MW, assembled complete at shipyards and towed to the site. These will require much less area and much smaller number of units for a given energy production than current 4-5 MW unit size land based designs. (6 refs.)

3.1.1(3) DESIGN STUDY AND ECONOMIC ASSESSMENT OF MULTI-UNIT OFFSHORE WIND ENERGY CONVERSION SYSTEMS APPLICATION. VOLUME I: EXECUTIVE SUMMARY; VOLUME 2: APPARATUS DESIGN AND COSTS; VOLUME 3: SYSTEM ANALYSIS; VOLUME 4: METEOROLOGICAL AND OCEANOGRAPHIC SURVEYS; FINAL REPORTS.

Kilar, L.A. (Westinghouse Electric Corp., Pittsburgh, PA.) Washington, DC. US Dept. of Energy. Westinghouse Report no: WASH-2330-78/4 (V.1) (V.2) (V.3) (V.4) June 1979 59 pp., 310 pp., 192 pp. & 344 pp.

Presents information concerning meteorological and oceanographic surveys of US offshore sites; apparatus designs and costs; and OWECS system economics and design.

3.1.1(4) OFFSHORE WIND ENERGY SYSTEMS FOR THE U.K.

Musgrove, P.J. (Reading Univ., U.K.) In International Conference on Future Energy Concepts, London, U.K. 10 January-1 February 1979 pp. 309-312. London, Institution of Electrical Engineers 1979 IEE Conf. Publ. no: 171.

There has been a growing awareness, in the U.K. and elsewhere, that offshore wind energy systems have much to commend them, despite the expense of construction. In the U.K. context one vital consequence of siting wind turbines offshore is that it greatly increases the potential contribution to energy needs. Offshore wind energy systems could in fact contribute more than one quarter of electricity needs; possibly much more if operation waters deeper than 20 metres prove economic. Considers some of the more important factors affecting the design of offshore wind energy systems. (16 refs.)

3.1.1(5) TECHNICAL AND ECONOMIC ASSESSMENT OF OFFSHORE WIND ENERGY CONVERSION SYSTEMS.
Kilar, L.A. and Chowaniec, C.R. (Westinghouse Electr. Corp., Pittsburgh, PA.) In IEEE Power Eng. Soc. Preparatory Summer Meeting, Vancouver, BC. 15-20 July 1979. New York, NY., USA IEEE 1979 Pap A-79-454-0 6 pp.
  Presents the principal results of a comprehensive assessment of offshore wind energy conversion systems. Designs and associated costs are given for the major components of multi-unit installations. Offshore energy costs are developed in terms of controlling environmental parameters, distance from shore, and equipment type.
See also similar paper by Kilar in Energy Technology: proceedings of 7th Energy Technology Conference, V. 2: Expanding Supplies and Conservation, sponsored by Govern. Inst. Inc., Washington 24-26 March 1980 pp. 1472-1483.

3.1.1(6) COASTAL ZONE WIND ENERGY. PART 1 - SYNOPTIC AND MESOSCALE CONTROLS AND DISTRIBUTIONS OF COASTAL WIND ENERGY.
Garstang, M. and others. (Virginia Univ., USA) Washington, DC., USA US Dept. of Energy March 1980 191 pp.
  Describes a method of determining coastal wind energy resources. Climatological data and a mesoscale numerical model are used to delineate the available wind energy along the Atlantic and Gulf coasts of the United States. The spatial distribution of this energy is dependent upon the locations of the observing sites in relation to the major synoptic weather features as well as the particular orientations of the coastline with respect to the wind.

3.1.1(7) OFFSHORE WIND ENERGY SYSTEMS.
Musgrove, P.J. Meteorol. Mag., 109 (1293) 1980 pp. 113-119.
  General review paper on wind energy systems design, costs and potential.

3.1.1(8) REQUIREMENTS FOR THE PROVISION OF OFFSHORE WIND DATA RELATING TO WIND ENERGY CONVERSION SYSTEMS.
Harris, R.I. Feltham, U.K., National Maritime Inst., February 1980 28 pp. NMI Report no: 75.
  Considers the effect of gusts and means of control of systems utilising data from wind instrumentation.

3.1.1(9) AN ASSESSMENT OF OFFSHORE SITING OF WIND TURBINE GENERATORS IN THE UNITED KINGDOM.
Simpson, P.B., Lindley, D. and Hardy, W.E. (Haywood Engineering, U.K.) Paper presented at 3rd International Conference on Future Energy

Concepts, London 27-30 January 1981 pp. 264-272. London, Institution of Electrical Engineers 1981.
  Reports on a technical and economic assessment of the generation of electricity using wind turbine generators located in shallow coastal waters of the U.K. Investment costs, including fixed construction costs, size dependent construction costs, and breakdown costs are analysed. Recurring costs are also considered, such as plant replacement and inspection and maintenance costs. Generating costs can be reduced by increasing rotor diameter, improving rotor performance, and developing a lighter support structure. (5 refs.)

3.1.1(10) OFFSHORE WIND-ENERGY-CONVERSION SYSTEMS.
Kilar, L.A., Stiller, P.H. and Ancona, D.F. (Westinghouse Electr. Corp., Pittsburgh, PA., USA) J. Energy, 5 (2) March-April 1981 pp. 79-83.
  Summarises the findings of a comprehensive assessment of offshore wind-energy-conversion systems (OWECS). Conceptual designs and associated costs are given for the major components of multi-unit installations which are then integrated into baseline systems. Through numerous studies, on-shore energy costs are developed in terms of controlling environmental parameters, distance from shore, and equipment type. (3 refs.) (Based on reports produced in 1979 and 1980 referred to earlier in this section of the bibliography.)

3.1.1(11) POWER TRANSMISSION FROM OFFSHORE WIND GENERATION SYSTEMS.
Franklin, P.J. and Gardner, G.E. (Central Electricity Generating Board, U.K.) Paper presented at 3rd International Conference on Future Energy Concepts, London 27-30 January 1981 pp. 327-330. London, Institution of Electrical Engineers 1981.
  Wind energy machines can be located in offshore coastal waters to take advantage of wind regimes, increase energy recovery, and reduce visual impact. Systems for enhancing generated power transmission from such offshore sites are discussed. The use of an aerogenerator in an interconnected arrangement is described. Synchronous generators are also considered for this application. (3 refs.)

3.1.1(12) REVIEW OF ELECTROCHEMICAL ENERGY CONVERSION AND STORAGE FOR OCEAN THERMAL AND WIND ENERGY SYSTEMS.
Landgrebe, A.R. and Donley, S.W. (US Dept of Energy, Washington, DC., USA) In proc. 16th Intersociety Energy Conversion Engineering Conference V. 2. Atlanta, GA., USA 9-14 August 1981. New York, NY., USA ASME 1981 pp. 2075-2080.
  Literature review on the application of electrochemical storage and manufacturing processes related to ocean thermal energy conversion (CTEC) and wind energy conversion (WEC). Storage system requirements and capabilities are estimated through the year 1995. Fuel and

capacity savings for the utility power sector are estimated for various battery storage/WEC systems. (11 refs.)

3.1.1(13) THE U.K. OFFSHORE WINDPOWER RESOURCE. Milborrow, D.J. and others. (Central Electricity Generating Board, U.K.) In Fourth International Symposium on Wind Energy Systems, Stockholm Sweden 21-24 September 1982 V. 1. Cranfield, U.K., BHRA Fluid Engng. 1982 Session D Paper D4 pp. 245-260.

An evaluation was made of the U.K. offshore power resource. The whole of the U.K. continental shelf starting 5 km. from the shore-line was examined and areas not excluded for technical or logistic reasons were designated as having potential for arrays of wind turbines. Annual average wind speeds at the 80 m. level were derived from high level (free stream) wind data and values assigned to each available area. Estimates of the energy generation capability of 100 m. diameter machines, spaced 1 km. apart in these areas were made and allowances were made for energy losses due to interactive effects in a cluster. The total resource from full development of the probable areas alone was estimated at 230 TWh/year, which is similar to the total U.K. electricity demands.

3.1.1(14) OFFSHORE WIND POWER IN DENMARK. Copenhagen, Denmark, Danske Elvaerkers Forening, March 1983 175 pp. Report no: EEV-83-01 (In Danish)

A survey is made of the wind power resources in Danish waters, and an upper limit of wind power production is calculated. A rough design is made of a wind turbine construction for offshore sites, and some technical problems are described. Two alternatives for a Danish offshore wind power production of 4 TWh/year are described with details of production method and plant. Based on this the costs of construction, operation, and maintenance are evaluated. An economic comparison is made between offshore and onshore wind power production, and between wind power production and conventional power production.

3.1.1(15) WATTS FROM THE WIND - WIND POWER PLANTS FOR SPECIAL APPLICATIONS. Anon. Siemens Rev. 50 (5) September-October 1983 pp. 8-13.

A survey of the current design criteria for wind turbines is given. A more detailed description is given of a large 3 mW Growian plant, for high wind use, located on the North Sea off Germany.

3.1.1(16) OFFSHORE SITING OF LARGE WIND ENERGY CONVERTER SYSTEMS IN THE GERMAN NORTH SEA AND BALTIC REGIONS. Pernpeintner, R. Modern Power Systems, 4 (6) June 1984 pp. 33-38.

Statistical data for wind speed and direction indicate that offshore wind conditions are better than those onshore for the siting of large wind energy converters. Considerations of water depth, wave height, seabed conditions, tidal range, sea currents, and wind characteristics are discussed. Plant component modifications necessary to facilitate marine operation are examined, and requirements for electric connections and power transmission lines are reviewed. (9 refs.)

3.1.1(17) OFFSHORE WIND-TURBINE SUB-STRUCTURE DESIGN. Swift, R.H. and Dixon, J.C. (Open University) In Wind Energy Conversion, proc. 5th BWEA Wind Energy Conf., Reading, U.K. 23-25 March 1983. P. Musgrove (ed.) pp. 334-341. Cambridge, U.K., Cambridge University Press 1984 375 pp. (Isbn 0-521-26250-X)

Design options for support structures for large offshore wind turbines are considered, with special attention to influence of the principal design parameters on the cost. A 90 m diameter teetered rotor compliant tower turbine is considered, mounted on a concrete gravity support structure. A range of possible shapes of the support structure are examined. Wave forces are calculated using stream function theory. (See also previous paper in this conference by these authors (pp. 326-333) on tower design for offshore structures.)

3.1.2 SIMULATION AND MODELLING

3.1.2(1) WIND ENERGY STATISTICS FOR LARGE ARRAYS OF WIND TURBINES (NEW ENGLAND AND CENTRAL U.S. REGIONS). Justus, C.G. (Georgia Inst. of Technol., Atlanta, USA.) In Sharing the Sun; Solar Technology in the Seventies, Joint Conference of International Solar Energy Society American Section and Solar Energy Society of Canada, Winnipeg, Manitoba 15-20 August 1976. Cape Canaveral, Fla. ISES/AS 1976 V. 7 pp. 268-288.

Performance characteristics are simulated for large dispersed arrays of 500 kW-1500 kW wind turbines producing power and feeding it directly into the New England or Central US utility distribution grids. Studies show that in good wind environments the 500 kW generators can average (on an annual basis) up to 240 kW mean power output, and the 1500 kW generators can average up to 350 kW mean power output. Better performance (averaging up to 470 kW) is obtained by an 1125 kW rated power unit designed to operate at lower wind speeds. The beneficial effect of operating widely dispersed arrays of wind turbines is that available power output can be increased and if winds are not blowing over one part of the array, they probably will over some other part of the array. (4 refs.)

3.1.2(2) AEROELASTIC WIND ENERGY CONVERTER. Ahmadi, G. (Pahlavi Univ., Shiraz, Iran) Energy Convers., 18 (2) 1978 pp. 115-120.

Describes the principle of aeroelastic wind energy conversion and an H-section model which works on the basis of torsional aeroelastic

instability. A mathematical formulation for
the prediction of the power coefficient of such
wind machines is presented. A small model is
constructed and tested in a wind tunnel. The
efficiency of the model was very low, but the
system can convert energy at a very low wind
speed. Furthermore, this wind energy converter is relatively simple and economical.
(18 refs.)

3.1.2(3) DATA ACQUISITION AND SIGNAL PROCESSING FOR A VERTICAL AXIS WIND ENERGY CONVERSION SYSTEM.

Stiefeld, B. and Tomlinson, R.N. (Sandia Lab.,
Albuquerque, New Mexico, USA) In proc. Seminar
on Test Solar Energy Materials and Systems,
Gaithersburg, MD. 22-24 May 1978. Mt. Prospect,
Ill., Inst. of Environ. Sci. 1978 pp. 241-244.

Describes the data acquisition and analysis
system developed to meet the needs of the 17-
metre vertical axis wind turbine at Kirtland
Air Force Base, New Mexico. The system employs
a minicomputer-based data acquisition system
with special peripheral equipment. Statistical
methods are described that are employed to
evaluate the performance of the system. The
objective of this programme is the development
of an instrumented outdoor wind laboratory with
the capability of evaluating various aspects of
wind turbine performance as well as analysis of
wind flow properties.

3.1.2(4) EFFICIENT USE OF WIND ENERGY BY USING STATIC SLIP RECOVERY SYSTEMS - A SIMULATOR STUDY.

Rajagopalan, V., Sankura Rao, K. and
Swamy, M.N.S. (Univ. of Quebec, Trois Rivieres,
Canada) In Energy '78: IEEE Reg. 5 Annual
Conference, Record of Conference Papers, Tulsa,
Oklahoma 16-18 April 1978. New York, NY., IEEE
(Cat. no: 78CH1283-1 Reg. 5) 1978 pp. 250-254.

Proposes a wind energy system using static
slip recovery scheme and a digital simulator,
WESS, which can simulate wind energy systems.
The Wind Energy System Simulator is capable of
simulating normal operating conditions of the
system as well as fault conditions, such as
commutation failure, firing control failure,
failure of an SCR, etc. Experimental results
on a 2 kW scheme together with results obtained
from simulator run are presented and compared.
(6 refs.)

3.1.2(5) ON THE FLUCTUATING POWER GENERATION OF LARGE WIND ENERGY CONVERTERS, WITH AND WITHOUT STORAGE FACILITIES.

Sorensen, B. (Univ. of Copenhagen, Niels Bohr
Inst., Denmark) Solar Energy, 20 (4) 1978
pp. 321-331.

Presents an analysis of the power fluctuations
and time duration patterns of large hypothetical
wind energy generators, using meteorological
data for Denmark. It is found that the fluctuations, relative to a load which varies through
the year in a manner similar to the actual load,
are no greater than the fluctuations relative
to a constant load. The addition of a hypothetical short-term storage, capable of delivering
the average power for 10-12 hr., makes the wind
energy system as dependable as one large nuclear
power plant. (19 refs.)

3.1.2(6) PLANS FOR WIND ENERGY SYSTEM SIMULATION.

Dreier, M.E. (Paragon Pac. Inc., El Segundo,
California, USA) In NASA Conf. Publ. CP
no: 2034 for a workshop: Wind Turbine
Structural Dynamics, NASA Lewis Research
Centre, Cleveland, Ohio 15-17 November 1977.
Washington, DC., NASA March 1978 pp. 261-264.

Introduces a digital computer code and a
special purpose hybrid computer. The digital
computer programme, the Root Perturbation method
or RPM, is a new implementation of the Floquet
procedure which circumvents numerical problems
associated with the extraction of Floquet
roots. The hybrid computer, the Wind Energy
Steam Time-domain simulator (WEST), yields
real-time loads and deformation information
essential to design and system stability
investigations.

3.1.2(7) ANALYSIS AND SIMULATION OF WIND ENERGY SYSTEMS.

Krause, P.C. Washington, DC., USA, NASA
Report no: NASA CR-162538 November 1979
60 pp.

Presents a simulation of two MOD-2 wind
energy systems. Results are given for the
power coefficient calculation, wind torque
equation, pitch control and torsional
dynamics of mechanical and aerodynamic
components of the system. The significance
of the wind fluctuation model in assessing
wind turbine dynamic performance is
considered.

3.1.2(8) INTERACTION IN LIMITED ARRAYS OF WINDMILLS: REVIEW OF EARLIER RESULTS FROM A SIMPLE MODEL AND A PRESENTATION OF THE CAPABILITIES OF A DYNAMIC PBI MODEL.

Crafoord, C. (Univ. of Stockholm, Sweden)
NTIS Report no: N80-11631 27 March 1979
57 pp.

Considers the problem of how closely packed
an array of windmills can be erected without
unduly interfering with each other. A general
technical background of the behaviour of a
single windmill in a homogenous flow, and a
list of important parameters for the efficient
performance of a group of windmills are
presented. A dynamic one dimensional
planetary model is used to determine wind
profiles behind a windmill unit. Preliminary
results indicate a variation in mean efficiency
of about 13% for a group of 80 windmill units.

3.1.2(9) SIMWEST: A SIMULATION MODEL FOR WIND AND PHOTOVOLTAIC ENERGY STORAGE SYSTEMS. VOLUME 1: CDC USER'S MANUAL. VOLUME 2: CDC PROGRAM DESCRIPTIONS.

Warren, A.W., Edsinger, R.W. and Burroughs, J.D.
(Boeing Computer Serv. Co. and NASA Lewis
Research Centre) Washington, DC., US Dept. of
Energy August 1979 486 pp. and 247 pp.
Report nos: DOE/NASA/0042-79/3 and 4; BCS 40262-1
and 2; CASA CR 159607 and CR 159608.

The User's Manual for CDC version of the
SIMWEST computer programmes developed by Boeing
Computer Services Company under NASA Contract
DEN3-42, an Expanded System Simulation Model for
Solar Energy Storage is presented. The SIMWEST
codes were originally developed for simulation
of wind energy storage systems, and an example
of the application of SIMWEST programme to a
100 kW wind energy storage system is given.
(See also earlier paper by same authors in
13th Intersociety Energy Conversion Conference
proc. 20-25 August 1978. IEEE 1978 V. 3.
pp. 2108-2114.)

3.1.2(10) SWECS TEST ACTIVITIES AT ROCKY FLATS.
Trenka, A.R. (Rockwell Intl.) Paper presented
at US Dept. of Energy Small Wind Turbine
Systems Research and Development Requirements
Conference, Boulder, USA 27 February-1 March
1979 V. 1 pp. 216-220.
  Reports on the testing of small wind energy
conversion systems under actual environmental
conditions at the Rocky Flats Test Centre.
Technical, performance and operational data
were to be collected and made available for
designers and engineers. Data acquisition and
component testing systems are described.

3.1.2(11) COMPARATIVE INVESTIGATIONS OF
OPERATING BEHAVIOUR OF WIND ENERGY INSTALLATIONS
OF 10 KW POWER CLASS.
Fries, S. and others. Statusber. Windenerg.
(Semin), Hamburg, Germany 24-26 June 1980.
Dusseldorf, Germany, VDI 1980 pp. 341-354.
(In German)
  Describes the testing of nine small wind power
plants in the North Frisian island of Pellworm
to determine the efficiency, reliability and
costs of each type. They were horizontal-axis
machines of different types and Darrieus verti-
cal-axis machines. Measurements and examples
of applications are described.

3.1.2(12) COMPUTER SIMULATION AND OPTIMAL
CONTROL OF WIND-DRIVEN POWER PLANTS.
Sherif, M.M. Salford, U.K., Salford Univ.
July 1980 211 pp. (Ph.D. Thesis)
  Reviews wind energy conversion systems and
wind turbines. Preliminary wind tunnel tests
of a horizontal axis wind turbine are described and a dynamic model of a wind energy
conversion system is developed and used as
the basis of analog and digital computer
models. These models were validated.
Optimum control of wind turbines is analysed
and a computer model of a wind energy conversion system with optimum control is
presented. The performance of optimum
control is discussed and compared with
sub-optimum control.

3.1.2(13) CONTROLLED VELOCITY TESTING OF
SMALL WIND ENERGY CONVERSION SYSTEMS. AN
EVALUATION OF A TECHNIQUE.
Balcerak, J.C. (Atomics International Div.,
Golden, CO., Rocky Flats Plant) Washington,

DC., USA, US Dept. of Energy November 1980
50 pp. (Rocky Flats Plant Rept. no: RFP-3189)
  Report on tests of a small wind energy
conversion system (SWECS) conducted at the
Department of Transportation Test Centre in
Pueblo, Colorado. The test machine was
mounted on a rail flatcar which was pushed
by a locomotive. The primary objective
was to determine the usefulness of SWECS controlled velocity testing (CVT) using this
method. Wind velocity profiles, acceler-
ation/deceleration forces, rotor yaw, power
output, rotor rpm, power coefficient, wake
measurements, and flow visualisation were
examined and results confirm the potential
benefit of this method as an addition to the
natural atmospheric testing done at the Rocky
Flats Wind Systems Test Centre.

3.1.2(14) A SIMULATION MODEL TO EVALUATE
THE PERFORMANCE OF WINDMILL SYSTEMS.
Wysk, A., Choi, B. and Tanchoco, J.M.A.
(Virginia Polytechnic Inst. and State University) Paper presented at American Society
of Agricultural Engineers National Energy
Symposium, Kansas City 29 September-1 October
1980 pp. 489-495.
  Presents a simulation model of a wind power
generation system including both design and
control variables as input. The effects of
these variables are determined through a
"one at a time search procedure", and an
optimal windmill system is developed using
the three design variable component capacities.
Component cost data is included in minimum cost
objective function, thus allowing the optimal
system to be configured. Windmill and battery
control, cooling and heating system control,
and performance are discussed. (5 refs.)

3.1.2(15) TRANSIENT BEHAVIOUR OF WIND ENERGY
SYSTEMS.
Sivasegaram, S. (Univ. of Peradeniya, Sri Lanka)
Wind Eng., 4 (2) 1980 pp. 53-63.
  Report on a study of the transient response
of wind energy systems to sudden changes in
wind speed and fluctuating wind speeds. It is
shown that analytical solutions are possible
in some instances and that results of general
validity can be obtained. The response time
is shown to be dependent upon the magnitude of
the change in wind speed and that the time
constant used on control theory for small per-
turbances will be often inadequate to describe
system behaviour. (2 refs.)

3.1.2(16) PROGRESS ON THE DEVELOPMENT OF INTER-
NATIONAL TESTING PRACTICES FOR WIND ENERGY
CONVERSION SYSTEMS.
Trenka, A.R. (Rockwell International) In proc.
5th Biennial Wind Energy Conference & Workshop,
Washington, DC., USA 5-7 October 1981 I.E. Vas
(ed.) V. 2. Palo Alto, USA, Solar Energy Res-
earch Institute 1981 Session 2A pp. 229-239.
Report nos: SERI/CP-635-1340 CONF-811043.
  Reports on efforts in the development of
internationally agreed test and reporting
practices for energy production, quality of

power, reliability, durability and safety. Cost effectiveness or economics, noise interference and impact on the environment are also assessed.

3.1.2(17) WIND-ENERGY CONVERSION SYSTEM SIMULATION PROGRAM.
Assarabowski, R.J. and Mankauskas, J.J. (United Technol. Research Centre, East Hertford, Conn., USA) J. Energy, 5 (2) March-April 1981 pp. 66-71.
Describes a programme which was developed to simulate the operation and performance of a wind-energy conversion system (WECS) for a non-utility potential user, and to perform a life-cycle economic analysis. The system consists of a wind turbine in conjunction with purchased utility power and an optional energy storage system. Because of the site-specific nature of WECS, the programme parameters are character-istic of the WECS site being simulated; e.g. the site's wind-velocity data, the non-utility owner's process energy requirements, the specific performance of the WECS installed equipment, and the owner's investment structure. (4 refs.)

3.1.2(18) CONTROL POLICIES FOR MAXIMISING ENERGY EXTRACTION FROM WIND TURBINES.
Casanova-Alcalde, V.H. and Freris, L.L. In papers presented at 4th International Symposium on Wind Energy Systems, Stockholm, Sweden 21-24 September 1982 V. 2. Cranfield, U.K., BHRA Fluid Engng. 1982 Session L Paper L3 pp. 233-246.
In a wind energy conversion system the energy extracted for a given wind speed is a complex function of the wind turbine characteristics, the selected matching and cut-in wind speeds, and the dynamics of the control system used. Some of these relationships are studied through the simulation of a number of wind turbines on an analogue computer.

3.1.2(19) FIELD MEASUREMENTS OF THE POWER OUTPUT OF SMALL WECS IN SERVICE.
Jensen, S.A. and Bjerregaard, E.T.D. (Danish Ship Research Lab.) In papers presented at 4th International Symposium on Wind Energy Systems, Stockholm, Sweden 21-24 September 1982 V. 1. Cranfield, U.K., BHRA Fluid Engng. Session F Paper F2 pp. 365-377.
The Dept. of Wind Engineering at the Danish Ship Research Laboratory, DSRL, is carrying out field measurements for the determination of the power curves of Wind Energy Conversion Systems in service. The power curves are used for the documentation of the power per-formance of the individual WECS and for the estimation of the energy production that can be obtained from the WECS in various wind conditions. The field measurements are carried out with the aid of a mobile test station which consists of easily erectable masts for wind measurements, equipment for measuring power, and a micro-processor which controls the data collection and stores the data for final analysis.

3.1.2(20) MOD-2 WIND TURBINE PROJECT ASSESSMENT AND CLUSTER TEST PLANS.
Gordon, L.H. (NASA Lewis Research Centre, USA) In proc. Workshop on Large Horizontal-Axis Wind Turbines, Cleveland, USA 28-30 July 1981 R.W. Thresher (ed.) Washington, DC., USA, NASA 1982 pp. 653-673. (NASA Conf. Publication 2230; DOE Publication CONF-810752 Report no: SERI/CP-635-1273)
Presents an assessment of the Mod-2 Wind Turbine project based on initial goals and 1981 results. The Mod-2 background, project flow, and a chronology of events/results lead-ing to Mod-2 acceptance is presented. After acceptance of the three operating turbines, NASA Lewis Research Centre will continue management of a two year test programme performed at the DOE Goodnoe Hills test site, expected to yield data necessary for the continued development and optimisation of wind energy systems. These test activities, their implementation, and the results to date are also presented.

3.1.2(21) PERFORMANCE ASSESSMENT AND COST EFFECTIVENESS OF WIND ENERGY CONVERSION SYSTEMS.
Miller, G. and others (New York University) J. Energy, 6 (2) March-April 1982 pp. 104-108.
Performance and cost characteristics of wind energy conversion systems are assessed. Both technological and cost-effectiveness consider-ations are discussed for these systems with emphasis on developing appropriate economic measures for the application of electric power generation. A cost-effectiveness analysis is developed for target costs per unit frontal area for such systems showing the influence of annual mean wind-speed, tower height, atmos-pheric boundary layer stability, capital-cost financing rates, and the cost escalation rate of the fossil fuel displaced.

3.1.2(22) SMALL WIND ENERGY CONVERSION SYSTEMS, APPLICATION AND PERFORMANCE.
Norton, J.H. (North Wind Power Co. Inc., USA) New York, USA, American Society Mechanical Engineers 7-10 March 1982 6 pp. ASME Paper no: 82-Pet-3.
Examines two basic applications of SWECS - remote and utility interface. The discussion of remote power briefly reviews the history of this type of application, describes the spec-ifications of the HR2 high reliability 2 kW wind system, and examines in detail two com-mercial applications in the petroleum industry - their operation, environment and performance. The discussion of utility interface briefly reviews the history of the L16 line interface 6 kW wind system, and discusses its potential application in small commercial windfarms, remote community power systems and dispersed industrial, residential and agricultural applications.

3.1.2(23) THE TEST FIELD PELLWORM FOR SMALL INTERMEDIATE WIND ENERGY CONVERSION SYSTEMS AT THE GERMAN COAST OF THE NORTH SEA.
Fries, S., Petersen, G. and Mengelkamp, H.T.

In papers presented at 4th International
Symposium on Wind Energy Systems, Stockholm,
Sweden 21-24 September 1982 V. 1. Cranfield,
U.K., BHRA Fluid Engng. 1982 Session F Paper F3
pp. 379-390.

The GKSS research centre at Geesthacht near
Hamburg has established a test field for small
and intermediate wind energy conversion systems
on the island of Pellworm, in the North Sea.
In the first phase of a comprehensive test
programme nine commercially available machines
or commercial prototypes of the 10 kW size are
operated at the test field.
(See also Report by same authors GK 55-82/E/48
1982 40 pp. (In German) from GKSS-Forschungs-
zentrum Geesthacht GmbH, Hamburg.)

3.1.2(24) TEST PROGRAM FOR WIND ENERGY
CONVERSION SYSTEM GROWIAN.
Koerber, F. Bundesministerium fuer Forschung
und Technologie, Bonn-Bad Godesberg, F.R.
Germany. Report no: BMFT-FB-T-82-072
June 1982 131 pp. (In German)
A test programme for the German large scale
wind energy converter GROWIAN is planned.
Special emphasis was laid on the action of wind
and the reaction of the machine in the field of
dynamics, load and control. For wind measurement a grid arrangement of anemometers was
designed covering entirely the rotor area in
25 m. steps. Rated power and efficiency of power
conversion will be measured with regard to rotor
aerodynamics and operation conditions. The
behaviour of the GROWIAN asynchronous generator
on the power supply line will be evaluated.
Another essential item is the analysis of
environmental impact of noise and TV interference.

3.1.2(25) WIND ENERGY CONVERSION SYSTEM ANALYSIS
MODEL (WECSAM) COMPUTER PROGRAM DOCUMENTATION.
Downey, W.T. and Hendrick, P.L. (Arthur D.
Little Inc., Cambridge, MA; Solar Energy
Research Inst., Golden, CO.) Washington, DC,
USA, Solar Energy R. I. Report no: SERI/SP-
19136-4 July 1982 78 pp.
Describes a computer-based wind energy conversion system analysis model (WECSAM) developed
to predict the technical and economic performance of wind energy conversion systems (WECS).
The model is written in CDC FORTRAN V. The
version described accesses a data base
containing wind resource data, application
loads, WECS performance characteristics,
utility rates, state taxes, and state
subsidies for a six state region (Minnesota,
Michigan, Wisconsin, Illinois, Ohio and
Indiana). The model is designed for analysis
at the county level. The computer model
includes a technical performance module and
an economic evaluation module. The modules
can be run separately or together. The model
can be run for any single user-selected county
within the region or looped automatically
through all counties within the region. In
addition, the model has a restart capability
that allows the user to modify any data-base
value written to a scratch file prior to the
technical or economic evaluation.

3.1.2(26) MATHEMATICAL PROGRAMMING MODELS FOR
THE ECONOMIC DESIGN AND ASSESSMENT OF WIND
ENERGY CONVERSION SYSTEMS.
Reinert, K.A. (Granville Corp.) Wind Engng.,
7 (1) 1983 pp. 43-59.
System reliability is one of the important
determinants of the economic and technical
feasibility of wind energy conversion systems.
The inclusion of storage facilities into these
systems greatly increases their reliability.
Recognising this, two inter-period mathematical
programming models are formulated which optimise the design of a wind energy conversion
system, locating wind turbines in a number of
distinct arrays, sizing a storage facility,
providing rules for the operation of the
storage facility, and ensuring that a pre-
specified level of demand is met.

3.1.2(27) METHODS OF REDUCING WIND POWER CHANGES
FROM LARGE WIND TURBINE ARRAYS.
Schlueter, R.A. and others. (Michigan State
University, East Lansing, Michigan, USA) IEEE
Trans Power Apparatus & Systems, 102 (6)
June 1983 pp. 1642-1650.
Presents analysis and simulation results to
demonstrate how to decrease a wind turbine
system's generation change over a 10-minute
period through selection of the wind turbine
generator model at each site and the siting
configuration. Wind generation change data
change caused by the passage of a thunder-
storm is analysed and presented to determine
the factors concerning the wind turbine model
and siting configurations. (4 refs.)

3.1.2(28) EFFECT OF SITE WIND CHARACTERISTICS
ON ENERGY PRODUCTION.
Pennell, W.T. and Wegley, H.L. (Pacific North-
West Lab.) In Alternative Energy Sources III,
V. 4 - Indirect Solar/Geothermal Energy, proc.
3rd Miami Intl. Conference on Alternative
Energy Sources, Miami Beach, USA 15-17 December
1980. T. Nejat (ed.) pp. 3-20. Washington, USA,
Hemisphere Publ. Corp. 1983.
Net energy production over a given period
can be estimated if both the performance
characteristics of the turbine and the wind
speed probability density function are known.
Simulations covering a range of probability
density functions and machine performance
characteristics showed that reasonable estimates of net energy production can be made
using simple, analytic probability density
functions. (See also paper by W.T. Pennell
in Intl. Wind Energy Symposium, 5th Annual
Energy Sources Technol. Conf. 7-10 March 1982
ASME pp. 265-274.)

3.1.2(29) DEVELOPMENT AND VERIFICATION OF MOD-2
AND MOD-OA SIMULATION MODELS.
Miller, A.H. and Formica, W.J. (Battelle Pacific
NorthWest Labs.) Pacific NorthWest Labs. Report
no: PNL-4864 August 1984 171 pp. (Dept. of
Energy Report no: DE 84017389/XAD)
Presents a method for generating annual energy
production estimates which includes computer
models that simulate the operation and power

output of the MOD-2 and MOD-OA wind turbines. The verification of the models, performed using data from actual wind turbine sites shows that they can be applied as tools in assessing the feasibility of wind energy in a particular situation.

3.1.2(30) DIRECTIONAL VARIATION OF WIND PROBABILITY AND WEIBULL SPEED PARAMETERS. Dixon, J.C. and Swift, R.H. (Open University, Faculty of Technology, Milton Keynes, U.K.) Atmos. Environ. 18 (10) October 1984 pp. 2041-2047. (See also Wind Engng. 8 (2) 1984 pp. 87-98.)

A three-parameter wind model is proposed. Two of these are the familiar Weibull characteristic speed and shape factor; the third is a measure of directionality. The model is essentially empirical, and is easily applied, giving the directional probability and the Weibull parameters to be used for any particular wind direction. The McWilliams wind model, in which beta is a dependent variable, and the Weibull model, where directionality is neglected, are each cases of the new model. The McWilliams model has previously been shown to be a fair predictor of ground-level winds. The models are here compared with geostrophic wind data. The directional variation of probability, characteristic speed and power density can be fitted well, although shape factor varies erratically. Thus, the suggested model should provide a broadly realistic representation for any altitude, with the advantage that overall characteristic speed, shape factor and directionality can be independently varied as required. (Refs.)

3.1.2(31) EXPERIMENTAL TECHNIQUE FOR MEASURE- MENT OF SLENDER BODY OSCILLATIONS. Nathan, G.K. and Balendra, T. (Faculty of Engng., National Univ. of Singapore) Wind Engng. 8 (2) 1984 pp. 78-86.

Describes an experimental set-up to measure the amplitude and frequency of oscillations of a slender-body model referred to three mutually orthogonal directions, using calibrated torsion bars. The experimental set-up is capable of measuring along wind, across wind and torsional oscillations either independently or in combin- ations. (12 refs.)

3.1.2(32) SIMULATION OF WINDS AS SEEN BY A ROTATING VERTICAL AXIS WIND TURBINE BLADE. George, R.L. Richland, USA, Battelle Pacific NorthWest Labs. Report no: PNL-4914 February 1984 52 pp.

Describes a technique which uses high speed turbulence wind data from a line of seven anemometers at a single level to simulate the wind seen by a rotating vertical axis wind turbine blade. Compares VAWT and horizontal axis wind turbine simulations and also studies direct VAWT blade wind measurements. (From author's abstract.)

3.2 Wind Turbines: Design and Testing

3.2(1) PRELIMINARY RESULTS OF THE LARGE EXPERI- MENTAL WIND TURBINE PHASE OF THE NATIONAL WIND ENERGY PROGRAM. Thomas, R.L. and Sholes, J.E. (Lewis Research Centre, Cleveland, Ohio, USA) In Front of Power Technology 8th Annual Conference, Oklahoma State University, Stillwater 1-2 October 1975 Paper 15 15 pp. Stillwater, USA, Oklahoma State University 1975.

A major phase of the wind energy programme is the development of reliable wind turbines for supplying cost-competitive electrical energy. The preliminary results of two projects in this phase of the programme are discussed. An experimental 100 kW wind turbine design and its status are reviewed and the results of two parallel design studies for determining the configurations and power levels for wind turbines with minimum energy costs given.

3.2(2) WIND ENERGY CONVERSION. VOLUME IV. DRIVE SYSTEM DYNAMICS. Martinez-Sanchez, M. and Labuszewski, T. (Massachusetts Inst. of Tech., Cambridge, MA) Washington, DC, USA, US Dept. of Energy Report no: COO-4131-T1(V. 4) September 1978 197 pp.

Describes the dynamics of the drive system and various approaches to power transmission. Effects on performance of using a constant rotor speed as opposed to a rotor speed varying with the wind speed are discussed for various rotor operating schedules and typical wind distributions. Dynamics of the com- bined rotor, alternator, and drive system are analysed and conditions which could lead to electro-dynamic instabilities and de- synchronisation are discussed. The dynamics of the drive system and important design conditions for various drive systems, such as location of the alternators, use of hydraulic drive systems and smoothing techniques are detailed.

3.2(3) EXECUTIVE SUMMARY MOD-1 WIND TURBINE GENERATOR ANALYSIS AND DESIGN REPORT. NTIS Report N80-11558 March 1979 61 pp.

Reports the activities leading to the design of a wind turbine generator having a nominal rating of 1.8 MW. Includes MOD-1 system description; structural dynamics; stability analysis; mechanical subassemblies design; power generation subsystem; and control and instrumentation.

3.2(4) DEVELOPMENT OF A 2 KW HIGH-RELIABILITY WIND TURBINE GENERATOR. Drake, W. and Clews, H. (Enertech Corp., VT.) Presented at U.S. Dept. of Energy Small Wind Turbine Systems R & D Requirements Conference, Boulder, USA 27 February-1 March 1979 V. 1 pp. 13-30.

Describes Enertech Corporation's programme to develop and fabricate prototypes for testing of a 2 kW high-reliability wind turbine generator. The machine is designed to produce 2 kW of electrical power on a 9 m/sec. wind, and should require not more than a day's service per year. Design, technical configuration, and economics of the generator are described.

3.2(5) FLUID DYNAMIC ASPECTS OF WIND ENERGY CONVERSION.
De Vries, O. (Advisory Group for Aerospace R & D, France) NTIS Report no: AD-A076 513 July 1979 150 pp.

Reviews the fluid dynamic aspects of wind energy conversion. Wind-driven turbines, including horizontal-axis and vertical-axis units are described. Inhomogeneous flow and turbulence effects on turbine performance are analysed.

3.2(6) GRUMMAN 8 KW WIND TURBINE.
Adler, F. (Grumman Energy Systems) Paper presented at US Dept. of Energy Small Wind Turbine Systems R & D Requirements Conference, Boulder, USA 27 February-1 March 1979 V. 1 pp. 106-120.

Outlines design requirements for a prototype 8 kW wind energy turbine machine by Grumman Energy Systems. The prototype must produce at least 8 kW of electrical power at a wind speed of 20 mph, able to withstand 165 mph winds and power generation must be usable AC electrical power. The system must be operational and repairable for 25 years and cost $750/kW produced. Design paths selected for complying with these specifications are summarised.

3.2(7) IMPROVING THE MECHANICAL LOAD MATCHING OF WIND ENERGY CONVERTERS.
Dixon, J.C. (Open University, U.K.) In proc. 1st British Wind Energy Association Wind Energy Workshop, Cranfield, U.K. April 1979. London, U.K., Multi-Science Publ. Co. Ltd. 1979 pp. 181-189.

The efficiency of the conventional design of windpump as an energy converter is about 0.05 due to three main factors: rotor aerodynamics, bad matching of pump to rotor with speed variation, and cyclic variations of torque required by the pump. A high-speed horizontal axis rotor or vertical axis rotor, although economically attractive, has a torque characteristic particularly unsuitable for use with a typical positive displacement pump. Some methods of overcoming these difficulties are compared.

3.2(8) RANDOM DATA ANALYSIS IN WTG TESTING.
Hansen, A.C. (Rockwell International, CO) Paper presented at US Dept. of Energy Small Wind Turbine Systems R & D Requirements Conference, Boulder, USA 27 February-1 March 1979 V. 1 pp. 221-232.

Describes three methods available for the analysis of wind turbine generator performance data at the Rocky Flats Wind Systems Test Centre in Colorado. Method-of-bins, frequency-matching,

and most-probable power methods are described. The method-of-bins is shown to be the most valid method for analysing continuously sampled, equal interval data. (3 refs.)

3.2(9) ROCKY FLATS SMALL WIND SYSTEMS TEST CENTRE ACTIVITIES. VOLUME II. CONTROLLED VELOCITY, VIBRATION AND DYNAMOMETER TESTING OF SMALL WIND ENERGY CONVERSION SYSTEMS. SECOND INTERIM REPORT.
Atomics International Div., Golden, CO, Rocky Flats Plant. Report no: RFP 3004 (V. 2) July 1979 44 pp.

Report on the controlled velocity, vibration and dynamometer testing performed on small wind energy conversion systems (SWECS) at Rocky Flats up to June 1979. Results of controlled velocity testing on wind machines and of vibration testing of five wind machines are included.

3.2(10) WIND SYSTEM DESIGN AND RESEARCH CONSIDERATIONS.
Thresher, R.W. (Oregon State University, USA) Paper presented at US Dept. of Energy Small Wind Turbine Systems R & D Requirements Conference, Boulder, USA 27 February-1 March 1979 V. 1 pp. 189-197.

Current wind turbine design efforts to reduce energy cost and provide safe, reliable turbine systems are identified. Considerable work is needed to establish design loads and loading cases, and to develop required analytical approaches to improve cost effective designs.

3.2(11) WORKSHOP ON LARGE WIND TURBINE DESIGN CHARACTERISTICS AND R & D REQUIREMENTS HELD AT CLEVELAND, USA ON 24-26 APRIL 1979.
Lieblein, S. Washington, DC, USA, US Dept. of Energy December 1979 464 pp. DOE Publication CONF-7904111; NASA-CP-2106.

Outlines the DOE Wind Energy Programme, a major phase of which is the research and development of large wind turbine systems that can be integrated into utility networks. Development status of horizontal and vertical axis machines in various countries is presented together with the design characteristics and operating experience of wind turbine blades using different materials.

3.2(12) AXIAL-FLOW WIND AIR TURBINE WITH STATOR AND ELECTRIC EDDY-CURRENT BRAKE.
Awano, S. (Nihon University, Japan) Trans. Japan Society Mechanical Engrs., (401) January 1980 pp. 57-66. (In Japanese)

Describes a new axial-flow wind turbine with an external diameter of 1.2m for high speed wind. One striking feature of the turbine is that it has a stator before the rotor to give a higher output at a lower rotational speed than that of the ordinary type without stator. Another feature is that it has an electric eddy-current brake to automatically hold steady the rotational speed of the generator, at a desired value even in violent storms. The brake also affords a new means for converting the wind energy directly to heat. The design, wind tunnel-test results and performance of the wind air-turbine and generator are described.

3.2(13) DESCRIPTION OF THE TWO DANISH 630 KW WIND TURBINES, NIBE, A AND NIBE, B AND SOME PRELIMINARY TEST RESULTS.
Pedersen, B.M. and Neilsen, P. (Denmark Tech. University) In proc. 3rd International Symposium on Wind Energy Systems, Lyngby, Denmark 26-29 August 1980. Cranfield, U.K., BHRA Fluid Engng. 1980 Paper E1 pp. 223-238.

Gives a brief description of the two demonstration wind turbines, NIBE-A and NIBE-B, of the Danish wind energy programme.

3.2(14) DEVELOPMENT OF A 4 KW WIND TURBINE GENERATOR.
Bottrell, G. and Sullivan, L.J. (Structural Composites Ind. Inc., USA) In IECEC '80, Energy to the 21st Century, proc. 15th Intersociety Energy Conversion Engng. Conference, Seattle, USA 18-22 August 1980 V. 1. New York, USA, American Institute Aeronaut. & Astronaut. Paper no: 809162, 1980 pp. 810-814.

SCI has been awarded a contract, funded by the US Department of Energy to develop a small wind energy conversion system. The system is to be rated at 4 kW and operated in a 10 mile per hour mean wind speed for remote residential applications. Results to 1979 are presented.

3.2(15) INSTALLATION AND CHECKOUT OF THE DOE/ NASA MOD-1 2000-KW WIND TURBINE GENERATOR.
Puthoff, R.L., Collins, J.L. and Wolf, R.A. (NASA Lewis Research Centre, Cleveland, Ohio, USA) In A Collection of Technical Papers from AIAA/Solar Energy Research Institute Wind Energy Conference, Boulder, USA 9-11 April 1980. New York, USA, American Institute Aeronaut. & Astronaut. 1980 Paper no: 80-0638 pp. 249-260.

Describes the wind turbine, the assembly and testing at Philadelphia, and the installation at Boone, North Carolina. The report concludes with performance data taken during the initial tests conducted on the machine.

3.2(16) LARGE WIND ENERGY CONVERTER: GROWIAN 3 MW.
Feustel, J.E., Helm, S. and Koerber, F. Washington, DC., USA, National Aeronautics and Space Administration, Final Report. Report no: NASA-TM-75404 November 1980 90 pp. (NTIS Report no: N81-20543/7) Translation into English from German paper originally published by Maschinenfabrik Augsberg-Nuremberg, ag., Nuremberg, Germany.

Summary of the final report on the projected application of large-scale wind turbines on the northern German coast. The designs of the tower, machinery housing, rotor, and rotor blades are described and various construction materials are examined. Rotor blade adjustment and auxiliary equipment are examined.

3.2(17) LARGE WIND TURBINES EARLY OPERATIONAL EXPERIENCE AND POTENTIAL FOR SUPPLYING ELECTRICITY.
Thomas, R.L. and Robbins, W.H. (NASA Lewis Research Centre, Cleveland, Ohio, USA) Presented at International Solar Energy Society American Section/Et Al 1980 Annual Conference, Phoenix, 2-6 June 1980 V. 3. 2 pp. 1486-1492.

Wind energy generation has the potential to save 6-7 q/yr of energy. Large wind turbines are being developed to harness this resource and the design and operation of these large turbines are described. First-generation technology turbines are in operation at selected utility sites. Second-generation machines are expected to generate electricity at less cost than conventional systems. (9 refs.)

3.2(18) A NEW WECS DESIGN: THE UNIFIED WIND DYNAMO.
Deibert, D.D. (Universal Technologies Corp., NJ.) Paper presented at 2nd Dept. of Energy/ Solar Energy Research Institute Wind Energy Innovative Systems Conference, Colorado Springs, USA 3-5 December 1980 pp. 35-46.

A radically new wind energy conversion system has been designed, manufactured, and successfully tested. The unified wind dynamo employs a specially designed, ultra low rpm alternator and is able to adapt to varied blade structural constraints.

3.2(19) PERFORMANCE ASSESSMENT OF A FLETTNER WIND TURBINE.
Crimi, P. (Avco Systems Div., Mass., USA) J. Energy, November-December 1980 4 (6) pp. 281-284.

Optimal aerodynamic configuration and power output as a function of wind speed are analysed, using the classical strip formulation. A limited performance analysis of a hybrid configuration is also included. Six factors were identified as having an impact on determining the cost of producing power with a Flettner wind turbine. These include relatively low power coefficient due to the low section lift-drag ratio; simple rotor and hub configurations considering design, construction, and material requirements; and no maximum operating wind speed, simplifying control requirements and increasing integrated output. (8 refs.)

3.2(20) VARIOUS SYTEMS FOR THE GENERATION OF ELECTRICITY USING UPPER ATMOSPHERIC WINDS.
Roberts, B.W. and Blackler, J. (University of Sydney, Australia) Paper presented at 2nd US Dept. of Energy/Solar Energy Research Institute Wind Energy Innovative Systems Conference, Colorado Springs, USA 3-5 December 1980 pp. 67-80.

Describes four basic wind energy system designs for generating electricity using upper atmospheric winds. The airship concept, the open-rotor type turbine and biplane design, the rotary wing concept with tail rotor, and ducted turbines with monoplanes or biplanes are discussed. Rotary wing theory was extended to justify a successful series of wind tunnel tests on a twin motor gyromill. This machine can operate as a helicopter and generate in an autorotative mode. (8 refs.)

3.2(21) WIND WHEEL TURBINE.
Frost, W. and Kaufman, J.W. Paper presented at 2nd U.S. Dept. of Energy/Solar Energy Research Institute Wind Energy Innovative Systems Conference, Colorado Springs, 3-5 December 1980 PL3L 26 pp.

Describes an analytical and experimental study of an innovative wind wheel turbine. The wind wheel apparatus is essentially a bladed wheel which is directly exposed to the wind on the upper half and exposed to wind through ducting on the lower half. These multiple ducts consist of a front concentrator and two side concentrators. The forced rotation of the wheel is then converted to electrical power.

3.2(22) AIRFOIL DATA FOR WIND TURBINES.
Snyder, M.H. (Wichita State Univ., USA) In Lecture Series on Wind Energy Conversion Devices, Rhode Saint Genese, Belgium 1-5 June 1981. Rhode Saint Genese, Belgium, Von Karman Inst. Fluid Dyn. 1981 18 pp. (Lecture Series 1981-8).

Reviews some of the problems encountered by wind turbine designers in applying aerodynamic characteristics of airfoil sections to the design of wind energy conversion systems. Also reviews studies which have been made of the effects of changing airfoils on wind turbine performance.

3.2(23) CONCENTRATOR SYSTEMS FOR WIND ENERGY, WITH EMPHASIS ON TIPVANES.
Van Holten, T. (Delft University of Technol., Netherlands) Wind Eng , 5 (1) 1981 pp. 29-45.

A general aerodynamic theory for wind energy concentrator systems is described and verified by wind tunnel experiments on the tipvane concentrator concept. Considerable reductions in the cost of wind energy may be achieved by the application of concentrator techniques. (10 refs.)

3.2(24) DEVELOPMENT AND MANUFACTURE OF LARGE WIND TURBINES.
Reijnders, H.B. Energiespectrum, 5 (7) July-August 1981 pp. 205-206. (In Dutch)

The Dutch National Research Programme on Wind Energy has provided a reliable measurement installation for deriving design data for wind turbines. This paper gives a detailed consideration of various cost aspects, with explanatory graphs.

3.2(25) EXPERIENCE AND ASSESSMENT OF THE DOE/ NASA MOD-1 2000 KW WIND TURBINE GENERATOR AT BOONE, NORTH CAROLINA.
Collins, J.L. and Poor, R.H. (NASA Lewis Research Centre; Gen. Electr. Co., USA) In proc. 5th Biennial Wind Energy Conference &

Workshop, Washington, DC., USA 5-7 October 1981 I.E. Vas (ed.) V. 1. Palo Alto, USA, Solar Energy Research Institute 1981 Session 1A pp. 125-142 (SERI/CP-635-1340)

The Mod-1 wind turbine was the first megawatt sized machine in the Federal Wind Energy Programme to produce electrical power from wind energy. Specific project objectives were: operational and performance data for a megawatt sized wind turbine in a utility operated application; demonstration of unattended, fail-safe operation; involvement of utility as user and operator; identification of maintenance needs for large wind turbines; involvement of industry in the design, fabrication and installation of the turbine; identify components/ subsystem modifications to reduce cost, improve reliability and increase performance and assess public reaction to large wind turbines.

3.2(26) EXPERIMENTS WITH A DIFFUSER AUGMENTED MODEL WIND TURBINE (DAWT).
Gilbert, B.L. and Foreman, K.M. (Grumman Aerospace Corp.) New York, USA, American Society Mechanical Engrs. 7-10 March 1981 9 pp. (ASME Paper no: 82-Pet-5)

A 3-bladed constant chord, untwisted turbine model was incorporated into a DAWT model. Objectives were to add real turbine characteristics such as swirl, and centrebodies effects, to the flow. Although this turbine model was not well matched to the diffuser, the model DAWT system increased the power output by more than four times that of the model turbine operating as conventional wind energy conversion systems. More than 3.4 times the power potential of an ideal wind turbine was measured.

3.2(27) HYDRAULIC WIND ENERGY CONVERSION SYSTEM.
(Jacobs Energy Research Inc. Audubon, MN., USA) Washington, DC, US Dept. of Energy Report no: DOE/R5/10236-2 July 1981 18 pp. (NTIS Report no: DE 81027122)

This design used a three bladed turbine, which drove a hydraulic pump. Energy is transmitted from the pump through a long hose and into a hydraulic motor, where the energy is used. This wind system was built and tested during the winter of 1980-1981. The power train included a five metre three bladed wind turbine, a gearbox, a 1.44 cubic inch displacement pump with a small supercharged gear pump attached.

3.2(28) THE MOD 2 WIND TURBINE DEVELOPMENT PROJECT.
Kennard, J. (NASA Lewis Research Centre, Cleveland, Ohio, USA) Paper presented at American Society of Mechanical Engineers/Et Al 3rd Annual Energy Symposium, Pennsylvania 31 March-1 April 1981 pp. 75-97.

Describes the Mod 2 project sponsored by DOE and NASA to develop intermediate and large wind energy conversion systems. A 36 month programme begun in the summer of 1977 - culminates earlier R & D work on three preceeding turbines Mod 0, Mod 0A and Mod 1. Mod 2's objective is to design a wind turbine to produce energy for less than 5.gnt/kwh based on 1980 cost forecasts.

3.2(29) ONE-ARMED MONSTER.
Scott, D. Popular Science, 218 (1) January 1981
pp. 83-87.

Report on the design of a 594 ft. tall 5 mW
single-blade wind turbine planned for construction in W. Germany. Advantages of the single-
blade design are simplified rotor-head construction, reduced bending loads during blade flapping,
and elimination of strength problems normally
associated with very large windmills. The
turbine drives an 800 kv ac generator through a
step-up gearbox, and voltage is held constant
through a wide range of wind speeds.

3.2(30) RIMMED POLYBLADE WIND TURBINE.
Smith, R.S. and Smith, D.J. (Wind Electric
Systems Inc., University of California,
Berkeley, USA) Sunworld, 5 (2) 1981 pp. 40-47.

The rimmed polyblade wind turbine is a
structurally simple wind-driven generator
system built to produce electric power
synchronised with that of a utility grid.
The design and operation of the turbine is
described. Performance, efficiency and
potential applications are surveyed.
(4 refs.)

3.2(31) PNEUMATIC ENERGY TRANSMITTAL IN WIND-
POWER SYSTEMS.
Kentfield, J.A. (University of Calgary,
Alberta, Canada) J. Energy, 5 (6) November-
December 1981 pp. 362-367.

Describes a simple, fixed-pitch moderate-
velocity-ratio wind turbine-termed a delta
turbine - suitable for the direct drive of
mechanical devices such as air compressors.
A pneumatic transmission to be used in conjunction with the delta turbines is also
examined; advantages of such transmissions
include available energy storage in the form
of stored compressed air, and thermally compressed air in solar-assisted systems.
(6 refs.)

3.2(32) ROTATIONAL DYNAMICS OF WIND TURBINE
GENERATORS.
Swansborough, R.H. and Ballard, L.J. (Era
Technology, U.K.) Paper presented at Future
Energy Concepts: proc. 3rd International
Conference, London 27-30 January 1981
sponsored by Institution of Electrical
Engineers pp. 357-360. London, IEE 1981.

Two contrasting types of wind turbine
generators are discussed; the fixed pitch,
induction generator and variable pitch,
synchronous unit. The rotational dynamic
performances of these turbines are very different, and their respective advantages are
best used in different applications. Turbine
aerodynamics and rotational properties are
analysed. (2 refs.)

3.2(33) SHROUDED WIND TURBINE RESEARCH IN
ISRAEL.
Igra, O. (University of Negev, Israel)
Ambient Energy, 2 (2) April 1981 pp. 85-96.

The shroud concept is analysed and experimental data revealing its power augmentation
capacities are shown. Technical improvements
to reduce cost while retaining aerodynamic
performance are outlined. A pilot plant,
built and tested reveals the usefulness of the
shroud for power augmentation
(14 refs.)

3.2(34) SOME INNOVATIVE CONCEPTS IN WIND
TURBINES OF THE AXIAL-FLOW, CROSS-FLOW, AND
COMBINED (DUAL) FLOW TYPES.
Ljungstrom, O. (Aeronaut Research Inst., Sweden)
In proc. 5th Biennial Wind Energy Conference &
Workshop, Washington, DC., USA 5-7 October 1981
I.E. Vas (ed.) V. 1. Palo Alto, USA, Solar
Energy Research Institute 1981 Session 1D
pp. 415-432. (SERI/CP-635-1340 CONF-811043)

Describes two concepts patented by the author
in Sweden: the swept blade turbine and the
catenary ribbon blade turbine. Augmentation of
turbine mass flow and power output by increasing
the effective sweep area of axial flow can be
achieved by moving the rotor hub. Some concepts and initial exploratory experiments are
described; new applications of the LDB concept
(L-blade and double blade) to straight bladed
CFT-VAWTS (eg Musgrove turbine, Giromill),
inclined shaft CFT with single and multiple
shafts, spiral blade concept, fluid ring bearing VAWT. Further, some dual flow turbine
concepts are discussed consisting of a large
primary turbine catching the wind energy and
transferring it directly to a much smaller
secondary turbine or brake turbine.

3.2(35) TECHNICAL AND ECONOMIC REVIEW OF SMALL
AND LARGE WIND SYSTEMS.
Hughes, W.L., Ramakumar, R.G. and
Lingelbach, D.D. (Oklahoma State University,
USA) Paper presented at Florida Solar Energy
Centre/Dept of Energy Solar Technology Assess-
ment Conference, Orlando, USA 29-30 January
1981 Section 8 11 pp.

Presents a technical and economic review of
small and large wind energy conversion systems.
System components and configurations, such as
the use of horizontal- or vertical-axis wind
blades, upwind or downwind rotors, and airflow
arrangements, are evaluated.

3.2(36) WIND-LOADING DEFINITION FOR THE
STRUCTURAL DESIGN OF WIND-TURBINE GENERATORS.
Kareem, A. and others. J. Energy, 5 (2) March-
April 1981 pp. 89-93.

The stochastic nature of atmospheric wind
must be considered in the design of a wind
turbine generator and the supporting structure.
Important statistical parameters of fluctuating
wind such as turbulence intensity, power
spectral density, and coherence functions are
identified. For the practical design of wind-
turbine generators, short-term extreme-wind-
loading descriptions for the stressing of
structural systems and long-term loading for
the fatigue analysis are discussed in time and
frequency domains.

3.2(37) WIND-TURBINE GENERATOR SYSTEMS.
McGraw, M.G. Electrical World, 95 (5) May 1981
pp. 97-110.

Surveys horizontal-axis and other turbine
configurations being researched. The Federal
and private sectors are engaged in developing
wind energy technology. Wind energy power
plant design and siting considerations are
identified.

3.2(38) DESIGNING A HIGH RELIABILITY WIND
MACHINE - HOW ONE COMPANY DID IT.
Wind Power Digest, (25) Autumn 1982 pp. 6-13.

Vermont's North Wind Power Co. decided to
develop a 2 kW wind machine for use in
remote areas where extreme weather conditions
can be expected. The design features incorporated were an upwind rotor, a three bladed
rotor, and direct drive, low speed generator.
Early tests conducted on the machine determined the effects of gyroscopic forces on
the rotor system, measured the power coefficient and output of the rotor in
relation to blade pitch, and investigated
vibrations on the blades during rotor tilt-
back. Design of the variable axis rotor
control system and the stub tower are
discussed.

3.2(39) EFFECT OF WIND SPEED FLUCTUATIONS ON
THE POWER OUTPUT OF THE WIND TURBINE IN
KALKUGNEN, SWEDEN.
Smedman, A. (Uppsala University, Sweden)
In papers presented at 4th International
Symposium on Wind Energy Systems, Stockholm,
Sweden 21-24 September 1981 V. 2. Cranfield,
U.K., BHRA Fluid Engng. 1982 Session K
Paper K4 pp. 181-194.

An extensive measurement programme, including turbulence measurements, has been carried
out at Kalkugnen wind energy test site. During
April 1980 measurements of the three components
of the wind and the power output from the wind
power plant were simultaneously made.
The fluctuations in power output and in wind
velocity have been analysed in the frequency
domain. A simple method has been developed
to calculate the variation of the power output
of the windmill at Kalkugnen when the spectrum
of the horizontal wind component is known.

3.2(40) SIZE EFFECTS IN DAWT INNOVATIVE WIND
ENERGY SYSTEM DESIGN.
Foreman, K.M. (Grumman Aerosp. Corp.) New York,
USA, American Society of Mechanical Engineers
1982 11 pp. (ASME Paper no: 82-WA/Sol-20)

Examines the effect of size on the estimated
weight and cost of an advanced wind energy conversion system, the diffuser-augmented wind
turbine (DAWT). Preliminary designs are described for three DAWT sizes (ratings) in all-
aluminium; ferrocement; and a hybrid fibreglass reinforced plastics diffuser shell on an
aluminium frame. Common design criteria are
employed in designs for these three materials
and installed cost estimates are made.

3.2(41) THEORETICAL DESIGN STUDY OF THE MSFC
WIND-WHEEL TURBINE.
Frost, W. and Kessel, P.A. (FWG Associates,
Tennessee, USA) NASA Report 3532 March 1982
42 pp.

Gives a mathematical evaluation of an innovative wind-wheel turbine conducted to assess
overall design features. The main parts of the
WWT are a bladed wheel, main housing, two forward ducts, side concentrators, elevated base,
and electrical/mechanical subsystems. Experimental tests and performance prediction
techniques are described. The turbine is
shown to be a viable system offering many
advantages compared with conventional wind
turbine systems.

3.2(42) DEVELOPMENT OF A WIND CONVERTER AND
INVESTIGATION OF ITS OPERATIONAL FUNCTION.
PART 1: TECHNICAL DESCRIPTION OF THE WIND
ENERGY CONVERTER; PART 2: AERODYNAMICS AND
CALCULATIONS OF LOADS; PART 3: DESIGN OF ROTOR
BLADE, PRODUCTION AND LOADING TESTS; PART 4:
TEST SET UP AND RESULTS OF MEASUREMENT.
Molly, J.P. and others. (Deutsche Forschung-
sanstalt fuer Hubschrauber und Vertikalflung-
technik, Stuttgart, F.R. Germany, Inst. fuer
Bauwesen- und Konstruktionsforschung) Final
Report April 1981 Report nos: BMFT-FB-T-82-204/
205/206/207 November 1982 60 pp., 62 pp.,
171 pp. and 48 pp. (In German with English
summary)

Describes the development of a 10 kW wind
energy converter by using as far as possible
standard serial production parts. Design
criteria and the description of the essential
machinery components of the MODA 10 wind
energy converter are given. For some special
load cases the safety calculations of the
important components are shown. The blade
control system which qualified for small wind
energy converters, is explained and weight
and cost of the MODA 10 considered. Blade
design power/blade pitch angle and critical
load conditions are assessed in Parts 2 and
3 and Part 4 includes tests on a 6 kW system
as well as the MOD 10 converter.

3.2(43) APPLICATION EXAMPLES FOR WIND TURBINE
SITING GUIDELINES. FINAL REPORT PREPARED BY
BATTELLE PACIFIC NORTHWEST LABS.
Wegley, H.L. and Pennell, W.T., EPRI March 1983
130 pp.

Describes the trial application by two
utilities of wind-turbine cluster-siting
guidelines. The application examples are
intended as an aid to other users of the siting
guidelines. In addition, the sensitivity of
wind-turbine economics to two types of wind-
turbine performance model is examined.

3.2(44) METHODS OF REDUCING WIND POWER CHANGES
FROM LARGE WIND TURBINE ARRAYS.
Schlueter, R.A. and others. (Michigan State
University, East Lansing, Michigan, USA)
IEEE Trans. Power Apparatus Systems, PAS-102
(6) June 1983 pp. 1642-1650.

Discusses methods of reducing the WECS
generation change through selection of the

wind turbine model for each site, selection of an appropriate siting configuration, and wind array controls. Detailed simulation results indicate more precisely how these factors can be exploited to minimise the WECS generation changes observed. (44 refs.)

3.2(45) MOD-2 WIND TURBINE DEVELOPMENT.
Gordon, L.H., Andrews, J.S. and Zimmerman, D.K. National Aeronautics and Space Administration, Cleveland, Ohio; Lewis Research Centre. US Dept. of Energy Report nos: DOE/NASA/20305-9 CONF-830631-14 1983 27 pp. (In Biennial Wind Energy Conference, Minneapolis, MN., USA 1 June 1983.)
Describes the development of the Mod-2 turbine, which was designed to achieve a cost of electricity that will be competitive with conventional electric power generation. The Mod-2 wind turbine system background, project flow, and a chronology of events and problem areas leading to Mod-2 acceptance are described. The role of the participating utility during site preparation, turbine erection and testing, remote operation, and routine operation and maintenance activity is reviewed. Discusses system performance, loads, and controls.

3.2(46) PERFORMANCE OF AN ANGULAR FLANGE AEROELASTIC WIND ENERGY CONVERTER.
Ahmadi, G. (Clarkson College of Technology, Dept. of Mechanical & Industrial Engineering, Potsdam, NY., USA) J. Energy, 7 (3) May-June 1983 pp. 285-288.
An angular flange H-section model of a torsional aeroelastic wind energy converter is constructed, and its performance under various conditions is investigated. The effects of the variations of the flange angle and the flange width on the performance of the model are studied. The weight of the pendulum also varied, and its effects on the power coefficient of the model are investigated. It is observed that the efficiency of energy conversion decreases with an increase in wind speed.

3.2(47) PRINCIPLES OF ENERGY EXTRACTION FROM A FREE STREAM BY MEANS OF WIND TURBINES.
Riegler, G. (Institute for Applied Systems Technology, Graz, Austria) Wind Engineering, 7 (2) 1983 pp. 115-126.
Equations for the energy extraction from a free stream are developed with particular stress on the necessary basic assumptions. It follows that the only usable form of energy in a free stream is its kinetic energy and in order to get a high energy output per turbine area it is necessary to have a high velocity reduction as well as a high mass flow rate.

3.2(48) WIND ENERGY CONCENTRATORS. THEORETICAL AND EXPERIMENTAL STUDY ON INCREASING THE POWER DENSITY OF WIND TURBINES IN VORTICES.
Greff, E. (RWTH, Inst. fuer Luft- und Raumfahrt, Aachen, West Germany) Fortschr. Ber VDI Z Reihe, 7 (73) 1983 273 pp. (In German)

Proposes the concept of a wind energy concentrator for a turbine operating in vortices as a means of increasing the power density of wind turbines. The vortex generators used in this case are delta-shaped surfaces with sharp leading edges. Water tunnel studies show a distinct improvement of the vortex stability through cambering of the delta surface. This allows the rotors to be fitted into the vortex cones without causing the latter to burst. Rotor design methods with various corrections for losses are rederived and are used for optimisation of the blade chord and blade twist distributions.

3.2(49) ELECTRICAL ENERGY FROM THE WIND.
Wilson, R.E. and Thresher, R.W. (Oregon State University, Corvallis, Oregon, USA) Mechanical Engng., 106 (1) January 1984 pp. 60-69.
Describes studies conducted to determine the wind energy resources in the United States. Wind turbine performance calculations are provided including horizontal axis and vertical axis wind turbines. Economic aspects of wind generators are discussed and the technical aspects important in considerations on the future of wind energy as a commercial source of electricity, are also reviewed.

3.2(50) DEVELOPMENTAL HISTORY OF HIGH RELIABILITY WIND TURBINES UTILISING SIDE TURNING ROTOR OVERSPEED CONTROL.
Sencenbaugh, J.R. (Sencenbaugh Wind Electr. Inc.) In Intl. Wind Energy Symposium 5th Annual Energy-Sources Technology Conference, New Orleans, USA 7-10 March 1982 pp. 101-106. New York, American Soc. Mech. Engrs. 1982.
Describes design of the assembly, which consists of propeller wood blade, 5 inch chord, and hub assembly, transmission alternator assembly, centre casting/slip ring housing and tail assembly. Operational history of the machines is noted and design modifications, design variants described. Used for battery charging in remote locations.

3.2(51) THE SOUTHERN CALIFORNIA EDISON WIND TURBINE GENERATOR TEST PROGRAM.
Wehrey, M.C. and Yinger, R.J. (Southern Calif. Edison Co.) In Intl. Wind Energy Symposium 5th Annual Energy-Sources Technology Conf., New Orleans, USA 7-10 March 1982 pp. 51-56. New York, American Soc. Mech. Engrs. 1982.
Edison's wind power test programme was developed for the Alcoa vertical axis wind turbine and the Bendix/Schachle horizontal axis machine. Lists the instrumentation and describes in more detail procedures involved in noise surveys. Gives details of the test site and test machines.

3.2(52) ON THE DEVELOPMENT OF INTERNATIONALLY AGREED UPON STANDARDS FOR TESTING AND EVALUATION OF WIND TURBINES.
Pedersen, B.M. (Denmark Tech. Univ.) In Intl. Wind Energy Symposium 5th Annual Energy-Sources Technology Conference, New Orleans, USA

7-10 March 1982 pp. 347-349. New York, American Soc. Mech. Engrs. 1982.

Outlines the work done by an expert group under the International Energy Agency (IEA), R & D Implementing Agreement, to create a set of recommended practices for wind turbine testing and evaluation. Examines the first parts of the recommendations dealing with power performance testing, evaluation of the cost of energy and methods and testing for the evaluation of the resistance of the structure to fatigue damage. Means of implementation and continuous revision of these recommendations are discussed.

3.2(53) TEST PERFORMANCE OF LARGE WIND TURBINES. Vachon, W.A. and Schiff, D. (Arthur D. Little Inc.) In Intl. Wind Energy Symposium 5th Annual Energy-Sources Technology Conf., New Orleans, USA 7-10 March 1982 pp. 137-158. New York, American Soc. Mech. Engrs. 1982.

A summary of recent test experiences and results from current, federally funded large wind turbine (WT) tests is presented. These results cover tests from four 200 kW MOD-OA WT's, three 2.5 mW MOD-2 WT's, and three 100 kW vertical axis wind turbines; all funded by the U.S. Department of Energy. Test results from these programmes covering overall system performance, component lifetime and reliability system availability, energy capture, and operations and maintenance requirements are discussed. In addition, key blade design modifications are discussed.

3.2(54) AN APPRAISAL OF STRAIGHT BLADED VERTICAL AND HORIZONTAL AXIS WINDMILLS. McAnulty, K. (A.E.R.E. Harwell, U.K.) In Wind Energy Conversion, proc. 5th BWEA Wind Energy Conf., Reading, U.K. 23-25 March 1983. P. Musgrove (ed.) pp. 245-252. Cambridge, U.K., Cambridge University Press 1984 375 pp. (Isbn 0-521-26250-X)

Performance in uniform wind, and in free air for both types of machine is broadly similar. Considers control constraints on operating tip speed, and contrasts the different stalling behaviour of each. Notes methods of pitch control. Studies rotor configuration, output shaft orientation and blade shape.

3.2(55) EVOLUTION OF WIND-TURBINES: AN HISTORICAL REVIEW. Fleming, P.D. and Probert, S.D. (Cranfield Inst. of Technology, School of Mechanical Engineering, U.K.) Appl. Energy 18 (3) November 1984 pp. 163-177.

The political and commercial forces leading to the harnessing of wind power and the spread of relevant technical knowledge are considered. A review is made of early developments, twentieth-century wind power utilisation incorporating both horizontal and vertical axis wind turbines, and future developments. (Refs.)

3.2(56) EXPERIENCE GAINED IN CONSTRUCTING AND COMMISSIONING TWO MEDIUM SIZED WIND TURBINES. Wilson, R.R. and Brown, A. (James Howden & Co.,

Glasgow, Scoltand) Proc. Inst. Mechanical Engineering Part A 198 (9) 1984 pp. 141-148.

Discusses design of 300 kW and 250 kW wind turbines for use in isolated sites. Outlines the design features of the two and describes the difficulties encountered and the lessons learned in constructing and commissioning the machines.

3.2(57) FLEXIBLE SAIL WIND-TURBINES: REVIEW OF PERTINENT THEORETICAL ANALYSES. Fleming, P.D. and Probert, S.D. (Cranfield Inst. of Technology, Bedford, U.K.) Appl. Energy (18) 2 1984 pp. 89-99. (See also paper by same authors in Applied Energy 17 (3) 1984 pp. 169-180)

Preliminary theoretical analyses of the behaviours of flexible-sail type wind turbines are virtually non-existent. Previous theoretical investigations about the pertinent aerodynamics of yacht sails are reviewed, and their possible modifications to describe the behaviours of flexible-sail wind-turbines are discussed. (30 refs.) (See also paper in Wind Energy Conversion proc. 5th BWEA Wind Energy Conf., Reading, U.K. 23-25 March 1983 pp. 350-357.)

3.2(58) METEOROLOGICAL AND PERFORMANCE MEASUREMENTS ON A 55 KW VESTAS AND A 30 KW SONEBJERG WIND TURBINE. EC WIND ENERGY PROGRAMME, ACTION 5. Jensen, S.A. and others. Luxembourg, Comm. European Communities. Report no: EUR 9041 EN 1984 76 pp.

A recording system has been designed for measurements of meteorological data and data concerning the performance of a wind turbine. Measurements were carried out on two commercially available Danish wind turbines: a 55 kW VESTAS wind turbine erected at a farm, and a 30 kW Sonebjerg wind turbine erected at a house. The same measuring programme carried out on both wind turbines included long-term measurements for the determination of wind statistics, power performance, and operation time; short-term measurements for the determination of power quality, skew wind and brake functions; and supervision measurements to record the behaviour of the wind turbine during an emergency.

3.2(59) 250 KW AND 3 MW WIND TURBINES ON BURGAR HILL, ORKNEY. Lindley, D. (Taylor Woodrow Construction Ltd., Wind Energy Group, Greenford, Middx., U.K.) Proc. Inst. Mechanical Engineering Part A 198 (9) 1984 pp. 149-160. (See also paper in Wind Energy Conversion 5th BWEA Wind Energy Conference, Reading, U.K. 23-25 March 1983 pp. 253-273. Cambridge Univ. Press 1983.)

Describes design and manufacture of a 20 m diameter wind turbine generator, rated at 250 kW, erected on Orkney in July 1983. It rotated for the first time in late July and was synchronised with the grid in August 1983. The commissioning tests have been completed and are to be followed by 12 months (+) of performance monitoring. This paper gives details of the machine design and specification, details of sensors and monitoring system, and early results from commissioning. (11 refs.)

3.2(60) WIND TURBINE RUNAWAY SPEEDS.
Milborrow, D.J. (C.E.G.B., U.K.) In Wind Energy
Conversion, proc. 5th BWEA Wind Energy
Conference, Reading, U.K. 23-25 March 1983.
P. Musgrove (ed.) pp. 358-366. Cambridge, U.K.,
Cambridge University Press 1984 375 pp.

In an accident with the MOD-2 wind turbine in
1981, the wind turbine was disconnected from a
load source and the rotor oversped. Analyses
this case to determine the maximum speed of the
wind turbine in the runaway condition and define
its relationship to wind speed and other
variables. Maximum tip speed (driving torque
zero) is reached at 276 m/s. Considers the
effect of aerofoil design, design lift coefficient and blade solidity and tabulates
estimated runaway speeds, and corresponding
minimum wind speed for two tip speed rotors.

3.2(61) WIND TURBINES - PERFORMANCE AND
CONSTRUCTION.
Buerskens, H.J. J. Klimaatbeheersing 13 (7)
July 1984 (In Dutch)

Reports on design aspects of wind turbines
based on research by the ECN Energy Centre and
BEOP Office for Energy Research Projects within
the framework of the NOW National Programme for
Wind Energy Development. Explains the capital
intensity of wind energy conversion processes
by the low energy density of wind forces to
large installations and the fact that wind turbines need storage systems or additional electricity generation and supply. Research is
focussed on maximising the output/cost ratio,
searching for the best locations and methods of
integration into the national grid.

3.2.1 VERTICAL AXIS TURBINES.

3.2.1(1) VERTICAL-AXIS WIND TURBINE
TECHNOLOGY WORKSHOP, 1976.
(ERDA, Wind Energy Conversion Branch, Oak
Ridge, Tenn., USA) Proc. of Vertical-Axis
Wind Turbine Technology Workshop, Albuquerque,
NM, 18-20 May 1976. Publ. by Sandia Lab.,
Albuquerque, NM. 1976 Report no: SAND76-5586.

Thirty papers deal with vertical-axis wind
turbine technology. Major topics covered are:
programme overviews; ERDA's Darrieus vertical-
axis wind turbines and important vertical-axis
wind turbine programmes.

3.2.1(2) WIND ENERGY - A DESIGN IDEA FOR A
SMALL VERTICAL-AXIS TURBINE.
Baker, W.J. Electron Power, 24 (3) March 1978
pp. 186-187.

There are many designs of wind-driven rotors,
mostly of the conventional horizontal-axis type
which have to be equipped with a method for
presenting the driving blades into the on-coming
wind. Vertical-axis rotors such as the 'Darrieus'
and 'Savonius' type do not require such methods.
A design for a small vertical-axis turbine is
presented. A hollow cylinder fitted with end
plates comprises the 'chassis' and this is
arranged so that it can rotate freely on a

vertical shaft. Wind vanes are mounted between
the end plates, hinged at their outer edges on
vertical spindles, so that they are free to
swing independently and automatically into
'driving' or 'feathering' positions depending
on wind direction but always in such a way as
to afford unidirectional rotation of the
assembly. By using a cylindrical form of
chassis, the air stream striking it is diverted
round its surface, causing an acceleration in
the velocity of the air impinging upon the
driving vanes. A substantial increase in
power output is achieved.

3.2.1(3) CHARATERISTICS OF FUTURE VERTICAL
AXIS WIND TURBINES.
Kadlec, E.G. Albuquerque, USA, Sandia Nat. Lab.
Report no: 79-1068 1979 17 pp.

Sandia Laboratories is developing Darrieus
vertical-axis wind turbine (VAWT) technology in
order to assess the practicality of wind-energy
systems for low-cost production and commercial
marketing by private industry. This report
describes the characteristics of current technology designs and assesses their cost effectiveness. Report claims that better aerodynamics and future structural requirements can
combine for potential energy cost reductions
of 35 to 40%.

3.2.1(4) DEVELOPMENT OF A 1-KW HIGH RELIABILITY
CYCLOTURBINE.
Zvara, J., Drees, H.M. and Noll, R.B. (Aerospace
Systems, Mass., USA) Paper presented at US Dept.
of Energy Small Wind Turbine Systems Research &
Development Requirements Conference, Boulder,
USA 27 February-1 March 1979 V. 1. pp. 6-12.

Under the DOE inititated wind energy R & D
programmes to develop prototype wind machines
in the 1, 8 and 40 kW ranges, a unique vertical
axis wind machine has been developed. Design
requirements for this prototype involve tolerance of extreme environmental conditions. The
configuration, construction and performance of
the system is examined.

3.2.1(5) DEVELOPMENT OF AN 8 KW VERTICAL AXIS
WIND TURBINE.
Stewart, T.D. (Aluminum Co. of America) Paper
presented at US Dept. of Energy Small Wind
Turbine Systems Research & Development Requirements Conference, Boulder, USA 27 February-
1 March 1979 V. 1. pp. 121-131.

A programme was initiated by Aluminum Co. of
America to develop a low-cost machine suitable
for farm and rural domestic applications. The
components and structural features of the prototype vertical axis turbine - three blades and
rotor tubes, base assembly, and tie-down
anchoring system - are described. The micro-
processor controller equipment is also discussed.

3.2.1(6) DEVELOPMENT OF VERTICAL AXIS WIND
TURBINES.
Shankar, P.N. (National Aeronaut. Lab.,
Bangalore, India) Indian Acad. Sci. Sect. C.,
proc. 2 (1) March 1979 pp. 49-61.

Describes the development of vertical axis
wind turbines based on the Darrieus rotor. A
performance analysis was developed which permits
the estimation of their characteristics. A
5 m. high wind turbine using curved wooden
blades was designed, fabricated and tested.
Both theory and initial tests confirmed the low
starting torque of the turbine. Wind tunnel
tests were performed on model Savonius rotors
to determine optimum starter bucket configurations. Finally a straight-bladed turbine
was designed and constructed. It is concluded
from present experience that Darrieus turbines
are likely to be useful in large systems used
to generate electrical power for national grids
but not for direct water pumping purposes.
(11 refs.)

3.2.1(7) 40 KW GIROMILL PROGRAM.
Duwe, W.D.(Valley Industries, St. Louis) Presented
at US Dept. of Energy Small Wind Turbine Systems
Research & Development Requirements Conference,
Boulder, USA 27 February-1 March 1979 V. 1.
pp. 132-141.

The McDonnell Aircraft Co. is contracted to
design, build and deliver a 40 kW vertical axis
windmill called a giromill. Specifications
include production of 40 kW in a 20 mph wind
with the rotor centre line at 75 ft. It is to
be designed for utility grid use, a 30-year
life span and an initial cost of $500/kW. The
selected configuration and control systems are
described.

3.2.1(8) USAF ACADEMY VERTICAL AXIS WIND
TURBINE DEVELOPMENT PROGRAM.
Kullgren, T.E., Wiedemeier, D.W. & Tinsley, J.T.
(US Air Force Academy, Colorado, USA) Paper
presented at US Dept of Energy Small Wind
Turbine Systems Research & Development Require-
ments Conference, Boulder, USA 27 February-
1 March 1979 V. 1. pp. 247-253.

A Darrieus-type machine with two fixed curved
blades was being developed. This machine's
features included a degree of portability and
ease of installation, variable speed operation
by alternator field control, a lightning
protection system, and segmented steel blades.
Preliminary field testing results are summarised.
(6 refs.)

3.2.1(9) WIND SYSTEM TEST PROCEDURES.
Atkins, R.E. (Virginia Polytechnic Inst. &
State University) Paper presented at US Dept.
of Energy Small Wind Turbine Systems Research
& Development Requirements Conference, Boulder,
USA 27 February-1 March 1979 V. 1. pp. 233-246.

Reviews test procedures used at the DOE/US
Sandia Labs vertical axis wind turbine test
facility. The method-of-bins as a technique
for determining the mean performance of a
wind turbine in a field environment is
explained and some complex test procedures
are detailed. (11 refs.)

3.2.1(10) CONTROL OF DISPERSED VERTICAL AXIS
WIND TURBINES.
Wan, Y., Dodd, C.W. and Evers, J.L. (Southern
Illinois University at Carbondale) In proc.
6th Annual UMR-DNR Conference & Exposition on
Energy, Missouri-Rolla, USA 16-18 October 1979.
J.D. Morgan (ed.) V. 6. Missouri-Rolla, USA,
Missouri-Rolla University, 1980 Session 4e
pp. 413-417.

Wind turbines used in a remote region as the
primary source of energy are called dispersed
wind energy systems. The use of a Vertical
Axis Wind Turbine in this mode of operation
requires special control features.
computer simulation examining the turbine's
output voltage under time varying wind
conditions evaluates three possible control
techniques.

3.2.1(11) DEVELOPMENT OF A 5.5 M IN DIAMETER
WIND ENERGY CONVERSION WITH A VERTICAL AXIS
(PHASE III).
Dekitsch, A. (Dornier Sys., Friedrichshafen,
Germany) Statusber. Windenerg. (Semin) Hamburg,
Germany 24-26 June 1980. Publ. by VDI, Dussel-
dorf, Germany 1980 pp. 305-327. (In German)

Describes tests of Darrieus and Savonius
rotor WECS over a prolonged period of time.
Some weak points were recognised making it
necessary to introduce mechanical and power
electronics improvements.

3.2.1(12) INVESTIGATIONS OF DYNAMIC LOADS OF
A WIND ENERGY CONVERTER WITH VERTICAL AXIS
AND OF ACCELERATION AND OVERLOAD CONTROL.
Auffarth, B. and others. Statusber. Windenerg.
(Semin), Hamburg, Germany 24-26 June 1980.
Dusseldorf, Germany, VDI 1980 pp. 289-303.
(In German)

Considers loads on wind energy converters
with a vertical axis and blades parallel to
the axis, and the possibilities of improved
acceleration and a safer overload control.
The WEC has a capacity of 2.5 kW, its blade
length is 3.5 m. and the diameter is 3.5 m.
The measurement and control systems are
described.

3.2.1(13) A LOW-SPEED VERTICAL-AEROFOIL,
VERTICAL-AXIS WIND TURBINE.
Inall, E.K. (Australian Natl. Univ.) Paper
presented at 2nd US Dept. of Energy/Solar
Energy Research Institute Wind Energy Innovative
Systems Conference, Colorado Springs, USA
3-5 December 1980 pp. 95-1. 1.

Most designs for wind energy turbines impose
high centrifugal forces on the rotors calling
for rotors made from expensive materials. An
alternative is to use a less efficient, slow
moving large rotor with lower stress and cheaper
material. Also these have much higher starting
torque, which is maintained over a range of
speeds and provides adequate output and simple
load control without special stalling provisions.

3.2.1(14) NEW CONCEPTS IN VERTICAL AXIS WIND
TURBINES (VAWT) AND APPLICATIONS TO LARGE
MULTI-MW SIZE, OFF-SHORE WIND TURBINE SYSTEMS.

Ljungstrom, O. (Aeronaut. Research Institute, Sweden) In : Collection of Technical Papers from AIAA/Solar Energy Research Institute Conference, Boulder, USA 9-11 April 1980. New York, USA, American Institute Aeronaut. & Astronaut. 1980 Paper no: 80-0620 pp. 286-298.

Presents some new concepts in cross-flow or vertical axis wind turbines with basic properties, test results and design studies of application to multi-mW VAWT, primarily for offshore siting in sheltered waters.

3.2.1(15) PRELIMINARY PERFORMANCE ANALYSIS FOR A FLEXROTOR INNOVATIVE WIND ENERGY SYSTEM.
Noll, R.B., Zvara, J. and Ham, N.D. (Aerospace Syst. Inc.; Massachusetts Institute of Technology) In 2nd Solar Energy Research Institute Wind energy Innovative Systems Conference, Boulder, USA 3-5 December 1980. Golden, USA, Solar Energy Research Institute 1980 Session 4 pp. 183-192. Report no: SERI/CP-635-938.

The FlexRotor is a three-bladed vertical axis wind energy system mounted on an unguyed, free-standing tower. The straight, untwisted blades and struts are made of commercially-available standard extrusions to simplify construction and reduce cost. High strength cables are used to support the blade to carry large tension forces and offset centrifugal loading. Control of overspeed is by an aeroelastically induced pitch angle produced through a canted hinge called the FlexHinge. Design features are presented and preliminary performance analysis results given showing the effects of size, blade swept area, rotor solidity, rotor speed and wind speeds.

3.2.1(16) TETHERED GYROTURBINE WIND ENERGY SYSTEM.
Noll, R.B. and Ham, N.D. (Aerospace Syst. Inc.; Massachusetts Institute of Technology) In 2nd Solar Energy Research Institute Wind Energy Innovative Systems Conference, Colorado Springs, USA 3-5 December 1980 V. II. Golden, USA, Solar Energy Research Institute 1980 Session 2 pp. 19-26. Report no: SERI/CP-635-1061.

Consists of a body/rotor system operating in an autogyro mode in the jet stream and tethered to ground anchors by high strength cables. The rotor produces sufficient power in the autogyro mode to sustain the weight of the entire system and to rotate an induction generator which produces power in the megawatt range. The GyroTurbine system and its operational modes are described, and its feasibility is discussed for systems of several sizes.

3.2.1(17) VERTICAL AXIS WIND ENERGY CONVERTER WITH STRAIGHT BLADES.
Reimerdes, H.G. (RWTH, Aachen, Germany) Paper presented at 3rd International Symposium on Wind Energy Systems, Lyngby, Copenhagen, Denmark 26-29 August 1980. Cranfield, U.K., BHRA Fluid Engng 1980 Paper G2 20 pp.

Deals with the rotor development of a vertical-axis 60 kW wind-turbine with straight rotor-blades. It is shown that this rotor-concept may be realised using simple construction techniques. The calculation was

carried out using normal methods, without any high numerical calculations. (6 refs.)

3.2.1(18) VERTICAL AXIS WIND TURBINE DESIGN TECHNOLOGY SEMINAR FOR INDUSTRY.
Johnston, S.F. Albuquerque, USA, Sandia Labs. Report nos: SAND-80-09847 CONF-800412-8 August 1980 339 pp.

The objective of the VAWT programme at Sandia National Laboratories is to develop economical, industry-produced, and commercially marketable wind energy systems. Emphasis was placed on technology transfer, on Sandia's technical developments and on defining the available analytic and design tools.

3.2.1(19) VERTICAL AXIS WIND TURBINE DESIGNED AERODYNAMICALLY AT TOKAI UNIVERSITY.
Kato, Y., Seki, K and Shimizu, Y. (Tokai University, Japan) Period. Polytech.-Mech. Engng., 25 (1) 1980 pp. 47-56.

Discusses the straight-wing type vertical axis turbine which is omnidirectional, i.e. able to convert wind energy from any direction. Design details and specifications are presented.

3.2.1(20) VERTICAL AXIS WIND TURBINE RESEARCH AT WEST VIRGINIA UNIVERSITY.
Walters, R.E., Migliore, P.G. and Wolfe, W.P. (West Virginia University, USA) Paper presented at 2nd US Dept. of Energy/Solar Energy Research Institute Wind Energy Innovative Systems Conference, Colorado Springs, USA 3-5 December 1980 20 pp.

Summarises developmental testing of the vertical axis wind turbine conducted at West Virginia University. Mathematical models are used to calculate induced downwash and outflows associated with turbine blades. A computer code was devised to calculate torques caused by various structural components of the turbine system. These calculations agree well with actual test data. (8 refs.)

3.2.1(21) WIND ENERGY CAPACITY OF A SINGLE AIRFOIL WITH VERTICAL AXIS ON A CIRCULAR TRACK.
Palmgren, D. and Otis, D.R. (University of Wisconsin, Madison, USA) In proc. 5th Intersociety Energy Conversion Eng. Conference, Energy to the 21st Century V. 1. Seattle, Washington 18-22 August 1980. Publ. by American Institute Aeronaut. & Astronaut., New York, NY. 1980 Paper 809478 pp. 840-845.

A vertical axis wind energy conversion system consists of a single vertical airfoil travelling at constant speed around a horizontal circular track. Describes a computer simulation to determine thrust, normal force and power coefficients and angle of attack for standard airfoils for airfoil speeds up to 10 times the wind speed. (7 refs.)

3.2.1(22) AERODYNAMIC INTERFERENCE BETWEEN TWO DARRIEUS WIND TURBINES.
Schatzke, P.R., Klimas, P.C. and Spahr, H.R.

(US Sandia Labs., New Mexico, USA) J. Energy, 5 (2) March-April 1981 pp. 84-88.

Calculations were made of the effect of aerodynamic intereference on the performance of two curved-blade vertical-axis wind turbines using a vortex/lifting-line aerodynamic model. The turbines have a tower-to-tower separation distance of 1.5 turbine diameters, with the line of turbine centres varying with respect to ambient wind direction. It was revealed that downwind turbine power decrement was significant only when the line of turbine centres was co-incident with ambient wind direction; the decrement increased with increasing tip-speed ratio. (7 refs.)

3.2.1(23) DESIGN ASPECTS OF SMALL STRAIGHT BLADED VERTICAL AXIS WIND TURBINES.
Bannister, W.S. and Gair, S. (Napier College, U.K.) In proc. International Colloquium on Wind Energy, Brighton, U.K. 27-28 August 1981. Cranfield, U.K., BHRA Fluid Engng. 1981 Session 4 pp. 237-242. (British Wind Energy Association)

3.2.1(24) DEVELOPING THE VARIABLE GEOMETRY VERTICAL AXIS WIND TURBINE FOR PRODUCTION.
Mays, I.D. and Holmes, B.A. (P.I. Specialist Engineers, U.K.) In 3rd International Conference on Future Energy Concepts, London 27-30 January 1981 pp. 283-287. London, Institution of Electrical Engineers 1981.

Describes a new type of windmill. The turbine was developed for cathodic protection for applications such as pipelines and tele-communications, but also has a wider market. Blade development and prototype construction are surveyed. The system can also be used for water pumping and electricity generation for commerical use. (7 refs.)

3.2.1(25) EFFECTS OF DYNAMIC STALL ON SWECS.
Noll, R.B. and Ham, N.D. (Aerospace Syst. Inc.; Massachusetts Institute of Technology) In proc. 5th Biennial Wind Energy Conference & Workshop, Washington, DC, USA 5-7 October 1981. I.E. Vas (ed.) V. 2. Palo Alto, USA, Solar Energy Research Institute 1981 Session 2B pp. 295-306. Report nos: SERI/CP-635-1340. CONF-811043.

A study was made of dynamic stall in order to define its influence on the aerofoil force and moment coefficients so that these effects can be included in the calculation of small wind energy conversion system (SWECS) loads and responses. A definition of a dynamic stall theory is made for use in SWECS design, and the theory is implemented in loads and dynamic response analyses. Sample calculations are made for a representative vertical-axis machine. It is shown that loads and moments on the blades may be underestimated if dynamic stall is not considered.

3.2.1(26) RECENT PROGRESS IN THE DEVELOPMENT OF THE MUSGROVE VERTICAL AXIS WIND TURBINE.
Musgrove, P.J. and Mays, I.D. (Reading University; Sir Robert McAlpine & Sons Ltd., U.K.)

In proc. 5th Biennial Wind Energy Conference & Workshop, Washington, DC, USA 5-7 October 1981. I.E. Vas (ed.) V. 1. Palo Alto, USA, Solar Energy Research Institute 1981 Session 1D pp. 445-456. (SERI/CP-635-1340 & CONF-811043)

The design of a 25 metre diameter test bed machine is described. It represents a one-quarter scale model of a 100 metre diameter, 4.4 mW rated machine suitable for off-shore arrays; and the requirement for long life with low maintenance in a marine environment has influenced the design. Economic studies of offshore wind energy system indicate that the optimum rotor diameter may be well in excess of 100 metres. Vertical axis wind turbines avoid the cyclically varying gravity loads which limit the size of horizontal axis wind turbines, and it is anticipated that subsequent development of the Musgrove wind turbine will include units substantially larger than 100 metres diameter.

3.2.1(27) ANALYTICAL AND EXPERIMENTAL INVESTIGATION OF A WOUND-ROTOR VARIABLE-SPEED, CONSTANT-FREQUENCY GENERATOR FOR SMALL WIND ENERGY SYSTEMS.
Higashi, K.K., Minges, G.P. and Price, G.D. (Atomics International Div., Golden, CO., Wind Energy Research Centre) Washington, DC, USA, US Department of Energy Report no: RFP-3488 October 1982 79 pp. (DE 83012273)

Report on research carried out by the Rocky Flats Wind Energy Research Centre. Testing confirmed the feasibility of this application and showed improved performance could be expected over a constant-speed, constant-frequency generator system. The most notable improvement was the ability to maintain a constant tip speed ratio near the maximum rotor performance coefficient over a wide range of wind speeds. Controlled start-up and shutdown could also reduce the high transient torques and concomitant inrush currents common to induction-generator systems.

3.2.1(28) WIND ENERGY CONVERTER WITH HIGH-SPEED VERTICAL AXIS ROTOR AND STRAIGHT ROTOR BLADES.
Zelck, G. (ERNO Raumfahrttechnik G.m.b.H., Bremen, F.R. Germany) Washington, DC, USA, NASA. Final Report no: BMFT-FB-T-82-201 November 1982 107 pp. (In German)

Documents the development of a wind energy converter with a vertical axis rotor and straight blades. The 2 blade rotor with rigid and rectangular wooden air-foils reaches the nominal output of 75 kva from 11.4 m/sec wind velocity onwards. The development activities are supported by wind tunnel and component tests and the final design selection based upon previous development work. Trade offs show that the design is more advantageous compared to other designs and the use of wood as a material for the rotary and horizontal blade supports is effective.

3.2.1(29) WIND ENERGY IN THE USA - PART II.
Taylor, D. (Open University, U.K.) Energy J -
New Zealand, 55 (2) February 1982 pp. 12-13.

The curved blade, Darrieus wind turbine has
achieved the most success of vertical axis wind
turbines under development in the U.S. Four
turbines comprised of 17 m. diameter x 25 m.
high rotor design with 2 extruded aluminium alloy
blades and rated at 100 kW have been installed
at Sandia Labs., New Mexico, and other sites.

3.2.1(30) AEROELASTIC ANALYSIS OF THE DARRIEUS
WIND TURBINE.
Meyer, E.E. and Smith, C.E. (Boeing Commercial
Airplane Co., Seattle, Washington, USA) J.
Energy, 7 (6) November-December 1983
pp. 491-497.

The flutter stability of a single Darrieus
wind turbine blade spinning in still air is
investigated. The blade is modelled as a thin,
uniform beam pinned to the rotor shaft, with
aerodynamic forces accounted for using strip
theory. The two most dangerous flutter modes
are characterised for a one-parameter family
of blades, and the flutter mechanism is shown
to be dominated by gyroscopic coupling between
motions in the plane of the blade and normal
to the plane of the blade.

3.2.1(31) AUGMENTATION OF POWER IN SLOW-RUNNING
VERTICAL AXIS WIND ROTORS USING MULTIPLE VANES.
Sivasegaram, S. and Sivapalan, S. (Dept. of
Mech. Engng., University of Peradeniya, Sri
Lanka) Wind Engng., 7 (1) 1983 pp. 12-18.

Presents a simple two-vane power augmentation
system for rotors of the Savonius-type. The
influence of important design parameters of the
augmenting system and that of wind direction
have been investigated and the system configuration giving maximum power augmentation
has been determined. It is shown that an
eighty percent increase in power output could
be achieved using a pair of vanes of moderate
size.

3.2.1(31) DARRIEUS WIND TURBINE: VARIABLE-SPEED
OPERATION.
Abbott, K.W. and others. (South Dakota State
University, Agricultural Engineering Dept.,
Brookings, SD., USA) American Society Agric-
ultural Engng. Trans. (Gen. Ed.) 27 (1)
January-February 1984 pp. 265-267, 272.

Objectives of the research project were to
determine the operating characteristics and
performance of the Darrieus wind turbine
operated in a variable-speed mode. The variable-speed system averages 32% efficiency
which is near the predicted optimal efficiency
and operated between rotational speeds of 60
and 180 r/min.

3.2.1(32) FEEDBACK CONTROL OF A DARRIEUS WIND
TURBINE AND OPTIMISATION OF THE PRODUCED
ENERGY.
Maurin, T., Henry, B. and Devos, F. (Institut
d'Electronique Fondamentale, University Paris-

Sud, Orsay, France) Rev. Sci. Instrum., USA
March 1984 55 (3) pp. 416-422.

Presents a microprocessor-driven control
system applied to the feedback control of a
Darrieus wind turbine. The use of a DC machine
as a generator to recover the energy and as a
motor to start the engine, allows simplified
power electronics. The control unit is built
to ensure four different functions: starting,
optimisation of the recoverable energy,
regulation of the speed, and braking. The
authors found that the electrical energy
recovery was much more efficient using the
proposed feedback system than without the
control unit. This system allows a better
characterisation of the wind turbine and a
regulation adapted to the wind statistics
observed in one given geographical location.
(A) (12 refs.)

3.2.1(33) NOISE AND VIBRATION MEASUREMENTS
OF 50 KW VERTICAL AXIS WIND TURBINE GEAR BOX.
Krishnappa, G. (National Research Council,
Engine Lab., Ottawa, Ontario, Canada) Noise
Control Engng. J., 22 (1) January-February
1984 pp. 18-24.

The spectra of noise and vibration signals
show strong peaks at the gear mesh frequency
and its harmonics even under light load
conditions. The second peak radiated from
the power house panels is significantly
higher than that radiated by the gear box
casing.

3.2.1(34) COMMERCIALISATION OF VERTICAL AXIS
WIND TURBINES AT DAF INDAL LTD.
Schienbein, L.A. and Roberts, G.D. (Daf Indal
Ltd., Canada) In Solar Energy Society of
Canada Sunfest Conference, Windsor, Ontario,
Canada 1-5 August 1983 5 pp. (See also paper
in Intl. Wind Energy Symposium 5th Annual
Energy Sources Technol. Conf. March 1982 ASME
pp. 35-44.)

The evolution of the wind turbine designs is
surveyed. Two 500 kW vertical axis machines
were recently installed by the firm, one in
California and one on Prince Edward Island,
reflecting a commitment to this emerging energy
technology. Wind farms and wind/diesel hybrids
are the major targeted markets for wind turbines.
The vertical axis units are designed to improve
cost-effectiveness and safety. (6 refs.)

3.2.1(35) ASPECTS OF THE DYNAMICS OF A
VERTICAL AXIS WIND TURBINE.
Ficemec, I.P. and Sala, A. (NEI Cranes Ltd.)
In Wind Energy Conversion, proc. 5th BWEA Wind
Energy Conference, Reading, U.K. 23-25 March
1983. P. Musgrove (ed.) pp. 100-107. Cambridge,
U.K., Cambridge University Press 1984 375 pp.
(Isbn 0-521-26250-X)

Describes a study of the intermodal coupling
type of interaction possible with the H type
rotor of a VAWT. A two degree of freedom
model is developed to represent cross arm tip
vertical displacement (flap detection) and
propeller type rotation of the blade. Results
of an analytical investigation of the system

characteristic equations are presented. The model is extended to include aerodynamic forces by using a simple streamtube model. Solutions for system frequencies and stability with reference to rotor speed are presented.

3.2.1(36) COMPARISON OF FINITE ELEMENT PRE= DICTIONS AND EXPERIMENTAL DATA FOR THE FORCED RESPONSE OF THE DOE 100 KW VERTICAL AXIS WIND TURBINE.
Lobitz, D.W. and Sullivan, W.N. Albuquerque, USA, Sandia Nat. Lab. Report no: SAND-82-2534 February 1984 18 pp.
 Several types of experimental data taken from the DOE 100 kW rotor are compared with pre- dictions. These data include parked rotor natural frequencies, very low wind centri- fugal and gravitational load response, and vibratory response from wind loads covering the rotor operational spectrum.

3.2.1(37) ELECTRICAL GENERATING USING A VERTICAL AXIS WIND TURBINE.
Clark, R.N. Trans. Am. Soc. Agric. Engrs. (Gen. Edn.) 27 (2) March-April 1984 pp. 577-580.
 A newly designed 100 kW vertical axis wind turbine was operated for one year at the USDA Conservation and Production Research Laboratory, Bushland, Texas. The turbine has an induction generator and supplies power to a sprinkler irrigation system with excess power being sold to the electric utility. The unit has obtained a peak efficiency of 43% at a windspeed of 7 m/s or 73% of theoretical maximum. Guy cables were enlarged to provide greater stiffness at the top of the turbine to reduce blade stress levels.

3.2.1(38) LABORATORY PERFORMANCE MEASUREMENTS OF A MODEL VERTICAL-AXIS WIND TURBINE.
Gair, S. and others. (Napier College) In Wind Energy Conversion, proc. 5th BWEA Wind Energy Conference, Reading, U.K. 23-25 March 1983. P. Musgrove (ed.) pp. 274-282. Cambridge, U.K., Cambridge University Press 1984 375 pp.
 Describes a low cost microcomputer data logging and software processing package suitable for laboratory and outdoor sites use in measur- ing wind turbine performance parameters. The model tested is a straight bladed vertical axis wind turbine, and an open jet wind tunnel is used for acceleration tests, to determine performance for a variety of configurations.

3.2.1(39) SIMPLIFIED DYNAMIC BEHAVIOUR OF A STRAIGHT BLADED VERTICAL AXIS WIND TURBINE.
Courtney, M.S. (City Univ., London, U.K.)
In Wind Energy Conversion, proc. 5th BWEA Wind Energy Conference, Reading, U.K. 23-25 March 1983. P. Musgrove (ed.) pp. 83-90. Cambridge, U.K., Cambridge University Press 1984 375 pp. (Isbn 0-521-26250-X)
 A two-degree-of-freedom model involving bending and twisting of the cross-arm is developed. The equations of motion are presented in a general form. The dynamic

behaviour of the model is presented firstly by examining the uncoupled equations of motion. Coupling terms are then introduced and are seen to cause dynamic instability under certain conditions. Mitigating factors including twist damping and blade offset are discussed.

3.2.2 HORIZONTAL AXIS TURBINES.

3.2.2(1) OPTIMUM CONFIGURATION OF ROTOR BLADES FOR HORIZONTAL WIND ENERGY CONVERTERS.
Weber, W. (University Stuttgart, Germany)
Z. Flugwiss, 23 (12) December 1975 pp. 443-447. (In German)
 Reports on design methods and results dealing with the aerodynamic problems of rotor blades configurations for wind energy converters. The aerodynamically optimal layout of rotor blades is calculated and results show some typical parameter influences on the layout and perform- ance of two-bladed wind energy converters.

3.2.2(2) DYNAMICS OF DRIVE SYSTEMS FOR WIND ENERGY CONVERSION.
Martinez-Sanchez, M. (MIT, Cambridge, Massachusetts) In proc. of a Workshop on Wind Turbine Structural Dynamics held at NASA Lewis Research Centre, Cleveland, Ohio 15-17 November 1977. Washington, DC., NASA Conf. Publ. CP 2034 March 1978 pp. 187-193.
 Calculations are performed to determine the dynamic effects of mechanical power transmission from the nacelle of a horizontal axis wind machine to the ground or to an intermediate level. It is found that resonances are likely at 2 or 4/rev, but they occur at lower power only, and seem correctable. (4 refs.)

3.2.2(3) WIND ENERGY CONVERSION. VOLUME II. AERODYNAMICS OF HORIZONTAL AXIS WIND TURBINES.
Miller, R.H. and others. (Massachusetts Inst. of Technology, Cambridge. Aeroelastic and Structures Research Lab.) US Dept. of Energy Report no: COO-4131-T1 (V. 2) September 1978 213 pp. (See also volumes 3 & 5 of same report series by same authors)
 Presents the aerodynamic theory of the wind turbine starting with the simple momentum theory based on uniform inflow and an infinite number of blades. The simple vortex theory is then developed. Following these swirl, non-uniform inflow, the effect of a finite number of blades and empirical correction theory for vortex ring condition are presented. The more complete vortex theory is presented which includes unsteady aerodynamic effects based on a semi-rigid wake. Methods of applying this theory for performance estimation are discussed as well as for the purpose of comput- ing time varying airloads due to windshear and tower interference.

3.2.2(4) DYNAMICS OF HORIZONTAL AXIS WIND TURBINES. WIND ENERGY CONVERSION.
Miller, R. and others. (Massachusetts Inst. of Technology, Cambridge. Aeroelastic and Struc-

tures Research Lab.) ASRL-TR-184-9
and COO-4131-TI(V. 3) September 1978 149 pp.

The dynamic analysis of horizontal axis turbines may be divided into two convenient areas:
the investigation of the aeroelastics and
response of a single blade on a rigid tower;
and the investigation of the mechanical stability
and vibrations of the rotor system on a flexible
tower. With a reasonable understanding of the
behaviour in these two areas the completely
coupled blade-tower aeroelastic system can be
better understood, and dynamic problems can be
better assessed. (See also paper by Dugundji
in Workshop on Wind Turbine Structural Dynamics
NASA CP 2034.)

3.2.2(5) WIND ENERGY CONVERSION. VOLUME V.
EXPERIMENTAL INVESTIGATION OF A HORIZONTAL AXIS
WIND TURBINE.
Dugundji, J., Larrabee, E.E. and Bauer, P.H.
(Massachusetts Inst. of Technology, Cambridge.
Aeroelastic and Structures Research Lab.)
US Dept. of Energy Report no: COO-4131-TI(V. 5)
September 1978 66 pp.

Presents results of some brief experiments
conducted on a wind turbine model of a rotor
system to verify the aerodynamic excitation
characteristics of wind turbines.

3.2.2(6) ON-LINE CONTROL OF A LARGE HORIZONTAL
AXIS WIND ENERGY CONVERSION SYSTEM AND ITS
PERFORMANCE IN A TURBULENT WIND ENVIRONMENT.
Kos, J.M. (United Technol. Corp., Hamilton
Stand, Windsor Locks, Conn., USA) In proc. 13th
Intersociety Energy Conversion Engng. Conference,
San Diego, California 20-25 August 1978.
Warrendale, Pa. SAE (Cat no: P-75) 1978. Also
New York, NY. IEEE (Cat no: 78-Ch1372-2 Energy)
V. 3 pp. 2064-2073.

Describes closed loop, shaft torque control
for controlling the power of a 2 megawatt,
variable pitch, horizontal axis wind turbine,
driving a synchronous generator connected to a
large power system. A method is presented
which provides a stable, responsive control
system by sensing shaft torque and rotor
rotational speed. The results show the dramatic
improvement in on-line system performance that
can be achieved with the responsive shaft
torque control described in this paper.

3.2.2(7) AUGMENTED HORIZONTAL AXIS WIND ENERGY
SYSTEMS ASSESSMENT. EXECUTIVE SUMMARY. FINAL
REPORT.
Arlington, USA, Tetra Tech. Inc., December 1979
94 pp. Report no: SERI/TR-98003-3.

3.2.2(8) THE DEVELOPMENT AND TESTING OF A
VARIABLE AXIS ROTOR CONTROL SYSTEM WITH 5
METRE ROTOR AND DIRECT-DRIVE ALTERNATOR.
Norton, J.H. (North Wind Power Co., VT.,
USA) Paper presented at US Dept. of Energy
Small Wind Turbine Systems Research & Develop-
ment Requirements Conference, Boulder, USA
27 February-1 March 1979 V. 1 pp. 31-64.

Describes the development of a prototype
1-2 kW high reliability wind energy system

for application in remote and extreme environments. It consists of a three-blade horizontal
axis, direct-drive system with fixed pitch
blades. Preliminary results of experimental
system operations are examined.

3.2.2(9) DEVELOPMENT OF A 40 KW HORIZONTAL
AXIS WIND TURBINE GENERATOR.
Bowes, M.A., Howes, H.E. and Perley, R.
(Kaman Aerospace Corp., Conn., USA) Paper
presented at US Dept. of Energy Small Wind
Turbine Systems Research & Development Require-
ments Conference, Boulder, USA 27 February-
1 March 1979 V. 1 pp. 142-158.

Describes work at Kaman Aerospace Corp.,
Bloomfield to design, fabricate, and test a
horizontal axis wind turbine generator capable
of producing 40 kW of electrical or mechanical
output power in a 20 mph wind. The rotor, drive
train, control, and tower subsystems are
described.

3.2.2(10) ROTOR AND CONTROL SYSTEM DESIGN FOR
THE 8 KW 10 METER DIAMETER DOE/WINDWORKS WECS.
Coleman, C. and Hunnicutt, W. Paper presented at
US Dept. of Energy Small Wind Turbine Systems
Research & Development Requirements Conference,
Boulder, USA 27 February-1 March 1979 V. 1
pp. 65-77.

The rotor is 10 m. in diameter with three
blades. It operates in a downwind configuration using positive pitching of the blades
for overspeed control. Other components of the
rotor and control system are described.

3.2.2(11) UTILITY OPERATIONAL EXPERIENCE ON
THE NASA/DOE MOD-OA 200 KW WIND TURBINE.
Glasgow, J.C. and Robbins, W.H. (NASA Lewis
Research Centre) In proc. 6th Energy Technology Conference: Energy Technology Achievements in Perspective, Washington, DC, USA
26-28 February 1979. R.F. Hill (ed.) Washington,
DC, USA, Gov. Inst. Inc. April 1979 Part 5
pp. 961-981.

The project is part of the Federal Wind
Energy Programme and is designed to obtain
early wind turbine operation and performance
data while gaining initial experience in the
operation of large, horizontal axis wind
turbines in typical utility environments.
This paper describes the machine and documents
the recent operational experience at Clayton,
NM.

3.2.2(12) AERODYNAMIC CHARACTERISTICS OF THE
TARP (TOROIDAL ACCELERATOR ROTOR PLATFORM)
WIND ENERGY CONVERSION SYSTEM.
Duffy, R.E., Jaran, C. and Ungermann, C.
(Rensselaer Polytechnic Inst., Troy, NY.)
New York State Energy Research and Development
Authority, Albany, USA. Final Report 1978-79
February 1980 72 pp. (NTIS Report no:
PB 81-140675)

The Toroidal Accelerator Rotor Platform
(TARP) is an innovative structure for use with
horizontal axis WECS. Its shape resembles
that of a horizontally-oriented wheel rim and

is intended to be built into or retrofitted onto structures built for other purposes, which could increase the use of WECS in urban areas. Variations of the basic TARP structure, about three feet in diameter, were tested in a wind tunnel to determine the optimum design. The model system produced up to 4.5 times the power which the rotor and generator extracted without the TARP. (See also paper by A.L. Weisbrich in 13th Intersociety Energy Conversion Annual Conference proc. V. 3. San Diego 20-25 August 1978.)

3.2.2(13) THE CONCEPTUAL DESIGN OF THE HIGH SPEED RADIAL WIND ROTOR FOR THE 52 M. DIAMETER HORIZONTAL AXIS VOITH WIND ENERGY CONVERTER.
Weber, W. (Voith Getriebe KG) In proc. 3rd International Symposium on Wind Energy Systems, Lyngby, Denmark 26-29 August 1980. Cranfield, U.K., BHRA Fluid Engng. 1980 Paper E2 pp. 239-251.

The Voith wind energy converter is a development of a horizontal axis turbine with two composite material blades. It is designed either to supply electric power with a 265 kW generator to an electric grid or to be an independent electricity source for specific tasks in urban or remote areas.

3.2.2(14) DEVELOPMENT STATUS OF THE TOROIDAL ACCELERATOR ROTOR PLATFORM TM (TARP) FOR WIND ENERGY CONVERSION.
Weisbrich, L. and Duffy, R.E. (Eneco., Conn.; Rensselaer Polytechnic Inst., USA) Paper presented at 2nd US Dept. of Energy/Solar Energy Research Institute Wind Energy Innovative Systems Conference, Colorado Springs, USA 3-5 December 1980 pp. 55-64.

To reduce energy costs a system should have a high performance capability for maximising energy recovery as well as a structure able to serve a variety of duties. The innovative toroidal accelerator rotor platform TM (TARP) flow augmentor design for wind energy conversion demonstrates these characteristics. Wind tunnel tests with TARP and a conventional horizontal axis wind turbine demonstrated power output amplification for the rotor in excess of 4.5 times its maximum identical free stream power output. (4 refs.)

3.2.2(15) LARGE WIND-TURBINE PROJECTS IN THE UNITED STATES WIND ENERGY PROGRAM.
Thomas, R.L. and Robbins, W.H. (NASA Lewis Research Centre, Cleveland, Ohio) J. Ind. Aerodyn., 5 (3-4) May 1980 pp. 323-335.

Significant advances have been made in the technology of large, horizontal-axis wind turbines since 1973. The capital costs for large, horizontal-axis wind turbines currently range from about $8000/kW for operational prototype units in the 200-kW class down to approximately $1000/kW for advanced design prototype units in the multi-megawatt range. Current designs of large wind turbines such as the 2500-kW Mod-2 are projected to be competitive for utility applications when produced in quantity, with capital costs of $600 to $700/kW.

3.2.2(16) SWEDEN TO BUILD LARGE-SCALE WIND TURBINE PROTOTYPES.
Hugosson, S. Intl. Power Generation, 3 (8) November 1980 pp. 37-39.

Two horizontal axis aerogenerators are to be erected in Sweden to obtain data and experience for Sweden's wind energy programme. One, at Maglarp, will be twin-bladed and of 77.6 m. diameter, with a capacity of 300 kW in a 14.2 m/s wind and an estimated annual output of 6-8 x 10SUB3 MWh. The other at Nasudden is to have a 75 m. diameter, twin-bladed rotor, rated at 2000 kW in a 12.5 m/s wind with the same output. Both feature blades of GRP-epoxy, a speed of 25 rpm, and blade pitch control.

3.2.2(17) TENSION-FIELD WIND MACHINE - A NEW CONCEPT IN LARGE-SCALE ENERGY PRODUCTION.
Bailey, D.Z. and Noll, R.B. (Bailey Engng.; Aerospace Syst. Inc.) In Collection of Technical Papers of the AIAA/SERI Wind Energy Conference, Boulder, USA 9-11 April 1980. New York, USA, American Institute Aeronaut. & Astronaut. 1980 Paper no: 80-0622 pp. 88-98.

Describes a unique wind energy conversion system for large-scale (megawatt range) energy production. The uniqueness of the system is the application of the tension-field concept of suspension bridges to provide the required structural strength for the WECS support and airfoil section. The machine typically has parallel, horizontal tensioned blades, spreaders to maintain rotational radius, and anchored end-bearings to accept the large axial tension. The basic configuration is called a horizontal cross-axis tension-field WECS or HCATF-WECS. Typical configurations, performance, cost effectiveness, and preliminary tests are discussed.

3.2.2(18) THE 25 M. EXPERIMENTAL HORIZONTAL AXIS WIND TURBINE (25 M HAT).
Schellens, F.J.C. (FDO Technische Adviseurs & V., Netherlands) In proc. 3rd Intl. Symposium on Wind Energy Systems, Lyngby, Denmark 26-29 August 1980. Cranfield, U.K., BHRA Fluid Engng. 1980 Paper H2 pp. 375-390.

Describes an experimental horizontal axis wind turbine which is part of the Netherlands research programme into the use of wind energy for electricity generation. Design information and component/systems information are included.

3.2.2(19) LINEAR STATIC AND DYNAMIC ANALYSIS FOR HINGED ROTOR BLADES OF 60 M. SPAN FOR A TWO BLADED HORIZONTAL AXIS WIND ENERGY CONVERTER.
Argyris, J.H. and others. (Stuttgart University, F.R. Germany, Inst. fuer Statik und Dynamik der Luft- und Raumfahrtkonstruktionen) Washington, DC., USA, National Aeronautics and Space Administration. Report no: ISD-291 1981 53 pp. (Presented at Von Karman Institute Lecture Series 1981-8 on Wind Energy Conversion Devices, Rhode Saint Genese, Belgium 1-5 June 1981.)

Report of an investigation into the linear static and dynamic behaviour of a rotor blade of a horizontal axis wind energy converter with flap and lag hinges and with coupling of flap and pitch. The linearised equations of motion are developed using a finite element method considering quasisteady aerodynamic forces. The complex eigenfrequencies are calculated. The time history response of the rotor blades is computed for cyclic gravitational loads at rated operation and for a global gust. The stresses at selected points along the blade and forces and moments acting on the tower are calculated from the structural deformation.

3.2.2(20) THE MOD-2 WIND TURBINE DEVELOPMENT PROJECT.
Linscott, B.S. and others. (NASA Lewis Research Centre) Report nos: DOE/NASA 20305-5; NASA-TM-82681, July 1981 24 pp.

The Mod-2 is a large (2.5 mW power rating) horizontal axis wind turbine designed for the generation of electrical power on utility networks. Three machines were built and are located in a cluster at Goodnow Hills, Washington. All technical aspects of the project are described: design approach, significant innovation features, the mechanical system, the electrical power system, the control system, and the safety system.

3.2.2(21) OPERATING EXPERIENCE WITH THE 200 KW MOD-OA WIND TURBINE GENERATORS.
Birchenough, A.G. and others. (NASA) In proc. 5th Biennial Wind Energy Conference & Workshop, Washington, DC., USA 5-7 October 1981 I.E. Vas (ed.). V. 1. Palo Alto, USA, Solar Energy Research Institute 1981 Session 1A pp. 107-118. Report nos: SERI/CP-635-1340 CONF-811043.

Documents 28,000 hours of Mod-OA operational experience. Characteristics of the wind energy generated, the machine performance, and the subsystem strengths and weaknesses are discussed. An assessment of the project success in fulfilling its goals and objectives is also presented.

3.2.2(22) AN OVERVIEW OF LARGE HORIZONTAL AXIS WIND TURBINE BLADES.
Faddoul, J.R. (NASA) In proc. 5th Biennial Wind Energy Conference & Workshop, Washington, DC., USA 5-7 October 1981. I.E. Vas (ed.) V. 3. Palo Alto, USA, Solar Energy Research Institute 1981 Session IVB pp. 113-143. Report nos: SERI/CP-635-1340 CONF-811043.

DOE/NASA wind energy programme has tried aluminium, steel, wood and fibreglass/epoxy for construction for operational wind turbine blades. In addition, polyurethane and cement have been investigated as potential low cost materials. In this report a review of the experience that has been gained from all of these blade materials in sizes ranging from 60 to 150 feet is presented. Fabrication processes are discussed for each blade material as well as cost and weight comparisons and/or projections.

3.2.2(23) UTRC 8-KW WIND-TURBINE TESTS.
Cheney, M.C. (United Technologies Research Center, Connecticut) J. Energy, 5 (2) March-April 1981 pp. 122-128.

The United Technologies Research Center (UTRC) 8 kW prototype turbine became operational in July 1979 and underwent testing in East Hartford, Connecticut until delivery to Rockwell International in February 1980. The prototype testing demonstrated the basic operation of the unique control concept of the UTRC Composite Bearingless Wind Turbine which utilises a hub-mounted pendulum employed to twist the graphite composite inboard region of the blade producing blade pitch variations. The tests also demonstrated the predicted performance of 9 kW at 20 mi/h (9 m/s) and the high-speed-stall control feature. (See also paper by same author in DOE Small Wind Turbine Systems R & D Requirements Conference February-March 1981 V. 1 pp. 78-105.)

3.2.2(24) CONTROL SYSTEM FOR THE 20 M. 250 KW W.T.G. FOR ORKNEY.
Cooper, B.J. (GEC Power Engng. Ltd.) In IEE Power Div. Colloquium on Energy Generating Systems for Wind Power, London 28 April 1982. London, U.K., Institution of Electrical Engineers 1982 Paper 2 pp. 2-1-2-6. IEE Colloquium Digest no: 1982/44.

Describes the basic philosophy and design of the control system for the 20 m. horizontal axis wind turbine generator to be erected at Burgar Hill, Orkney by the Wind Energy Group (Taylor Woodrow, British Aerospace & GEC).

3.2.2(25) DESIGN, CONSTRUCTION AND TEST PROCEDURES FOR HAWT ROTORS.
Bryant, P.J. (Missouri University at Kansas City) In proc. 1982 Wind and Solar Energy Technology Conference, Kansas City, USA 5-7 April 1982. Columbia, USA, Missouri-Columbia University 1982 pp. 15-20.

Presents a review of some procedures by which a Horizontal Axis Wind Turbine (HAWT) may be evaluated and describes a research project to analyse the dynamic stability of HAWT rotors. Controlled Velocity Test procedures are presented and calibration data for the power output of a Small Wind Energy Conversion System (SWECS) as derived by the CVT method are given.

3.2.2(26) THE EFFECT OF ROTOR DIAMETER TOWER HEIGHT AND GENERATOR RATING ON THE ECONOMICS OF WIND ENERGY.
Wilson, R., Jamieson, P. and McLeish, D. (James Howden & Co. Ltd.) In proc. 4th British Wind Energy Association Wind Energy Conference, Cranfield, U.K. 24-26 March 1982. P.J. Musgrove (ed.). Cranfield, U.K., BHRA Fluid Engng. 1982 pp. 275-282.

Presents a study showing the effects of variations in running speed, blade setting angle, blade diameter, tower height and generator rating. The design considered is for a 25 m. diameter, horizontal axis, fixed pitch wind turbine for use on isolated sites

experiencing medium to high wind regimes.
Analyses based on a computer model of the
aerodynamic performance coupled with detailed
information on the capital costs are shown.

3.2.2(27) INTRODUCTION TO WIND ENERGY.
Lysen, E.H. (Steering Committee Wind Energy
for Developing Countries (SWD)) 1982 Publication no: SWD 82-1 310 pp. (Available from
SWD, P.O. Box 85, 3800 AB Amersfoorst,
Netherlands.)
    Sets out the basic theory of selected
aspects of wind energy conversion in relation
to horizontal axis machines of moderate size
driving generators and pumps. Among the
topics treated are the analysis of wind
regimes, the aerodynamics of the rotor, basic
forms of pump and generator and the corresponding coupling theories, the general matching of turbines to wind regimes, rotor stressing, safety systems and elementary system
economics.

3.2.2(28) A REVIEW OF RESONANCE RESPONSE IN
LARGE, HORIZONTAL-AXIS WIND TURBINES.
Sullivan, T.L. (NASA Lewis Research Centre,
Cleveland, Ohio, USA) Solar Energy, 29 (5)
1982 pp. 377-383.
    Field operation of the DOE Mod-0 and Mod-1
wind turbines has provided valuable information
concerning resonance response in large, two-
bladed, horizontal-axis wind turbines. Operational experience has shown that one/rev excitation exists in the drive train and that high
aerodynamic damping prevents resonance
response of the blade flat modes. Teetering
the hub substantially reduces the chord blade
response to odd harmonic excitation. It has
also been found that present analytical techniques can accurately predict wind turbine
natural frequencies. (12 refs.)

3.2.2(29) DO CENTRE-BODIES IMPROVE THE PERFORM-
ANCE OF HORIZONTAL-AXIS SAIL-TYPE WIND-TURBINES?
Fleming, P.D. and others. (Cranfield Inst. of
Technology, School of Mechanical Engineering,
Cranfield, Bedfordshire, U.K.) Appl. Energy,
14 (2) 1983 pp. 123-130.
    The power harnessing advantages of using
hub-fairings - either rotating or stationary,
with or without stationary after-bodies - have
been measured experimentally. For the flexible
sail wind-turbines tested, stationary rather
than rotating hub-fairings produced greater
power augmentations. Employing a stationary
after-body is also desirable for the turbine
in order to increase the rate of harnessing of
wind energy.

3.2.2(30) THE EFFECTS OF TOWER SHADOW ON THE
DYNAMICS OF A HORIZONTAL AXIS WIND TURBINE.
Powles, S.R.J. (Cavendish Laboratory,
Cambridge University, U.K.) Wind Engng.,
7 (1) 1983 pp. 26-35.
    The philosophy behind the design of hori-
zontal-axis wind turbines has recently
changed to using compliant rather than

rigid systems. These are advantageous from
a weight and cost consideration, but the
effects of the aerodynamic shadow on the
rotor blades must also be considered. Wind
speed measurements of the shadow wake behind
a tubular steel tower are presented, and the
effects of this and other types of shadow are
discussed. Computed predictions of the blade
motion, as the blade passes through the tower
shadow, are presented and discussed for the
articulated wooden blades of the Cambridge
5 m. two bladed research turbines. Initial
results indicate that it is not disadvantageous
to use solid towers instead of lattice towers
for these wind turbines.

3.2.2(31) STUDIES ON THE DYNAMIC CHARACTERISTICS
OF THE HORIZONTAL SHAFT WIND TURBINE.
Shimizu, Y. and others. (Mie University, Japan)
J. Energy, 7 (6) November-December 1983 pp. 589-
595.
    The rotor surface of the turbine is continu-
ously and periodically inclined in the steady
flow, and the instantaneous torque and angular
speed are measured along with the coefficients
of power and torque of the wind turbine. These
results are compared with those obtained when
the rotor surface is statically inclined. In
addition, the velocity distributions around the
wind turbine are measured by hot wire anemometers.

3.2.2(32) WHIRL FLUTTER ANALYSIS OF A HORIZONTAL
AXIS WIND TURBINE WITH A TWO-BLADED TEETERING
ROTOR.
Janetzke, D.C. and Kaza, R.V. Solar Energy, 31
(2) 1983 pp. 173-182.
    An investigation is presented to explore the
possibility of whirl flutter and to find the
effect of pitch-flap coupling on teetering
motion of the DOE/NADA Mod-2 wind turbine.
The equations of motion are derived for an
idealised five-degree-of-freedom mathematical
model of a horizontal-axis wind turbine with
a two-blade teetering rotor. The model accounts
for the out-of-plane bending motion of each
blade, the teetering motion of the rotor, and
both the pitching and yawing motions of the
rotor support. Results show that the Mod-2
design is free from whirl flutter. Selected
results are presented indicating the effect
of variations in rotor-support damping, rotor-
support stiffness and on pitching, yawing,
teetering and blade-bending motions.

3.2.2(33) EVALUATION OF A HORIZONTAL SAVONIUS.
Farsaie, A., Smith, G.L. and Brill, B. (University of Maryland, Agricultural Engineering
Dept., College Park, Md., USA) American Society
Agricultural Engng. Trans. (Gen. Ed.) 27 (1)
January-February 1984 pp. 241-244, 247.
    The Savonius rotor generally consists of two
semi-cylinders offset to form a 'S' shape and
sandwiched between circular end discs. Horizontal and vertical Savonius wind energy conversion systems were modelled and wind tunnel
tests were conducted. Roof pitch, wind
incidence angle, and rotor positioning were
varied. The effect of system orientation and

wind direction on power output were studied. The horizontal Savonius rotor demonstrated up to 99% increase in overall power output over the vertical design.

3.2.2(34) DEVELOPMENT OF THE WTS-4 WIND TURBINE DESIGN.
Standard, H. and others. (United Technol. Corp.) In Intl. Wind Energy Symposium 5th Annual Energy-Sources Technology Conference, New Orleans, USA 7-10 March 1982 pp. 1-22. New York, American Soc. Mech. Engrs. 1982.
Describes design of the 4 mW WTS-4 downwind horizontal axis wind turbine, including the rotor nacelle, controller and tower. Trade-offs considered are noted and design selections include a teetered rotor and soft tower, allowing easier yaw drive design and total pitch management. Filament wound fibreglass blades are used.

3.2.2(35) APPLICATION OF THE VORTEX THEORY TO HIGH-SPEED HORIZONTAL-AXIS WIND TURBINES.
Maekawa, H. (Kagoshima Univ., Faculty of Engineering, Kagoshima, Japan) Bull JSME 27 (229) July 1984 pp. 1460-1466.
The turbine blade can be replaced by a bound vortex system, and when the energy loss per unit time is a minimum, the trailing vortices move backwards with a constant velocity and build a helical vortex sheet. A velocity potential function which represents the vortex system is obtained. Torque, power and resistance are computed for a turbine which has two to six blades.

3.2.2(36) ON LIFTING LINE ANALYSIS OF HORIZONTAL AXIS WIND TURBINES.
Politis, G.K. and Loukakis, T.A. (Athens Nat. Tech. Univ.) Wind Engng. 8 (1) 1984 pp. 23-35.
Lifting line correction factors are introduced and compared with those of Prandtl and Goldstein. Lifting line and strip theory formulations are applied for the calculation of performance for two wind turbines.

3.2.2(37) ON THE DESIGN OF HORIZONTAL AXIS TWO-BLADED HINGED WIND TURBINES.
Hohenemser, K.H. and Swift, A.H.P. (Washington Univ. and Texas Univ.) ASME Trans. J. Solar Energy Engng. 106 (2) May 1984 pp. 171-176.
Discusses and evaluates various hinge configurations for two-bladed rotors and shows why the conventional teeter hinge leads to non-uniform blade angular velocity in the bladetip path plane. A solution calling for the suppression of blade flapping by passive blade cyclic pitch variation produced by a strong negative pitch flap coupling, was found to be practical for upwind tail vane stabilising two-bladed wind turbines.

3.2.2(38) WIND ENERGY CONVERTER GROWIAN II.
Braun, D. and others. Bundesministerium fuer Forschung und Technologie, Bonn-Bad Godesberg, F.R. Germany. Report no: BMFT-FB-T-84-063

April 1984 148 pp. (In German) (U.S. Dept. of Energy Report no: DE 84751901)
The objective of the programme was to investigate advanced concepts for multi-mW-wind energy conversion systems in the rotor class of 135 m diameter. The preferred turbine was a variable-speed horizontal-axis downwind machine with a one-bladed teetering rotor and a guyed soft steel tower. A 1 : 3 scaled demonstrator with a rotor diameter of 48 m was built and erected at a site close to Bremerhaven.

3.2.3 TORNADO TYPE WIND TURBINES.

3.2.3(1) TORNADO-TYPE WIND ENERGY SYSTEM.
Yen, J.T. (Grumman Aerosp. Corp., Bethpage, NY, USA) In 10th Intersociety Energy Conversion Energy Conference Record, Univ. of Del., Newark 18-22 August 1975 Paper 759148 pp. 987-994. New York, NY, IEEE 1975. (Also ASME Paper 76-WA/Ener-2 1976 11 pp.)
Describes a new type of wind energy system. Data from wind tunnel tests of a small model are presented and compared with several analyses. Applying these to larger systems it is found that for the same sized turbine and the same wind speed, power outputs of this design may be 100 to 1000 times that of the conventional design. (9 refs.)

3.2.3(2) EFFECT OF LATENT HEAT FROM MOIST AIR ON THE VORTEX IN A TORNADO-TYPE WIND ENERGY GENERATOR SYSTEM.
Chen, J.M. (National Tsing Hua Univ., Hsinchu, Taiwan) J. Ind. Aerodyn., 5 (1-2) October 1979 pp. 53-60.
Presents a local similarity method for the equations that describe a vortex which is generated by a stationary tower with a partly opening top, directed by vertical vanes, and which is affected by the latent heat of condensation from moist air creating buoyant forces.

3.2.3(3) INVESTIGATIONS OF THE TORNADO WIND-ENERGY SYSTEM. FINAL REPORT, SEPTEMBER 1978-APRIL 1980.
Yen, J.T. (Grumman Aerosp. Corp., Bethpage, NY, Research Dept.) Washington, DC, US Dept. of Energy Report no: DOE/ET/20022-T1 May 1980 149 pp.
Tornado-type Wind Energy Systems (TWECS) use the core of a vortex in a hollow tower to provide a low pressure exhaust reservoir for a wind turbine. Wind tunnel data are presented, and previous and planned test programmes described, including the evolution of tower design from the spiral-cross-sectional design through the passive omnidirectional design to the fixed-vane omnidirectional design. A larger model with a 15 ft. height tower and a 30 in. diameter turbine is described. Details on design, configuration, and instrumentation, plus details on the test-rig for testing the 30 in. turbine under a straight-flow environment are given. Results on turbine efficiency, tip-speed-ratio, and turbine disk loading

coefficient are presented and discussed and a new cost analysis is presented.

3.2.3(4) TORNADO-TYPE WIND ENERGY SYSTEM: OUTLINE OF RECENT DEVELOPMENTS.
Yen, J.T. (Gurmman Aerosp. Corp.) In SERI 2nd Wind Energy Innovative Systems Conference, Colorado Springs, USA 3-5 December 1980 V. II. Golden, USA, Solar Energy Research Institute 1980 Session 2 pp. 9-18. Report no: SERI/CP-635-1061.
An outline of recent developments in the Tornado-type Wind Energy System (TWES) is presented. Small models up to 1.1 m. in tower height and 0.1 m. in turbine diameter were tested in wind tunnels. Data obtained from the Grumman Tunnel show a best power coefficient based on the tower frontal area of around 14%, while the Langley test shows a maximum value of around 10%. A cost analysis based on a 10% power coefficient was carried out. It is concluded that the TWES has a good commercial potential, reinforcing the conclusions reached in previous cost analyses. (See earlier papers.)

3.2.3(5) NUMERICAL MODEL FOR THE FLOW WITHIN THE TOWER OF A TORNADO-TYPE WIND ENERGY SYSTEM.
Ayad, S.S. (Solar Energy Research Institute, Golden, Colo., USA) ASME Trans. J. Solar Energy Engng., 103 (4) November 1981 pp. 299-305.
Describes the use of a two-equation turbulence model to predict numerically the flow within the tower of a tornado-type wind energy system. Calculations are carried out for a tower in a uniform flow. Both cases of closed-bottom tower and simulated turbine flow with a variety of turbine-to-tower diameter ratios and turbine flow rates are considered. Calculated values of pressure for closed-bottom tower are compared with experimental values. (11 refs.)

3.2.3(6) A PARAMETRIC STUDY OF TORNADO-TYPE WIND ENERGY SYSTEMS.
Ayad, S.S. In proc. 5th Biennial Wind Energy Conference & Workshop, Washington, DC, USA 5-7 October 1981: I.E. Vas (ed.) V. 1. Palo Alto, USA, Solar Energy Research Institute 1981 Session 1D pp. 487-499. Report nos: SERI/CP-635-1340 CONF-811043.
The tornado-type wind energy system uses the pressure drop created by an intense vortex. The vortex is generated in a tower mounted at the turbine exit. The tower serves as a low pressure exhaust for the turbine. The paper provides results to show effects of embedding the tower in an atmospheric boundary layer, varying the tower height to diameter ratio, and varying tower diameter using the same system geometry and approach flow conditions. The results indicate a reduction of 28% in power output caused by atmospheric boundary layer effects using a power low profile, a minimum height/diameter ratio of 1.0 to avoid asymmetric vortex decay, and decreasing improvements in system performance with increasing system size.

3.2.3(7) PERFORMANCE OF TORNADO WIND ENERGY CONVERSION SYSTEMS.
Volk, T. (New York University, NY, USA) J. Energy, 6 (5) September-October 1982 pp. 348-350.
Describes an experimental investigation performed to examine the flow characteristics and power production capabilities of the Tornado Wind Energy Conversion System (TWECS). Results indicated that the confined vortex in the tower of TWECS rotates approximately as a solid body and only supplements total power production, most of which comes from the tower acting as a bluff body. (6 refs.)

3.2.3(8) TORNADO WIND ENERGY CONVERSION SYSTEM (TWECS) EVALUATION PROGRAM.
Miller, G. and others. (New York University, NY, USA) NTIS Report no: PB 83-227207 September 1982 136 pp.
This study has three major components: the testing and optimisation of TWECS wind tunnel models, the development of a realistic theoretical model of TWECS performance and the construction and field testing of a prototype TWECS. Wind tunnel testing with optimised TWECS models resulted in low power coefficients (below 3%). Similar results were also experienced under atmospheric tests with a large field prototype (30 feet high and 15 feet in diameter), and the effects of forced vortices considered.

3.2.3(9) NUMERICAL STUDY OF THE PERFORMANCE OF TORNADO-TYPE WIND ENERGY SYSTEMS.
Ayad, S.S. J. Energy, 7 (2) February 1983 pp. 134-140.
The tornado-type wind energy system was proposed to utilise the pressure drop created by an intense vortex in a tower. The tower serves as a low pressure exhaust for the turbine. Describes a numerical solution of the tower flow, using the two equation (k-e) turbulence model, for the small size system tested by Yen, with 0.127 m. diameter tower standing in a uniform wind flow. A comparison of the results with the measured pressure values verified the model. The same numerical model is used for a system with a tower that is completely embedded in an atmospheric boundary layer. The results show, for a tower in a boundary layer of the one-seventh power-law profile, about a 28% reduction of the values of power coefficient from those for a tower in a uniform stream. Numerical predictions are made for systems with tower diameters between 0.5 and 8 m. using the same system geometry and approach flow conditions, showing that the scale effect is negligible for systems of 4 m. tower diameter or larger. (A)

3.2.3(10) PERFORMANCE OF TORNADO-TYPE WIND TURBINES WITH RADIAL INFLOW SUPPLY.
Hsu, C.T. and Ide, H. (Iowa State University, Ames, Iowa, USA) J. Energy, 7 (6) November-December 1983 pp. 452-453.
It is shown that the radial inflow supply is necessary for intensifying a vortex in the wind

collecting tower and, consequently, for enhancing the power efficiencies of the wind turbines.

3.2.3(11) POWER COEFFICIENT OF TORNADO-TYPE WIND TURBINES.

Rangwalla, A.A. and Hsu, C.T. (Iowa State University, Ames, Iowa, USA) J. Energy, 7 (6) November-December 1983 pp. 735-737.

The present work is a continuation of So's analysis but the floating parameter, which determines the mass flow through the turbine, is allowed to vary, like a conventional windmill in which the actual mass flow through the turbine is not determined a priori by the actuator disk analysis.

3.2.3(12) PREDICTION OF THE MAXIMUM POWER COEFFICIENT OF A TORNADO-TYPE WIND-ENERGY SYSTEM.

Dick, E. (Ghent State University, Ghent, Belgium) J. Wind Eng. Ind. Aerodyn., 12 (2) July 1983 pp. 101-108.

Makes a theoretical prediction for the power coefficient of a tornado-type wind-energy system. The vortex model developed is simple, but is shown to be consistent with the dynamics of the tornado system. The predicted power output, although still favourable, is considerably lower than was estimated by the inventor of the system.

3.2.3(13) RESEARCH RESULTS FOR THE TORNADO WIND-ENERGY SYSTEM: ANALYSIS AND CONCLUSIONS.

Jacobs, E.W. (Solar Energy Research Institute, Golden, Colo., USA) Washington, DC, Solar Energy Research Institute Report no: SERI/TP-211-1889 January 1983 14 pp.

The Tornado Wind Energy System (TWES) concept utilises a wind driven vortex confined by a hollow tower to create a low pressure core intended to serve as a turbine exhaust reservoir. Numerous experimental and analytical research efforts have investigated the potential of the TWES as a wind energy conversion system (WECS). This paper summarises and analyses much of the research to date on the TWES. A simplified cost analysis incorporating these research results is also included. Based on these analyses, the TWES does not show significant promise of improving on either the performance or the cost of energy attainable by conventional WECS.

3.2.3(14) EXTRACTION OF POWER FROM A TORNADO WIND ENERGY SYSTEM (TWES).

Windrich, J. and Fricke, J. (Phys. Inst., Wurzburg Univ., Germany) Solar Energy (USA) 32 (6) 1984 pp. 765-770.

Vortex flow and power extraction from a vortex were studied both experimentally and theoretically by an analytical approach. The measurements were performed with a small scale wind tunnel model of logarithmic spiral shape. Power estimates were derived from measurements with grid simulated turbines. The performance of the system can be improved if interaction between vortex and top stream is avoided and

if advantage is taken of the axial pressure gradient within the tower flow. (11 refs.)

3.2.4 OSCILLATORY WIND ENERGY CONVERSION SYSTEMS.

3.2.4(1) OSCILLATORY WIND ENERGY CONVERTOR.

Ahmadi, G. (Shiraz University, Iran) Wind Engng., 3 (3) 1979 pp. 207-215.

The principle of wind energy conversion through the oscillation of an aerodynamically unstable system is reviewed, and a nonlinear theory prediction of the power coefficient for such a device is developed. The test results for a small mathematical model are presented and compared with the theoretical prediction, and reasonable agreement is observed. (18 refs.)

3.2.4(2) AN OSCILLATORY KINETIC ENERGY CONVERTOR: A NEW DEVICE FOR PUMPING.

Clayton, B.R. and Filby, P. (University College, London, U.K.) Ambient Energy, 1 (4) October 1980 pp. 197-208.

A device for converting kinetic energy of wind or water streams to the fluctuating motion of a mechanism is described. Large models could produce enough power to support isolated installations, such as radio or meteorological buoys in coastal waters, or pumping sets for irrigation. Model flow is inconsistent and is related to those found in vertical axis wind-turbine rotors. A description of flow is given using a visualisation method and mathematical model.

3.2.4(3) EXPERIMENTS ON AN OSCILLATING AEROFOIL AND APPLICATIONS TO WIND-ENERGY CONVERTERS.

Ly, K.H. and Chasteau, V.A.L. (University of Auckland, New Zealand) J. Energy, 5 (2) March-April 1981 pp. 116-121.

Presents forces measured on a two-dimensional aerofoil oscillating in pitch and combined pitch-heave. Amplitudes of pitch, heave, Reynolds numbers and frequencies are reported. The application of this data to the prediction of Darrieus turbine performance, by approximating blade-wind interaction to a pure pitch oscillation, showed some siginficant differences from steady aerofoil predictions. (14 refs.)

3.2.4(4) OSCILLATING KINETIC ENERGY CONVERTERS.

Clayton, B.R. and Filby, P. (University College, London, U.K.) Paper presented at 3rd International Conference on Future Energy Concepts, London 27-30 January 1981 pp. 273-276. London, Institution of Electrical Engineers 1981.

Conversion of kinetic energy from wind using a device in which oscillatory rather than rotational motion is used is possible. A number of such devices have been built and tested. The design, performance, and operation of such a converter is described. The motion of the hydrofoil converter is complex and consists of surge, sway, and rotational components optimised by a series of linkages. (9 refs.)

3.2.4(5) THE WINGMILL: AN OSCILLATING-WING WINDMILL.

McKinney, W. and Delaurier, J. (University of Toronto, Canada) J. Energy, 5 (2) March-April 1981 pp. 109-115.

Describes the analytical and experimental investigation of a windmill utilising a harmonically oscillating wing to extract wind energy. The wing's span is horizontally aligned and the airfoil is a chordwise-rigid symmetrical section. The whole wing oscillates in vertical translation and angle of attack, with prescribed phasing between the two motions. Theoretical and experimental analyses show the wingmill to be capable of efficiencies comparable to rotary designs. (12 refs.)

3.2.4(6) DEVELOPMENT OF AN OSCILLATING-VANE CONCEPT AS AN INNOVATIVE WIND-ENERGY-CONVERSION SYSTEM.

Bielawa, R.L. (Solar Energy Research Institute, Golden, Colo., USA) Washington, DC, US Dept. of Energy. Solar Energy Research Inst. Report no: SERI/TR-98085-2 March 1982 114 pp. (US Dept. of Energy Report no: DE 82012870) (See also Solar Energy Research Inst. Report no: SERI/CP-635-1061.)

Report on the investigation of an oscillating vane wind energy conversion system, incorporating the bending-torsion flutter characteristics of a cantilevered wing. The system employs the high fatigue strength characteristics of a composite material. Results are presented of experimental and analytic studies to provide technical data upon the concept. Two variants of the concepts are examined: single-vane and split-vane configurations. The experimental results consist of the mechanical power generation and dynamic response and stress characteristics for a model having a span of approximately one metre.

3.2.5 ROTOR DESIGN, AERODYNAMICS, MATERIALS.

3.2.5(1) WIND ENERGY CONCENTRATORS.

Loth, J.L. (West Virginia University, Morgantown, USA) UMR-MEC (Univ. of Mo, Rolla-Mo Energy Council) 2nd Annual Conference on Energy, University of Mo, Rolla 7-9 October 1975 pp. 93-107. North Hollywood, California, USA, West Period Co. 1976.

Presents two alternatives to the shrouded propeller wind energy concentrator. Operation is based on generating a low pressure area, with high local wind velocity, around the windmill rotor. The two types of wind energy concentrators are considered and the performance parameters such as the power concentration ratio and the associated area ratio have been determined theoretically. Some preliminary experimental data are included. (10 refs.)

3.2.5(2) PREDICTION OF THE OPTIMUM PERFORMANCE OF VENTURI-TYPE WIND-ENERGY CONCENTRATORS.

Kentfield, J.A.C. (University of Calgary, Alberta, Canada) In 5th Annual UMR DNR Conference on Energy, University of Mo-Rolla 10-12 October 1978. University of Mo-Rolla, Ext. Div. 1978 pp. 57-64.

Presents a simple theoretical analysis which allows the optimum performance to be predicted for venturi-type wind-energy concentrators. Results are presented in convenient parametric form in such a manner that, for example, the effect on optimum performance of reducing the outlet area of the exit diffuser, of reducing the bulk and cost of a venturi-type concentrator, can be established directly from the curves. The most important parameters affecting concentrator performance are: the effectiveness of the exit diffuser, the area ratio of that diffuser and the magnitude of the base-drag coefficient applicable at the diffuser exit plane. (5 refs.)

3.2.5(3) WIND ENERGY CONVERSION. VOLUME VI. NONLINEAR RESPONSE OF WIND TURBINE ROTOR.

Chopra, I. (Massachusetts Inst. of Tech., Cambridge. Aeroelastic and Structures Research Lab.) Washington, DC, US Dept. of Energy September 1978 231 pp. DOE Report no: COO-4131-T1 (V. 6).

Using Lagrange's equations the nonlinear equations of motor for a rigid rotor restrained by three flexible springs representing the flapping, lagging and feathering motions, are derived, for arbitrary angular rotations. These are reduced to a consistent set of nonlinear equations. Analysis is divided into three parts: Part A consists of forced response of two-degree flapping-lagging rotor under the excitation of pure gravitational field; in Part B, the effect of aerodynamic forces on the dynamic response of two-degree flapping-lagging rotor is investigated; in Part C, the effect of third degree of motion, feathering, is considered.

3.2.5(4) WIND ENERGY CONVERSION. VOLUME IX. AERODYNAMICS OF WIND TURBINE WITH TOWER DISTURBANCES.

Chung, S.Y. (Massachusetts Inst. of Tech., Cambridge. Aeroelastic and Structures Research Lab.) Washington, DC, US Dept. of Energy September 1978 103 pp. DOE Report no: COO-4131-T1 (V. 9).

Lifting line theory, the counterpart of Prandtl's lifting line theory for rotating wing, is employed for the overall performance analysis of a horizontal axis wind turbine rotor operating in a uniform flow. The wake system is modelled by non-rigid wake which includes the radial expansion and the axial retardation of trailing vortices. For non-uniform flows which are caused by the ground, the tower reflection, or the tower shadow, the unsteady airloads acting on the turbine blade are computed, using lifting line theory and a non-rigid wake model. An equation which gives the wind profile in the tower shadow region is developed. Also, the equations to determine pitch angle control are derived to minimise the flapping moment variations or the thrust variations due to the non-uniform flow over a rotation.

3.2.5(5) WIND ENERGY CONVERSION. VOLUME VIII. FREE WAKE ANALYSIS OF WIND TURBINE AERODYNAMICS.
Gohard, J.C. (Massachusetts Inst. of Tech., Cambridge. Aeroelastic and Structures Research Lab.) MIT Report no: ASRL-TR-184-14 September 1978 294 pp. DOE Report no: COO-4131-T1 (V. 8).
Presents the theory for determining blade and rotor/tower vibration and dynamic stability characteristics as well as the basic dynamic operating loads. Starting with a simple concept of equivalent hinged rotors, the equations of motion for the blade including pitch, flap and lag motions are developed.

3.2.5(6) WIND ENERGY CONVERSION. VOLUME VII. EFFECTS OF TOWER MOTION ON THE DYNAMIC RESPONSE OF WINDMILL ROTOR.
Sheu, D.L. (Massachusetts Inst. of Tech., Cambridge. Aeroelastic and Structures Research Lab.) MIT-ASRL Report September 1978 61 pp. DOE Report no: COO-4131-T1 (V. 7).
Study of the effects of tower motion on the dynamic response of a windmill rotor. In the analysis the blade lagging and side tower motion are taken into consideration. The equations of motion for the system are a set of linear ordinary differential equations having periodic coefficients. The periodic coefficients of the equations of motion for a three bladed rotor are eliminated by using the multiblade coordinate transformation method. For a two bladed rotor, the equations of motion are solved by using the harmonic balance method.

3.2.5(7) AEROELASTIC BEHAVIOUR OF LARGE DARRIEUS-TYPE WIND ENERGY CONVERTERS DERIVED FROM THE BEHAVIOUR OF A 5.5 M. ROTOR.
Vollan, A.J. (Dornier Syst., Friedrichshafen, Germany) In 2nd International Symposium on Wind Energy Systems, Amsterdam, Netherlands 3-6 October 1978. Cranfield, U.K., BHRA Fluid Engng. 1978 Paper C5 pp. 67-88.
An investigation into the conditions of similarity for Darrieus rotors of different sizes and specially shaped blades proved that the rotors are similar with regard to their aeroelastic behaviour. However, scaling up a small rotor will lead to a very heavy structure and the control of the bending stresses would create problems. Therefore it is necessary to reduce the structural complexity of larger wind energy converters which results in increased danger of aeroelastic instability. (8 refs.)

3.2.5(8) WIND ENERGY CONVERSION. VOLUME X. AEROELASTIC STABILITY OF WIND TURBINE ROTOR BLADES.
Wendell, J. (Massachusetts Inst. of Tech., Cambridge. Aeroelastic and Structures Research Lab.) MIT-ASRL Report September 1978 94 pp. DOE Report no: COO-4131-T1 (V. 10).

The nonlinear equations of motion of a general wind turbine rotor blade are derived from first principles. The twisted, tapered blade may be preconed cut of the plane of rotation, and its root may be offset from the axis of rotation by a small amount. The equations are applicable to studies of forced response or of aeroelastic flutter, however, neither gravity forcing, nor wind shear and gust forcing are included. The equations are applied to study the aeroelastic stability of the NASA-ERDA 100 kW wind turbine, and solved using the Galerkin method. The numerical results are used in conjunction with a mathematical comparison to prove the validity of an equivalent hinge model developed by the Wind Energy Conversion Project at the Massachusetts Institute of Technology.

3.2.5(9) EVALUATION OF FEASIBILITY OF PRE-STRESSED CONCRETE FOR USE IN WIND TURBINE BLADES.
Lieblein, S. and others. (NASA) Dept of Energy Report no: DOE/NASA 5906-79/1 September 1979 119 pp. (NASA CR-159725).
A baseline blade design was achieved for the DOE/NASA Mod-0 100 kW experimental wind turbine that met aerodynamic and structural requirements. Calculated blade weight and cost were 4900 lb. and around $18,000, compared to 2000 lb. and around $200,000 for a Mod-0 aluminium blade, but significant cost reductions were indicated for volume production. Casting of a model blade section showed no fabrication problems. Dynamic analysis revealed that adverse rotor-tower interactions can be significant with heavy rotor blades. Design options are discussed.

3.2.5(10) STATIC AND DYNAMIC INVESTIGATIONS USING A WINDMILL MODEL.
Argyris, J.H. and others. (Stuttgart University, Stuttgart, Fed. Rep. Germany) June 1979 69 pp. (In German)
A scale model of a wind rotor was constructed, then used to develop and test a data acquisition and transmission system. With this system experimental data are collected from the operating model and displayed on a screen. Comparison between experimental data and computed model results were used to check applied static and dynamic analyses of the rotor model.

3.2.5(11) WIND TUNNEL TESTS ON A 3 M. DIAMETER MUSGROVE WINDMILL.
Willmer, A.C. (British Aerospace, U.K.) Ambient Energy, 1 (1) January 1980 pp. 21-26.
A 3 m. diameter model of a two-bladed Musgrove vertical-axis windmill was tested in a wind tunnel at Fulton, U.K.. Tunnel constraints were minimised by using a low flow blockage. Results demonstrate the good performance of this type of windmill. The measured performance showed excellent agreement with predictions from a simple mathematical model. However, maximum loads measured on the windmill are not well predicted by the model. To reconcile measurement and prediction, large induced crossflows must be postulated at some blade rotational positions. (4 refs.)

3.2.5(12) APPLICATION OF THE FINITE ELEMENT
METHOD TO THE STUDY OF THE BEHAVIOUR OF BLADES
OF INDUSTRIAL WIND ENERGY SYSTEMS.
Sorel, J. and others. (French Atomic Energy
Commission; AEROWATT Co.) In proc. 3rd Intl.
Symposium on Wind Energy Systems, Lyngby,
Denmark 26-29 August 1980. Cranfield, U.K.,
BHRA Fluid Engng. 1980 Paper D3 pp. 193-206.

The development of new materials and the
improvement of manufacturing techniques have
made possible high power (100 kW) wind energy
systems with large diameter (18 m.) propellers.
Those erected on windy sites are often exposed
to heavy mechanical overloads due to severe
climatic conditions. The blade attachment is
thus critical and led to an economic way of
strain distribution. The blade footing has
then been computed by the finite element method
for strain mapping providing vital information
concerned with turbine design.

3.2.5(13) CONSTRUCTION AND INVESTIGATION OF
OPERATING BEHAVIOUR OF A WIND ENERGY CONVERTER
OF MODULAR DESIGN.
Molly, J.P. Statusber Windenerg. (Semin),
Hamburg, Germany 24-26 June 1980. Dusseldorf,
Germany, VDI 1980 pp. 371-388. (In German)

A 10 kW modular wind power plant is operated
at the joint Voith/DFVLR wind power testing
ground in Schnittligen/Boehmenkirch, Germany,
at which an extensive electronic data acquisition
facility has been set up. Eight rotor blades
of different types and materials have been manufactured. The purpose was reduction of costs.
The blades are 5.5 m. long and the tower is
10 m. high.

3.2.5(14) EXCHANGE OF NATIONAL EXPERIENCE IN
THE FIELD OF NEW ENERGY SOURCES - IN PARTICULAR
SOLAR, WIND AND GEOTHERMAL ENERGY: CONCENTRATOR
SYSTEMS FOR WIND ENERGY, WITH EMPHASIS ON
TIPVANES.
van Holten, T. (Technische Hogeschool, Delft,
Netherlands, Dept. of Aerospace Engineering)
Geneva, Switzerland, Economic Commission for
Europe (UN) 30 October 1980 34 pp. (NTIS PB 81-
239576)

Summarises theory of concentrator-systems
and discusses experimental verification of its
theory describing tipvane augmentation, mass-
flow augmentation by cross-wind forces, and
the power required to maintain a vortex ring.
Practical development problems include aeroelastic instabilities, turbine blade shape and
viscous interference between turbine blade and
tipvane. Performance and cost are discussed.

3.2.5(15) LOW COST COMPOSITE MATERIALS FOR
WIND ENERGY CONVERSION SYSTEMS.
Weingart, O. (Structural Composites Industries,
California, USA) Paper presented at Inter-
national Solar Energy Society American Section/
Et Al 1980 Annual Conference, Phoenix 2-6 June
1980 V. 3.2 pp. 1502-1506.

Describes the development of a new composite
filament winding technology for low cost
production of wind turbine generators. A
method called the TFT process is employed in

fabricating wind turbine blades. The process
yields the tapered cross-section, tapered wall
thickness, and highly axial orientation
characteristics needed for such wind energy
applications. The design and operation of a
4 kW conversion system fabricated from
composite materials are described.

3.2.5(16) NUMERICAL CALCULATION OF STEADY
INVISCID FULL POTENTIAL COMPRESSIBLE FLOW
ABOUT WIND TURBINE BLADES.
Dulikravich, D.S. (NASA Lewis Research Centre,
Ohio) NTIS Report no: N80-18497/1 1980 11 pp.

Describes an exact nonlinear mathematic model
that accounts for three-dimensional cascade
effects about the inner portions of wind turbine
rotor blades. Potential compressible flow about
the tip regions of the blades can also be computed. A periodic computation mesh was generated for determining exact boundary conditions.

3.2.5(17) A PRELIMINARY STUDY OF A TETHERED
WIND ENERGY SYSTEM INCLUDING THE EFFECT OF A
DUCTED ROTOR.
Sivier, K.R. (Illinois University at Urbana-
Champaign, USA) In 2nd Wind Energy
Innovative Systems Conference, Colorado Springs,
USA 3-5 December 1980 V. II. Golden, USA,
Solar Energy Research Institute 1980 Session 3
pp. 55-66. (Solar Energy Research Institute
Report no: SERI/CP-635-1061)

Altitudes from 150 m. to 3050 m. were investigated. The basic study involved systems with
free (unducted) rotors and preliminary estimates
were made of the effects of using a ducted rotor.
Because of the increase in mean windspeed with
altitude, the rotor diameter and the weight
of the basic wind energy system decrease with
altitude. However, the aerostat volume and
the total TWES weight exhibit minimums at
moderate (600 to 1200 m.) altitudes.

3.2.5(18) STRESS ANALYSIS AND TEST PHILOSOPHY
FOR WIND ENERGY CONVERTER BLADES.
Bansemir, H. and Pfeifer, K. (Messerschmitt-
Boelkow-Blohm G.m.b.H., Munich, F.R. Germany.
Unternehmensbereich Raumfahrt) Washington, DC,
NASA Report no: NBB-UD-300-80-0 1980 28 pp.
(Presented at 4th Meeting of Experts on Rotor
Blade Technology with Special Respect to
Fatigue Design Problems, Stockholm 21-22 April
1980)

Presents a description and analysis of
projected wind energy converter blades for a
5 mW plant. The calculation of laminate cross-
section properties as well as the overall
behaviour of the blade is considered with special
attention given to the bolted area and to the
compression stresses in the blade. Tests
performed in order to obtain basic data of the
materials used in the design are discussed.
Special components such as the bolted area are
tested by static and dynamic loads. Finally
the strains in the entire blade under load are
measured.

3.2.5(19) TESTS PERFORMED ON THE 2 MW TVIND
WECS.
Jensen, S.A. and Bjerregaard, E.T.D. (Danish
Ship Research Lab., Denmark) Paper presented
at 3rd International Symposium on Wind Energy
Systems, Lyngby, Copenhagen, Denmark
26-29 August 1980. Cranfield, U.K., BHRA Fluid
Engng. 1980 Paper H3 pp. 391-400.
   Measurements on the 2 mW WECS, built by the
Tvind Schools in Denmark, have been carried
out. A brief description is given of the
instrumentation. Measurements were made with
strain guages mounted at various positions
along the rotor blades. The rotor shaft torque
and rpm were measured, enabling the mechanical
power efficiency to be investigated.

3.2.5(20) TURBULENCE AND WIND TURBINE PERFORMANCE.
Hansen, A.C. (Rockwell International, USA)
ASCE proc. J. Transportation Engng., 106 (TE6)
November 1980 pp. 675-683.
   Research activities at Rockwell International for
U.S. Department of Energy have begun to determine
the quantitative significance of turbulence to
wind-turbine utilisation. Limitations and
advantages of the analysis methods are highlighted. Wind energy conversion system development subcontracts have resulted in improved
understanding of the importance of turbulence
in system design. In nine designs, turbulence
(gust) considerations have been second only to
maximum survival wind speed, in importance to
structural design.

3.2.5(21) ADVANCED AND INNOVATIVE WIND ENERGY -
ENERGY CONCEPT DEVELOPMENT: DYNAMIC INDUCER
SYSTEM. RESEARCH REPORT.
Lissaman, P.B.S., Zalay, A.D. and Hibbs, B.H.
(Solar Energy Research Institute, Golden,
Colorado; AeroVironment Inc., Pasadena, CA.)
Washington, DC, US Dept. of Energy May 1981
89 pp. (Solar Energy Research Institute Report
no: SERI/TR-8085-1-T2) (An executive summary
is also available in Report no: SERI/TR-8085-
1-T1 16 pp.)
   Demonstrates the performance benefits of the
dynamic inducer tip vane system. Tow-tests
conducted on a three-bladed, 3.6 metre diameter
rotor have shown that a dynamic inducer can
achieve a power coefficient (based upon power
blade swept area) of 0.5, which exceeds that of
a plain rotor by about 35%. Wind tunnel tests
of a model dynamic inducer achieved a power
coefficient of 0.62 which exceeded that of a
plain rotor by about 70%. The dynamic inducer
substantially improves the performance of
conventional rotors and indications are that
higher power coefficients can be achieved
through additional aerodynamic optimisation.

3.2.5(22) ATMOSPHERIC TESTING OF A TWO BLADE
FURL CONTROLLED WIND TURBINE WITH PASSIVE CYCLIC
PITCH VARIATION.
Hohenhemser, K.H. and Swift, A.H.P. In proc. 5th
Biennial Wind Energy Conference & Workshop,
Washington, DC, USA 5-7 October 1981. I.E. Vas
(ed.) V. 1. Palo Alto, USA, Solar Energy Res-
earch Institute 1981 Session 1D pp. 457-467.
Report nos: SERI/CP-635-1340 CONF-811043.

Assesses whether a two-bladed furl controlled
wind rotor with passive cyclic pitch variation
has the potential of cost effective wind energy
conversion. Passive cyclic pitch variation
was achieved by letting the blade pair freely
oscillate about a common pivot with which the
blades formed a small prelag angle. This simple
two-bladed rotor was found to be capable of
rapid yaw rates suitable for rotor speed control
by yawing, without imposing vibratory hub
moments and without producing appreciable out-of-
plane blade excursions. During the first phase
of testing the 7.6 m. diameter rotor was automatically furled when 228 rpm at 10 kW rotor
power was exceeded. Unfurling was performed
manually. During the second phase of testing
fully automatic furl control systems were used
which limited rotor speed and torque quite
accurately to predetermined values. It was
found that passive cycle pitch variation and
the associated rapid wind direction following
did not degrade and possibly enhanced aerodynamic efficiency.

3.2.5(23) CALCULATION OF NATURAL MODES OF
VIBRATION FOR ROTOR BLADES BY THE FINITE
ELEMENT METHOD.
Kiessling, F. and Ludwig, D. Gottingen,
F. R. Germany, DFVLR 1981 65 pp. (DFVLR-
Forschungsbericht 81-07) (In German)
   The mass and stiffness matrices for a
rotating blade are established by the
finite element method, based on the Lagrange
function for combined flapwise bending,
chordwise bending and torsion of twisted non-
uniform rotor blades. Eigen value analyses
are performed for a nonrotating homogeneous
beam and for the rotor blade of a wind energy
converter.

3.2.5(24) AN OVERVIEW OF FATIGUE FAILURES AT
THE ROCKY FLATS WIND SYSTEM TEST CENTRE.
Waldon, C.A. (Rockwell International, CO, USA)
Paper presented at 16th ASME Intersociety
Energy Conversion Engineering Conference,
Atlanta, GA, USA 9-14 August 1981 V. 2
pp. 2047-2052.
   Identifies problems associated with small wind
energy conversion systems. The structural
integrity of specially designed small wind energy
conversion system components is analysed.

3.2.5(25) SOME OBSERVATIONS ON RESULTS OF ROTOR
PERFORMANCE TESTING.
Balcerak, J.C. (Rockwell Inc., USA) In proc.
5th Biennial Wind Energy Conference & Workshop,
Washington, DC, USA 5-7 October 1981. I.E. Vas
(ed.) V. 2. Palo Alto, USA, Solar Energy Research Institute 1981 Session 3B pp. 587-595.
Report nos: SERI/CP-635-1340 CONF-811043.
   Discusses a sample of the performance test
results of several small wind energy conversion
systems. These results show some apparent
anomalies. Sampling indicates that poor system
performance can stem from several sources of
the overall system, but that in the main, the
efficiency of the rotor in extracting energy
from the wind needs improving and considers
various design options.

3.2.5(26) CATALOG OF LOW-REYNOLDS-NUMBER AIRFOIL DATA FOR WIND-TURBINE APPLICATIONS.
Miley, S.J. (Rockwell International, Golden, CO, Rocky Flats Plant; Texas A and M University, College Station, Dept. of Aerospace Engineering) Rockwell Intl. Rocky Flats Plant Report no: RFP-3387 February 1982 623 pp. (DE 82 021712)
A literature survey was performed to acquire airfoil data at low Reynolds numbers which would be applicable to small wind energy conversion systems. Each set of data includes airfoil coordinates, lift, drag and pitching moment characteristics in both graphical and tabular form. A discussion of airfoil behaviour and the effects of Reynolds number, surface roughness and turbulence is given.

3.2.5(27) THE SPECTRUM OF WIND SPEED FLUCTUATIONS ENCOUNTERED BY A ROTATING BLADE OF A WIND ENERGY CONVERSION SYSTEM.
Connell, J.R. (Battelle Pacific NorthWest Lab., Washington, USA) Solar Energy, 29 (5) 1982 pp. 363-345.
A point on a rotating wind turbine blade encounters turbulence with characteristics different from turbulence measured by a stationary anemometer. The rotational sampling effect is quantified using measurements of wind velocity at circular arrays of anemometers and measurements by laser and hotwire anemometers traversing cross-wind circular paths and a simple theoretical model of the spectrum of turbulence which is rotationally sampled is developed. (9 refs.)

3.2.5(28) CONSTRUCTION, TESTING AND DEVELOPMENT OF LARGE WIND ENERGY FACILTIES.
Windheim, R. and Cuntze, R. (NASA) NASA Report no: NAS 1.15:76933 JUEL-SPEZ-138 September 1982 412 pp. (Transl. into English of proceedings of Seminar on Energy Research, F. R. Germany Report Juel-Spez-138 1981 415 pp.
Discusses the building of large rotor blades and control of oscillations in large facilities and concludes that the technical problems in the design of large rotor blades and control of oscillations can be solved.

3.2.5(29) ANALYSIS OF SOUND PRODUCED BY WIND TURBINES IN TURBULENT FLOW.
Juvet, P.J.D. (Aeronautical Research Inst. of Sweden) M.S. Thesis. Massachusetts Institute of Technology Report no: FFA-TN-1983-44 October 1983 114 pp.
An analytical method developed for helicopters and propellers was used to calculate the far field sound spectrum produced by the interaction of wind turbines with atmospheric turbulence. A linearised airfoil response to a two-dimensional gust was used to calculate the unsteady blade loading. The calculated sound spectrum presents a region of peaks and valleys at low harmonics, due to blade to blade correlation, degenerating into a smooth spectrum at higher frequencies.
(From author's abstract)

3.2.5(30) BLADE- AND ROTOR LOADS FOR VESTAS 15.
Rasmussen, F. (Risoe National Lab., Roskilde, Denmark) Risoe National Lab. Report no: RISO-M-2392 June 1983 50 pp. (In Danish) (DE 83 751370)
Contains a comparison of measurements and calculations of aerodynamic blade- and rotor forces for a stall-controlled VESTAS 15 windmill. The investigation focuses on the situation in skew wind, as the blade forces are found as a function of the three basic parameters: windspeed, yaw angle and blade position. Lift- and drag coefficients for the blade profiles are calculated iteratively from measurements of blade root bending moment in two directions as a function of windspeed under axial airflow.

3.2.5(31) METHOD FOR ESTIMATING THE AERODYNAMIC COEFFICIENTS OF WIND TURBINE BLADES AT HIGH ANGLES OF ATTACK.
Beans, E. and Jakubowski, G.S. (University of Toledo, Ohio, USA) J. Energy, 7 (6) November-December 1983 pp. 747-749.
Presents a method for estimating the coefficients in the post-stall region. The method agrees with experimental data obtained by the authors and from other sources.

3.2.5(32) OPTIMUM PROPELLER WIND TURBINES.
Sanderson, R.J. and Archer, R.D. (University of New South Wales, Kensington, NSW, Australia) J. Energy, 7 (6) November-December 1983 pp. 695-701.
The Prandtl-Betz-Theodorsen theory of heavily loaded airscrews has been adapted to the design of propeller windmills which are to be optimised for maximum power coefficient. It is shown that the simpler, light-loading, constant-area wake assumption can generate significantly different 'optimum' performance and geometry, and that it is therefore not appropriate to the design of propeller wind turbines. Design curves for optimum power coefficient are presented and an example of the design of a typical two-blade optimum rotor is given.

3.2.5(33) PERFORMANCE OF PROPELLER WIND TURBINES.
Wortman, A. (California State University at Fullerton, Calif., USA) J. Energy, 7 (6) November-December 1983 pp. 640-643.
Exhibits a new computational technique which yields performance directly when tangential speed ratio and section aerodynamic characteristics are specified. The off-design performance of the finite drag/lift was far better than that of their zero drag counterparts, except in a $\pm$ 20% region about the design conditions. Tolerance to off-design operation increased with decreasing tip speed ratios so that the annual energy capture for tip speed ratios between 2 and 4 was about 87% of the ideal turbine value.

3.2.5(34) THEORETICAL IMPROVEMENT OF WIND TURBINE OUTPUT BY THE ADDITION OF TIP DEVICES.
Sundar, R.M. and Sullivan, J.P. (Purdue University, West Lafayette, Indiana, USA)

J. Energy, 7 (6) November-December 1983
pp. 749-750.
Presents a method of analysis and some comparisons of the aerodynamic performance of wind turbines with and without tip devices.

3.2.5(35) INVESTIGATION OF THE AERODYNAMICS OF HORIZONTAL-AXIS WIND TURBINE ROTORS. FINAL TECHNICAL REPORT.
Cromack, D.E. and others. (Massachusetts Univ., Amherst, Dept. of Mechanical Engineering)
U.S. Dept. of Energy Report no: DOE/CS/8900-T1 September 1981 37 pp. (DE 84005245/XAD)
Includes a description of the data acquisition system for the Wind Rotor Test Facility as well as an extensive review of the literature on the aerodynamic theories developed to date. This review is the background to the formulation of a synthesised rotor performance theory and computer code.

3.2.5(36) AERODYNAMIC ANALYSIS OF A HORIZONTAL AXIS WIND TURBINE BY USE OF HELICAL VORTEX THEORY. VOLUME 1. THEORY.
Jeng, D.R. and others. (Toledo Univ., Ohio, Dept. of Mechanical Engineering) U.S. Dept. of Energy Report no : DOE/NASA-0005-1; NASA-CR-163054 December 1982 88 pp. (DE 84016035)
The method assumes that a helical vortex emanates from each blade element. Collectively these vortices form a vortex system that extends infinitely far downstream of the blade. The wind turbine performance of a two-bladed rotor is determined and compared to existing experimental data and to corresponding values computed from the widely used PROP code. It was found that the present method compared favourably with experimental data especially for low wind velocities.

3.2.5(37) EXAMINATION, EVALUATION, AND REPAIR OF LAMINATED WOOD BLADES AFTER SERVICE ON THE MOD-OA WIND TURBINE.
Faddoul, J.R. (NASA Lewis Research Center, Ohio) In American Wind Energy Association Wind Energy Expo National Conference, San Francisco 17-19 October 1983 pp. 317-357.
As part of the NASA development effort for large horizontal-axis wind turbines, four rotors were fabricated for the 200 kW machines using an epoxy-impregnated laminated wood material. Stud failure on a Hawaii blade set was caused by improper installation and inadequate corrosion protection; no structural damage in the laminated wood was detected. A crack in a Rhode Island set was attributed to a manufacturing process problem; again no significant structural problems were observed. (5 refs.)

3.2.5(38) WAKE STRUCTURE MEASUREMENT AT THE MOD-2 CLUSTER TEST FACILITY AT GOODNOE HILLS.
Lissaman, P.B.S. and others. Columbus, USA, Battelle Mem. Inst. March 1983 49 pp. (PNL-4572) (See also similar paper by Baker, R.W. and Walker, S.N. Oregon State Univ. DOE/BP-182 1982 91 pp.)

Investigates the rate of decay of wake velocity deficit with downwind distance in various meteorological conditions. A radio sonde suspended from a tethered balloon was used for wake-wind speed measurements. Isolations of the high wind speed velocity ratios showed velocity deficits of about 50 percent at 3 diameters and 5 percent at 5 diameters downstream.

3.2.5(39) AN ANALYSIS OF THE AERODYNAMIC FORCES ON A VARIABLE GEOMETRY VERTICAL AXIS WIND TURBINE.
Anderson, M.B. (Sir Robert McAlpine & Sons Ltd.) In Wind Energy Conversion, proc. 5th BWEA Wind Energy Conference, Reading, U.K. 23-25 March 1983. P. Musgrove (ed.) pp. 224-234. Cambridge, U.K., Cambridge University Press 1984 375 pp. (Isbn : 0-521-26250-X)
A theoretical model is developed, based on the two-disc multiple streamtube approach, to predict the aerodynamic forces on a variable geometry vertical axis wind turbine. The model includes the effect of the lift and drag produced by inclined struts, interference losses and parasitic drag.

3.2.5(40) DEVELOPMENT OF A WIND ENERGY CONVERTER WITH SINGLE BLADE ROTOR.
Hipp, K. (Boewe Maschinenfabrik GmbH, Augsberg, F.R. Germany, Bereich Technik Maschinenbau) Final Report for June 1981 Report no: BMFT-FB-T-84-114 June 1984 43 pp. (In German)
Wind energy converters with high tip speed ratio, 12 m diameter rotors, and a capacity of up to 50 kW in a 8.5/msec wind speed were developed. The concept of a cost effective plant as a high speed turbine with a supercritically running one blade rotor, gust balance out, automatic blade adjustment to ensure good starting qualities, proves to be a success.

3.2.5(41) EFFECT OF VORTEX GENERATORS ON THE POWER CONVERSION PERFORMANCE AND STRUCTURAL DYNAMIC LOADS OF THE MOD-2 WIND TURBINE.
Sullivan, T.L. (NASA Lewis Research Center, Cleveland, Ohio) NASA Tech. Memo 83680 June 1984 26 pp.
Improved performance came at the cost of a small increase in cyclic blade loads in below rated power conditions. Cyclic blade loads were found to correlate well with the change in wind speed during one rotor revolution.

3.2.5(42) INTERPRETATION AND ANALYSIS OF FIELD TEST DATA.
Anderson, M.B. (Sir Robert McAlpine & Sons Ltd.) In Wind Energy Conversion, proc. 5th BWEA Wind Energy Conference, Reading, U.K. 23-25 March 1983. P. Musgrove (ed.) pp. 283-295. Cambridge, U.K., Cambridge Univ. Press 1984 375 pp. (Isbn: 0-521-26250-X)
Examines the three sources of error associated with anemometer test data i.e. instantaneous velocity differences, spatial averaging and flow modification associated with wind turbines.

Discusses siting of an anemometer in relation to a turbine and examines ways to increase the correlation coefficient. Outlines methods to apply correction factors. Determines the aerodynamic power output, versus wind speed relationship of a 6 m diameter turbine from experimental results using the method of 'bias'.

3.2.5(43) MEASUREMENT OF THE WAKE OF AN AEROPOWER SL1000 WIND TURBINE.
Hansen, A.C. (Rockwell International, USA)
Wind Technol. J., 3 (1-2) 1984 pp. 2-7.
   Measurements were taken in August 1980 at the Department of Transportation Rail Test Facility in Pueblo, Colorado to quantitatively measure the nature and extent of the wake around the horizontal axis wind machine and to provide data necessary to validate rotor wake models.

3.2.5(44) NOTE ON A SIMPLIFIED APPROACH TO DESIGN POINT PERFORMANCE ANALYSIS OF HAWT ROTOS.
Shepherd, D.G. (Cornell University)
Wind Engng., 8 (2) 1984 pp. 122-130.
   Method is based on the use of the optimised analysis for determining the blade geometry, thus requiring only fixed values of the axial induction factor and corresponding optimised rotational induction factors. Power output is obtained by using a blade-element torque relationship which allows for drag and includes the tip-loss effect by direct application of the customary loss factor without requiring recalculation of induction factors.

3.2.5(45) A NUMERICAL SIMULATION OF THE RESPONSE OF A LARGE HORIZONTAL AXIS WIND TURBINE TO REAL WIND DATA.
Powles, S.J.R. and others. (Cambridge Univ. and Sir Robert McAlpine & Sons Ltd.) In Wind Energy Conversion, proc. 5th BWEA Wind Energy Conference, Reading, U.K. 23-25 March 1983. P. Musgrove (ed) pp. 119-128. Cambridge, U.K., Cambridge University Press 1984 375 pp. (Isbn: 0-521-26250-X)
   Blade element theory is used in calculation of steady state aerodynamic blade loading, and various methods of applying tip loss correction may be used. Describes procedures to obtain the blade dynamics input.

3.2.5(46) PERFORMANCE TESTING OF A SAVONIUS WINDMILL ROTOR IN SHEAR FLOWS.
Mojola, O.O. and Onasanya, O.E. (Dept. of Mech. Engineering, Ife University, Nigeria)
Wind Eng. (GB), 8 (2) 1984 pp. 109-121.
   Assesses the effects of flow shear and/or instability on the power-producing performance of a Savonius windmill rotor. Measurements are made of the speed, torque and power of the rotor at a number of streamwise stations for each of four values of the bucket overlap ratio. All pertinent details such as flow velocity profiles and the wind shear are also determined.

3.2.5(47) THE SCALING LAWS APPLIED TO WIND TURBINE DESIGN.
Peterson, H. Wind Engng., 8 (2) 1984 pp. 99-108.
   Considers derivation of the equations neglecting the variation of the Reynolds number. Studies variation of the rotational speed, rotor radius, blade incidence, and discusses optimisation of a stall controlled wind turbine to different wind climates by variation of the generator size.

3.2.5(48) SURVEY OF MATERIALS SUITABLE FOR USE IN MW SIZED AEROGENERATORS.
Wyatt, L.M. and others. (Fulmer Research Labs.) In Wind Energy Conversion, proc. 5th BWEA Wind Energy Conference, Reading, U.K. 23-25 March 1983. P. Musgrove (ed.) pp. 235-244. Cambridge, U.K., Cambridge University Press 1984 375 pp.
   Outlines typical stresses encountered in horizontal and vertical axis wind turbines, suggesting the modifications possible. Examines assessment of fatigue performance of a welded steel spar and also considers the effect of offshore siting on steel. Manufacturing procedures for G.R.F.P. and laminated wood are considered, and requirements for fatigue testing are studied.

3.2.5(49) THEORETICAL STUDY ON THE FLOW ABOUT SAVONIUS ROTOR.
Ogawa, T. (Hyogo University Tech. Educ.)
ASME Trans. J. Fluids Engng., 106 (1) March 1984 pp. 85-91.
   Presents a method for the two-dimensional analysis of the separated flow about Savonius rotors. Calculations are performed by combing the singularity method and the discrete vortex method. The method is applied to the simulation of flows about both stationary and rotating rotors. Torque and power coefficients are computed and compared with the experimental results presented by Sheldahl et al.

3.2.5(50) THERMAL-STRESS ANALYSIS FOR WOOD COMPOSITE BLADE.
Fu, K.C. and Harb, B. (Toledo University, Ohio) Report 5 December 1982-4 April 1984 Report no: NAS 1.26:173830 20 July 1984 117 pp. (NASA-CR-173830)
   The thermal-stress induced by solar insolation on a wood composite blade of a MOD-OA wind turbine was investigated. The temperature distribution throughout the blade was analysed and the thermal-stress distribution of the blades caused by the temperature distribution determined.

3.2.5(51) TURBULENCE INDUCED LOADS IN A WIND TURBINE ROTOR.
Garrad, A.D. and Hassan, U. In Wind Energy Conversion, proc. 5th BWEA Wind Energy Conf., Reading, U.K. 23-25 March 1983. P. Musgrove (ed.) pp. 108-118. Cambridge, U.K., Cambridge University Press 1984 375 pp. (Isbn: 0-521-26250-X)
   Discusses dynamic analysis of a two bladed teetered horizontal axis wind turbine rotor excited by a turbulent wind. Presents the equation of motion, and lists the dimension-

less parameters used. Transforms the frequency domain equations obtained into spectral form and discusses the output spectra.

3.2.5(52) WIND TURBINE DYNAMIC BLADE LOADS DUE TO WIND GUSTS AND WIND DIRECTION CHANGES. Stoddard, W. (Pioneer Wind Power) Wind Technol. J., 3 (1-2) 1984 pp. 8-12.

A simple method was derived for the calculation of wind turbine blade root bending moments due to aerodynamic, inertial, and coupled effects. The method was applied to the calculation of blade dynamic loading caused by wind direction changes and steady crossflow wind. Equilibrium blade trajectories and root flapping and lead lag bending moments were found for various values of steady wind crossflow. These allowed the calculation of tower yawing and pitching reaction moments.

4 Institutional Incentives and Controls on Wind Energy Conversion Systems

4.1 State and Government Incentives

4.1(1) COMMERCIALISATION ANALYSIS OF LARGE WIND ENERGY CONVERSION SYSTEMS. FINAL REPORT.
Boyd, D.W., Buckley, O.E. and Haas, S.M. (Decision Focus Inc., Palo Alto, CA, USA)
US Dept.of Energy Report no: DOE/ET/23119-T1 June 1980 311 pp.

Describes a framework that can be used to evaluate potential new federal incentives to facilitate the market acceptance of utility-scale wind energy conversion systems. The insights gained from applying this framework to evaluate a variety of hypothetical federal incentives are discussed. The heart of the evaluation framework is an explicit represent-ation of the decisions made by utility pur-chasers, suppliers, and government agencies with respect to the utilisation and fabri-cation of large wind energy conversion systems. The demand-side and supply-side aspects of the commercialisation model are described and the model's structure is explained.

4.1(2) CURRENT STATE INCENTIVE PROGRAMS FOR SMALL WIND ENERGY CONVERSION SYSTEMS.
Kornreich, T.R. and Devine, D..(Science Appli-cations Inc., McLean, VA) Washington, DC
Report no: RFP-3128/05480/3533/80-6 July 1980 81 pp.

Written primarily for the use of US state governments, focuses on the current status of state incentive programmes and regulations throughout the United States at December 1979. Based on the examination of all state incentive programmes in effect during this reporting period, they are divided into six categories: income taxes; property taxes; sales and use taxes; loans; grants; and others. The effect of state programmes on SWECS usage is also examined.

4.1(3) ECONOMIC INCENTIVES TO WIND SYSTEM COMMERCIALISATION.
Lotker, M. (Synectics Group Inc., USA) In Collection of Tech. Papers of the AIAA/SERI Wind Energy Conference, Boulder, USA 9-11 April 1980. New York, USA, American Institute Aero-naut. & Astronaut. 1980 Paper no: 80-0617 pp. 61-72.

Reports on the impact of US Government financial incentives on the implementation possibilities for wind energy conversion systems. These incentives need to be broad and flexible for the variety of potential WECS users to be persuaded to invest.

4.1(4) FINANCING OPTIONS FOR WIND POWER DEVELOPMENT.
Balma, M. and others. (SRI International, California, USA) Paper presented at Inter-national Solar Energy Society American Section/ Et Al 1980 Annual Conference, Phoenix 2-6 June 1980 V 3.2 pp. 1282-1285.

Identifies financing options that can ease the commercialisation of wind energy conversion systems. The economic characteristics of wind power are matched with potential purchaser requirements in light of new federal regulations and tax policies. Key markets for utility scale wind machines include utilities having high fuel costs, such as those in the SouthWestern US. Private investors and businesses can reduce the risk involved in wind energy ventures through leasing agreements, warranties, perform-ance guarantees, and proper management of tax benefits.

4.1(5) GENERAL LEGAL PRINCIPLES OF MUNICIPAL SECURITIES FINANCING.
Paper presented at US Dept. of Energy Local Alternative Energy Futures Conference, Austin, USA 11-13 December 1980 pp. 13-35.

Considers the general powers and limitations associated with the authorisation, issuance, and sale of municipal bonds in financing the development of alternative energy resources. Municipal financing is primarily applied to the development of solar and wind facilities. Basic bonds such as industrial development and revenue bonds are described. Exemptions and federal taxation requirements are also considered.

4.1(6) SWECS QUALIFICATIONS CRITERIA FOR STATE TAX INCENTIVE PROGRAMS.
Guerrero, J.V. and Alford, C. (Rockwell Intl.) In a Collection of Technical Papers from AIAA/ SERI Wind Energy Conference, Boulder, USA 9-11 April 1980. New York, USA, American Insti-tute Aeronaut. & Astronaut. 1980 Paper no: 80-0650 pp. 205-208.

Discusses how US state governments can stimu-late greater public interest and investment in wind energy by promoting a belief, based on qualifying criteria, that SWECS are a reliable, economic and reasonably safe renewable energy alternative.

4.1(7) COST OF FEDERAL TAX CREDIT PROGRAMS TO DEVELOP THE MARKET FOR INDUSTRIAL SOLAR AND WIND ENERGY TECHNOLOGIES. FINAL REPORT TO LAWRENCE LIVERMORE LABORATORY, UNIVERSITY OF CALIFORNIA. VOLUMES 1 & 2.
Downey, W.T. and others. (Arthur D. Little Inc., Cambridge, MA) Washington, DC, US Dept. of Energy Report no: UCRL-15446 (V. 1 & V. 2) 12 November 1981 138 pp. & 99 pp.

In 1978, the US Congress passed the Energy Tax Act which, in addition to the 10% investment tax credit then available, added a 10% tax credit (increased to 15% by the Crude Oil Windfall Profits Tax Act of 1980) to specifically encourage industrial investment in equipment which either generated energy from renewable sources or conserved energy. This study aims to estimate the amount of renewable energy investment these tax credits will cause, if they were retained until 1981.

4.1(8) PRELIMINARY PLAN SUMMARY FOR IMPLEMENTING SECTIONS 201 AND 210 OF THE PUBLIC UTILITY REGULATORY POLICIES ACT OF 1978.
Seilander, S. (United Power Association, Minn, USA) Paper presented at International Solar Energy Society American Section/Et Al Wind Power Energy Alternatives for MidWest Conference, Minnesota 3-4 April 1981 pp. 53-55.

Presents a preliminary plan for implementing sections 201 and 210 of the Public Utility Regulatory Policies Act of 1978 (PURPA), with respect to the interconnection of small power production facilities and cogeneration facilities with major utilities. Rules and regulations covering the interconnected operations for the purpose of ensuring personnel safety and public safety and of preventing the degradation of the energy supply and distribution system are discussed. Commercial rates for purchasing power acceptable to both qualified small power producers and cogenerators, and for selling power to these facilities are reviewed.

4.1(9) THE PUBLIC UTILITY REGULATORY POLICIES ACT OF 1978 (PURPA).
O'Sullivan, J. Presented at 2nd International Solar Energy Society American Section/Et Al Wind Power Energy Alternatives for MidWest Conference, Minnesota 3-4 April 1981 pp. 49-51.

Reviews the considerations that led to the enactment of the Public Utility Regulatory Policies Act of 1978, which deals with cogeneration and small power production. In the past, utilities have refused to buy power from small power producers or failed to offer a fair price for their power. However, under Section 210 of PURPA, utilities must buy power from small power producers at fair rates. The major benefits of these regulations, and the rules governing the eligibility of small power producers are discussed.

4.1(10) PURPA:A NEW LAW HELPS MAKE SMALL-SCALE POWER PRODUCTION PROFITABLE.
Gipe, P. (Centre for Alternative Resources, PA) Sierra, 66 (6) November-December 1981 pp. 52-55.

The US Public Utility Regulatory Policies Act 1978 (PURPA) requires utilities to buy excess electricity - at reasonable rates - from small power producers. This eliminates barriers to greater use of small wind generators as small power producers are exempt from federal restrictions, and windmill users are assured backup power at non-discriminatory rates. Such producers are also guaranteed a market for their excess energy, helping to make wind energy conversion systems ecologically feasible.

4.1(11) RENEWABLE POWER SPARKS FINANCIAL INTEREST.
Norman, C. Science, 212 (4502) 16 June 1981 pp. 1479-1481.

Decentralised electricity production from renewable resources is getting a boost from provisions of the Public Utility Regulatory Policies Act of 1978 (PURPA) and from tax credits. However, much controversy is surrounding the growing use of wind energy, solar energy and other sources for electricity production. Article claims that electric utilities are being thwarted of their exclusive rights to generate and sell power by these legal inducements, which are being taken advantage of by individuals and industry.

4.1(12) THE WISCONSIN ANEMOMETER LOAN PROGRAM.
Hershberg, E. (Wisconsin Div. of State Energy) Paper presented at 2nd International Solar Energy Society American Section/Et Al Wind Power Energy Alternative for MidWest Conference, Minnesota 3-4 April 1981 pp. 79-82.

The Wisconsin Anemometer Loan Programme is designed to assist Wisconsin residents who are interested in using wind energy, to set up plant. It is a detailed site analysis programme that will provide participants with information concerning the wind energy available at a particular site, the expected output of commercially available wind energy conversion systems, and an approximate assessment of the relative economics of using wind energy at that site.

4.1(13) SMALL POWER PRODUCTION AND WIND ENERGY: REGULATORY ACTIONS UNDER PURPA. SEIDB INFORMATION MODULE 1012.
Lornell, R. and Schaller, D.A. (Solar Energy Research Institute, Golden, CO) Solar Energy Research Institute Report no: SERI/SP-635-794 January 1982 66 pp. (DE 83019250)

Provides electric utilities and their regulatory authorities with a clear perspective of the small power producer rules under the US Public Utility Regulatory Policies Act (PURPA) of 1978 as it may affect their decisions on the integration of small wind systems to electric distribution networks. Sections 201 and 210 of PURPA, and the federal rulemaking which implements this statute, create several responsibilities for electric utilities and their respective regulatory agencies. Principal among these are the requirements for interconnection with qualified small power producers and the purchase of electric power from these producers at a rate which reflects the utilities' avoided cost. Also covers three issues key to wind energy development under the small power producer rules: interconnection, capacity displacement, and power purchase rates. It serves as a technical guide to these issues and includes a summary of selected utility and state commission decisions.

4.1(14) BUSINESS ENERGY INVESTMENT CREDIT FOR SOLAR AND WIND ENERGY GENERAL ACCOUNTING OFFICE, WASHINGTON, DC. RESOURCES COMMUNITY ECONOMIC DIV.
General Accounting Office Report nos: GAO/RCED-83-8 & B-209008 7 March 1983 55 pp.
(PB 83 193235)
The Business Energy Investment Credit, which provides a 15% tax credit to owners of new solar and wind energy equipment, is scheduled to expire on 31 December 1985. Reports on GAO's analysis of the credit's effect on the economics of our projects which were to employ solar and wind energy systems; a summary of the views of a range of organisations on the desirability of extending the credit past its currently scheduled expiration date, and discusses the possible revenue loss to the Treasury resulting from an extension.

4.1(15) FINANCING WIND TURBINE/WIND FARM INSTALLATIONS.
Lotker, M. (Synetics Group Inc.) In International Wind Energy Symposium 5th Annual Energy-Sources Technology Conference, New Orleans, USA 7-10 March 1982 pp. 183-195. New York, American Society Mechanical Engrs. 1982.
Examines several U.S. laws which affect the commercial development of wind power including the Public Utilities Regulatory Act, the Crude Oil Windfall Profits Tax Act, the Economic Recovery Tax Act and the Wind Energy Systems Act. Discusses use of the incentives available and studies the Small Power Producer sector with an examination of the economic benefits of the Limited Partnership ownership option with tax sheltered financing. Gives example calculations of cash flows and income.

4.2 Legal and Environmental Implications of WECS

4.2(1) SOME LEGAL-INSTITUTIONAL IMPLICATIONS OF OFFSHORE WIND ENERGY CONVERSION SYSTEMS.
Mayo, L.H. (George Washington University, Washington, DC, USA) In Sharing the Sun: Solar Technology in the Seventies, Joint Conference of the International Solar Energy

Society American Section and Solar Energy Society of Canada Inc., Winnipeg, Manitoba 15-20 August 1976. Cape Canaveral, Fal., ISES/AS 1976 V. 7 pp. 195-214.
Discusses potentially conflicting National-International and Federal-State jurisdictional claims, competing sea area uses, land use management, utility policy and regulation, and incentives to utilise alternative energy technologies. The existing array of Federal energy status appears to provide adequate legal framework for research, development and demonstration programmes but specific legislation to eliminate or ameliorate certain constraints on the implementation may be advisable if significant private sector participation is to be encouraged.

4.2(2) LEGAL-INSTITUTIONAL IMPLICATIONS OF WIND ENERGY CONVERSION SYSTEMS (WECS)
George Washington University, Programme of Policy Stud. in Sci. and Technol., Washington, DC, USA. Final Report prepared for NSF, RANN (Research Appl. to National Needs) NSF/RA-770204, Washington, DC September 1977. Available from GPO, Washington, DC, US Government Printing Office 1977 320 pp.
Discusses the interplay between the technical, economic, social and legal aspects of wind energy conversion systems, and describes the applications of wind energy, the restrictions to the implementation of land-based WECS, including problems with public utilities, zoning, housing codes and FCC regulations, and legal issues concerning offshore and land based WECS.

4.2(3) PROGRAMMATIC ENVIRONMENTAL ASSESSMENT: WIND-ENERGY-CONVERSION SYSTEMS (WEC) PROGRAM. ISSUES-IDENTIFICATION REPORT.
(Energy and Environmental Analysis Inc., Arlington, VA, USA) US Dept. of Energy Report no: DOE/ET/20510-T2 2 March 1979 30 pp. (Report no: DE 81 027410)
Describes environmental issues that may be associated with deployment of wind energy conversion systems. These issues will be further elaborated and qualified to form the core of the impacts section of the WECS programmatic environmental assessment. The identified issues may be modified or refined and new issues added as analytical efforts continue and new information is made available.

4.2(4) WIND ENERGY: LEGAL ISSUES AND INSTITUTIONAL BARRIERS.
Coit, L. Washington, DC, USA, US Dept of Energy June 1979 31 pp. (Solar Energy Research Institute Report no: SERI/TR-62-241)
Before the potential of wind energy can be realised, large scale commercialisation will have to occur, but standing in the way of commercial development are various institutional and legal barriers. These include possible conflicts with existing zoning and other land-use planning schemes, the question of guaranteeing access to the wind, possible tort and environmental law issues raised by WECS

operation, and the critical problem of generating financial incentives. Implications of each of these issues are discussed and solutions proposed.

4.2(5) LEGAL AND INSTITUTIONAL OBSTACLES TO THE DEVELOPMENT OF LARGE SCALE WIND ENERGY CONVERSION SYSTEMS IN THE STATE OF NEW HAMPSHIRE.
Smukler, L.M. (Franklin Pierce Law Centre, New Hampshire, USA) Paper presented at International Solar Energy Society American Section/ Et Al 1980 Annual Conference, Phoenix 2-6 June 1980 V. 3.2 pp. 1517-1521.

Analyses the legal and institutional obstacles that may confront a developer attempting to site a wind energy conversion system in New Hampshire. Issues related to land use, environmental impacts, regulations pertaining to interference with broadcasting signals, marketing of wind-generated electricity and criteria for using state or private land are considered. Few of these obstacles are seen as serious enough to impede wind energy development, if it can be seen to be economically viable.

4.2(6) LEGISLATIVE, REGULATORY AND INSTITUTIONAL BARRIERS TO ELECTRIC UTILITY PARTICIPATION IN ENERGY CONSERVATION AND RENEWABLE RESOURCE DEVELOPMENT PROGRAMS.
Benson, C.C. (Dallas Power & Light Co., Texas, USA) Paper presented at American Solar Energy Society Improving Energy Efficiency in Buildings Conference, Santa Cruz, USA 10-22 August 1980 Paper 5.7 14 pp.

US Electric and Gas utilities face significant legislative, regulatory and institutional barriers to cost-justified participation in conservation and renewable resource development activities. Disincentives and prohibitions created by federal laws and regulations promulgated by the US Teasury, US Internal Revenue Service and other agencies in support of these laws are also examined.

4.2(7) ACQUISITION OF WIND RIGHTS FOR WIND ENERGY DEVELOPMENT.
Noun, R.J. (Solar Energy Research Institute, Golden, CO, USA) Solar Energy Research Inst. Report nos: SERI/TP-211-1421 CONF-811043 November 1981 12 pp. (Biennial Wind Energy Conference and Workshop, Washington, DC, USA 5 October 1981)

Identifying suitable sites for large wind machine clusters, or wind farms, requires more than finding a location with an adequate wind resource. Consideration must also be given to the question of how land-use policies and regulations will affect the siting of wind system installations. In particular, the issue of acquiring wind rights, or guaranteed access to the wind resource for electric power generation is vital to the development of wind energy. Several methods for acquiring and preserving access to the wind resource and for dealing with related land-use issues are examined.

4.2(8) INSTITUTIONAL, FINANCIAL AND LEGAL ISSUES.
In proc. 5th Biennial Wind Energy Conference & Workshop, Washington, DC, USA 5-7 October 1981. I.E. Vas (ed.) V. 3. Palo Alto, USA, Solar Energy Research Institute 1981 Session IVA pp. 1-60. Report nos: SERI/CP-635-1340 CONF-811043.

Includes papers on making the most of Federal Tax Laws for WECS development; planning the wind farm; state and local concerns; protecting small wind energy conversion systems from unnecessary regulation; lessons for the Federal, state and local level; residential wind prospecting and the acquisition of wind rights for wind energy development.

4.2(9) LAND-USE IMPLICATIONS OF WIND ENERGY CONVERSION SYSTEMS (WECS)
Noun, R.J. (Solar Energy Research Institute, Golden, CO, USA) Paper presented at 2nd International Solar Energy Society American Section/ Et Al Wind Power Energy Alternatives for Mid-West Conference, Minnesota 3-4 April 1981 pp. 69-73.

Examines legal, social and environmental considerations in siting large-scale wind energy projects. Federal studies have indicated that most potentially adverse land-use impacts of wind energy conversion systems will be site-specific, and that these impacts can be minimised or avoided by careful planning, siting and design. Application of land-use regulations to WECS siting is discussed. Wind rights, state and local regulation of privately owned land, and regulations governing federally-owned land in the USA are reviewed.

4.2(10) MEASUREMENT AND ASSESSMENT OF THE NOISE PRODUCED BY SMALL WIND ENERGY SYSTEMS.
Hansen, A.C. and Martin, D.L. (Rockwell International) In proc. 5th Biennial Wind Energy Conference & Workshop, Washington, DC, USA 5-7 October 1981. I.E. Vas (ed.) V. 3. Palo Alto, USA, Solar Energy Research Institute 1981 Session IVC pp. 189-203. Report nos: SERI/CP-635-1340 CONF-811043.

Noise measurement can be accomplished easily and with little or no ambiguity, but valid prediction of community response to noise levels is somewhat more difficult. A method for systematically assessing probable response to small wind energy conversion system noise is proposed. The method uses the measured SWECS noise in conjunction with a Rayleigh wind distribution to estimate the daily average noise level. The resulting noise level is then assessed using a synthesis of noise annoyance predictors to estimate the percentage of population who would be highly annoyed by the noise.

4.2(11) UTILITY SITING OF WECS: A PRELIMINARY LEGAL/REGULATORY ASSESSMENT.
Noun, R.J., Lotker, M. and Friesema, H.P. (Solar Energy Research Institute, Golden, CO, USA) Solar Energy Research Institute Report no: SERI/TR-744-778 May 1981 100 pp.

Experiences of several utilities in dealing with the legal and regulatory issues raised in siting wind energy installations are examined and recommendations given for how utilities can deal with these issues.

4.2(12) A PRELIMINARY ESTIMATION OF THE EXPECTED NOISE LEVELS FROM THE SWEDISH WIND ENERGY CONVERSION SYSTEM (WECS) PROTO-TYPES MAGLARP AND NAESUDDEN.
Soedeqvist, S. (Aeronautical Research Institute of Sweden, Stockholm) Report no: FFA-TN-1982-01 13 April 1982 36 pp.
Predicts noise levels for horizontal axis turbines whose diameter is 75 m. and electric power output 2 to 3 mW. Noise criteria in the context of Swedish regulations and the effects of wind dependent background noise are considered.

4.2(13) SIDE EFFECTS OF RENEWABLE ENERGY RESOURCES.
Medsker, L. (National Audubon Society Environmental Policy Research Dept. Report 15 December 1982 74 pp.
Describes environmental impacts from various aspects of renewable energy source development. Side-effects can include large land requirements; disruption of wildlife habitats; increased materials demand for construction and manufacturing: health and safety problems associated with construction, operation and maintenance of energy facilities; potential air and water pollution from the release of working fluids and other chemicals, and use of non-renewable energy for manufacturing construction materials and for transporting resources or equipment. Includes wind energy.

4.2(14) SOCIOECONOMIC ASPECTS OF A VALUE ANALYSIS OF WIND ENERGY-ILLUSTRATED FOR THE NETHERLANDS.
Dub, W. and Pape, H. (University of Regensburg, W. Germany) Energy Sources, 6 (3) 1982 pp. 245-250.
Discusses socioeconomic aspects of an environmental impact analysis of wind energy. Factors considered include electricity production and delivery costs, pollution, and resource availability. Large-scale wind turbines are likely to become a contributing factor for future electricity generation, as they cause no pollution or depletion of resources. (13 refs.)

4.2(15) WIND FARMS: LOCAL REGULATION AND ASSESSMENT.
Odland, R. (Robert Odland Assocs.) In proc. 1982 Wind and Solar Energy Technology Conference, Kansas City, USA 5-7 April 1982. Columbia, USA, Missouri-Columbia University 1982 pp. 237-241.
Riverside County, California and the US Bureau of Land Management have commissioned a study of the impacts of wind energy development and the regulatory approaches to guide such development in San Gorgonio Pass area in Riverside County, California. A number of environmental and safety issues were identified including safety, noise, electronic interference, and aesthetic impacts.

4.2(16) WINDPOWER AND THE LANDSCAPE.
Bergsjo, A., Nilsson, K. and Skarback, E. (Sveriges Lantbruksuniv., Sweden) National Swedish Board for Energy Source Development Report 1982 188 pp. (In Swedish)
Research was initiated to analyse the influence of possible wind power grouped stations on the visual character of the landscape. Principles of locating wind power units and planning measures to avoid future environmental conflicts are discussed.

4.2(17) CLIMATE, ENERGY AND TRANSPORTATION.
Critchfield, H.J. (Western Washington University) International J. Env. Studies, 20 1983 pp. 149-157.
Climatic processes offer the potential for harnessing wind energy, solar energy and other energy forms. In turn, the products of energy consumption disrupt energy and mass budgets in the environment. Attempts to reduce environmental impacts or to modify climate deliberately for improvement of energy management require further expenditures of energy; these are likely to generate a new series of effects. (48 refs.)

4.2(18) THE ENVIRONMENTAL IMPACT OF THE USE OF LARGE WIND TURBINES.
Manning, P.T. (Environmental Section, CEGB, U.K.) Wind Engng., 7 (1) 1983 pp. 1-11.
Experience with the operation of wind turbines has indicated that they are not entirely environmentally benign. This paper reviews those aspects of wind turbines which could produce adverse environmental effect; although it specifically addresses horizontal-axis land based machines, much of it is generally applicable. Visual impact is potentially important but difficult to quantify. TV interference, if it occurs, can generally be ameliorated at a relatively modest cost. Noise is potentially a serious and intractable problem, but further research and experience is required before its true significance can be assessed. Other impacts are unlikely to be significant provided some care is taken in the design and siting of machines.

4.2(19) TECHNICAL, ECONOMIC AND LEGAL ASPECTS OF WIND ENERGY UTILISATION.
Obermair, G.M. and Jarass, L. (University of Regensburg, Dept. of Physics, Regensburg, W. Germany) Wind Engng., 7 (2) 1983 pp. 99-103.
Discusses technical, safety, economic and legal problems which are associated with renewable sources of energy when combined with a more conventional form of energy distribution such as the elctric grid. Presents a review based on studies and reports from international organisations and national and public sector industries in many industrialised countries.

4.2(20) ENERGY AND OUR FUTURE ENVIRONMENT.
Ireland, F.E. Energy World (Bull Inst. Energy)
(110) January 1984 pp. 23-24.
  Review of the Institute of Energy (U.K.)
Annual Conference in November 1983. Includes
paper by H.G. Tollard on wind energy which draws
attention to the environmental difficulties
experienced in using the simple idea of wind
energy.

4.2(21) MEASUREMENTS AND OBSERVATION OF NOISE
FROM A 4.2 MEGAWATT (WTS-4) WIND TURBINE
GENERATOR.
Shepherd, K.P. and Hubbard, H.H. Hampton, USA,
Bionetics Corp. May 1983 33 pp. (NASA/CR-66124)
  Noise measurements and calculations are being
made for large wind turbine generators to
develop a data base for use in designing and
siting such systems for community acceptance.
Paper presents the results of these exploratory
measurements for power output conditions in
the range 1.0 to 4.2 mW. Data include noise
levels, spectra, radiation patterns, effects of
distance, and the associated perception thresh-
olds for use in the further development of
acceptance criteria for this type of machine.

4.2(22) MEASUREMENTS OF TELEVISION INTERFERENCE
CAUSED BY A VERTICAL-AXIS WIND MACHINE.
Sengupta, D.L. and others (SERI, Golden, USA)
Solar Energy Research Institute Report no :
SERI/STR-215-1881 January 1983 92 pp. (See also
SERI/STR-215-1879 by same authors.)
  Describes measurements made to assess the
electromagnetic interference to television
reception caused by the 17 m Darrieus wind turbine at Albuquerque, New Mexico. Unacceptable
interference can be observed on some channels
at sites up to 33 m in the forward and backward
regions. With the antenna beam directed at the
wind turbine, interference was observed at all
sites on some or all TV channels. Uses a model
developed to analyse such interference from a
vertical axis wind turbine to provide an estimate for the equivalent scattering area of the
Darrieus, in terms of the wavelength. The wave-
forms for a Darrieus appear more complex than
those for a horizontal axis wind turbine.

4.2(23) METHOD FOR PREDICTING IMPULSIVE NOISE
GENERATED BY WIND TURBINE ROTORS.
Viterna, L.A. (Lewis Research Center, Cleveland)
NASA Lewis Research Center 1982 7 pp.
(NASA/TM-82794)
  Noises can be controlled by proper choice of
rotor design parameters such as rotor location
with respect to the supporting tower, tower geometry and tip speed. A method was developed to
calculate the impulsive noise generated when the
wind turbine blade experiences air forces that
are periodic functions of the rotational
frequency. This phenomenon can occur when the
blades operate in the wake of the support tower
and the nonuniform velocity field near the
ground due to wind shear.

4.2(24) NEGOTIATING LAND CONTRACTS FOR WIND
FARMS.
Walters, M.L. (Harrang Swanson Long & Watkinson,
Oregon) In American Wind Energy Association
Wind Energy Expo National Conference, San
Francisco 17-19 October 1983 pp. 125.
  Enumerates factors to consider in devising a
clear, comprehensive, and enforceable agreement
between a landowner and a wind farm developer.
Types of contracts, leasing, options, and purchase rights are explained and easements and
covenants discussed. The provisions of a
typical contract are outlined. Covenants of
title, period of contract, payment, rights
and obligations of parties, liabilities and
tax provisions are also included.

4.2(25) PROTECTING WIND ACCESS : A PRELIMINARY
ASSESSMENT.
Noun, R.J. (Solar Energy Research Institute,
Golden, USA) In ASME Wind Energy Symposium
6th Annual Energy-Sources Technology Conference
& Exhibition, Houston, USA 30 January-3 February
1983 pp. 607-617. New York, American Society
Mechanical Engrs. 1983.
  Considers the question of how land-use policies
and regulations will affect the siting of wind
system installations. In particular, the issue
of acquiring 'wind rights', or guaranteed access
to the wind resource for electric power generation is vital to the development of wind energy.

4.2(26) THE ROLE OF INSURANCE IN WIND FARMS.
Johnson, R.T. (Reed Stenhouse, San Francisco)
In American Wind Energy Association Wind Energy
Expo National Conference, San Francisco
17-19 October 1983 pp. 231-238.
  Proper insurance coverage can offer the needed
security to influence the financial community
to support wind energy system development.
Individual investors would also be encouraged
to underwrite wind farm construction. Some
insurance companies will even consider the loss
of earnings coverage if there is a lack of
wind which prevents normal performance of the
system.

4.2(27) SECURITIES LAW ISSUES OF WIND FARM
DEVELOPMENT.
Lyons, R.A. (Niesar Moody Hill Kregstein &
Hamilton, California) In American Wind Energy
Association Wind Energy Expo National Conference,
San Francisco 17-19 October 1983 pp. 239-253.
  Violation of securities laws can result in
legal action that may impair the viability of a
successful wind farm. Securities issues that
have a practical bearing on the planning for
wind energy projects are highlighted. The
designation of a wind turbine as a security
under current definitions and interpretations
is explained. Purchase transactions, private
offering exemptions from federal securities
laws, and intrastate exemptions are covered.

4.2(28) ACOUSTIC TESTS OF THE MOD-0/5A WIND
TURBINE ROTOR WITH TWO DIFFERENT AILERONS.
Shepherd, K.P. and Hubbard, H.H. (Bionetics

Corp., Hampton, Virginia) Report no: NASA-CR-1724279 August 1984 31 pp.

Data were obtained for a wide range of aileron deflection angles and for limited ranges of wind velocity and power output. Noise levels increased as deflection angles-increased and were higher in the upwind than in the downwind direction. The plain aileron gave out a howling noise at deflection angles for which flow induced cavity resonances were significant.

4.2(29) THE LEGAL ASPECTS OF USING WIND ENERGY SYSTEMS.
Hottin, F. Cahiers CSTB 248 (1923) April 1984 12 pp. (In French)

Notes the problems of safety and nuisance which can be associated with the installation of wind energy systems. Reviews studies and legislation carried out abroad and formulates observations concerning legislation governing the use of wind energy in France, and what solutions might be considered.

4.3 Safety Aspects of WECS

4.3(1) SAFETY OF WIND ENERGY CONVERSION SYSTEMS (WECS): PRELIMINARY STUDY.
Eggwertz, S. and others. Stockholm, Sweden, Swedish Aeronaut. Research Institute Report no: FFA-AU-2126 November 1979 136 pp.

Considers large land based horizontal axis wind turbines. An overall description of the system, statistical information concerning loads and strength of materials considered, and a discussion of geometrical tolerances are included. Formulae and procedures to be employed in the risk analysis are presented, followed by comments on acceptable risk levels. Critical events, safety systems, inspection and repairs are discussed.

4.3(2) STUDY OF WIND ENERGY CONVERSION SYSTEMS (WECS) IN A FARM AREA AND WECS SAFETY LIMIT REQUIREMENTS. MINUTES FROM EXPERT MEETING IEA RESEARCH AND DEVELOPMENT WECS, ANNEX ONE, SUBTASK A1.
Eggwertz, S. Stockholm, Sweden, Swedish Aeronaut. Research Institute Report no: FFA-TN-HU-2218 June 1980 114 pp.

Includes the description of two 2500 kW windmill prototypes, safety studies performed in several countries, and a contribution concerning fault tree analysis and load case recommendations. The introduction of safety zones, the crack detection system and operation during icing conditions are discussed.

4.3(3) ESTABLISHING REGULATIONS AND STANDARDS FOR WIND GENERATION SYSTEMS.
Galli, R. and Bianchi, E. In proc. International Colloquium on Wind Energy, Brighton, U.K. 27-28 August 1981. Cranfield, U.K., BHRA Fluid Engng. 1981 Session 2 pp. 107-112.

Exploitation of wind energy gives rise to the designing of more sophisticated and larger units with related problems of design and installation. The need for safety requires the establishment of regulations safeguarding both the manufacturers and the users. This proposal recommends the setting up of a committee defining the required regulations.

4.3(4) HAZARDS, SAFETY AND SMALL WIND ENERGY CONVERSION SYSTEMS.
Bass, L. and Weis, P. (Kairos Co., Cupertino, California, USA) J. Prod. Liability, 4 (2) 1981 pp. 203-214.

A generic hazard analysis and construction of a safety profile can shed light on the risks involved in using a product even though there is little empirical performance data available. Work funded by the US Department of Energy is drawn on to provide basic information for the development of voluntary consensus standards on the safety of small wind energy conversion systems. By listing hazards, their causes, and the relative frequency and severity of occurrence, areas of major concern were identified.

4.3(5) HEALTH AND SAFETY ISSUES OF ALTERNATIVE ENERGY SYSTEMS.
Watson, A.P., Etnier, E.L. and Walsh, P.J. (Oak Ridge National Laboratory) Paper presented at American Nuclear Society Alternative Energy Sources for Electrical Power Conference, MA, USA 4-7 October 1981 18 pp.

Occupational public health and safety implications are explored for alternative energy sources including wind energy. (80 refs.)

4.3(6) SAFETY AND WIND ENERGY CONVERSION SYSTEMS WITH HORIZONTAL AXIS (HAWECS)
Eggwertz, S. and others. Stockholm, Sweden, Swedish Aeronaut. Research Institute Report no: FFA/TN/HU-2229 19 March 1981 170 pp.

Hazards imposed by a WECS on the general public and on the operator personnel by complete collapse, by separation of fractured parts, or by pieces of ice (breaking) were calculated to provide a manual for safety evaluations. Land based large scale turbine systems with horizontal axes situated in areas with sparse population are considered. Blade material is assumed to be steel, aluminium alloy or fibre reinforced plastic, the tower of steel or reinforced concrete. Primary structure, function and failure modes are identified. Statistical information on loads and load combinations, strength of materials and geometric deviation are provided. A simplified method of risk analysis is described. The aims and functions of a safety system are reviewed, considering the effects of inspection and repair. In the event of a fracture occurring the probability of being hit is evaluated, and a risk zone is established.

4.3(7) SMALL WIND ENERGY CONVERSION SYSTEMS (SWECS) SAFETY DATA.
Hickey, M. and Bass, L. (Rockwell International)

In proc. 5th Biennial Wind Energy Conference & Workshop, Washington, DC, USA 5-7 October 1981 I.E. Vas (ed.) V. 2. Palo Alto, USA, Solar Energy Research Institute 1981 Session 2A pp. 193-201.

Data on accidents and safety incidents experienced with existing SWECS installations were collected. Case-study incidents were used to identify specific failure modes. Reasons for failure and the components involved were characterised and a generic hazards analysis presented. From this analysis, a comprehensive set of recommendations was developed which included design requirements to be considered, improvements in manufacturing and installation practices, and test procedures. Existing standards literature was reviewed for material relevant to SWECS and together with the recommendations will be used as input in the development of consensus safety standards for SWECS.

4.3(8) AN ANALYSIS OF LOCAL REGULATION OF WIND ENERGY CONVERSION SYSTEMS (WECS).
Moorehead, M.A. (Dept. of Energy) In American Wind Energy Association Wind Energy Expo National Conference, San Francisco 17-19 October 1983 pp. 135-146.

A review of over 60 wind energy ordinances shows the diversity of siting standards imposed by local bodies. While local agencies use different regulatory tools, the purpose of most standards is to promote safe, environmentally acceptable system installation and operation.

4.3(9) STUDY OF PRODUCT LIABILITY INSURANCE ISSUES RELATED TO SMALL WIND ENERGY CONVERSION SYSTEMS.
Bass, L. and others. (Hans W. Winholds Co., Cupertino, CA, USA) US Dept. of Energy Report no: RFP-3178-80/21 November 1980 155 pp.

This study was sponsored by the Department of Energy to analyse the products' safety and liability status of the SWECS industry and to propose strategies for addressing their needs. Eight strategies were identified which would reduce the manufacturers' risk, redistribute the economic burden of products liability, and redefine the concept of acceptable risk. (See also paper in ISES-AS/Et Al 1980 Annual Conference, Phoenix 2-6 June V. 3.2 pp. 1522-1525 by J. Noun.)

4.3(10) PRODUCT LIABILITY INSURANCE AND SMALL WIND ENERGY CONVERSION SYSTEMS.
Bass, L. (Kairos Co., Cupertino, CA, USA) J. Product Liability, 4 (3-4) 1981 pp. 265-273.

A number of wind energy manufacturers have experienced difficulty in obtaining products liability insurance at reasonable costs. Such insurance is available for all but the smallest and newest manufacuturers, but can be quite expensive. Costs of products liability insurance varies considerably, depending on both the manufacturer's performance and the insurer's perception of exposure.

5 Applications of Wind Energy Conversion Systems

5.1 Economics and Marketing of WECS for Various Applications

5.1(1) BARRIERS TO SMALL SCALE ENERGY TECHNOLOGIES.
Hocking, J.D., Isaacs, N.P. and Jeffs, L.H. (Earth Research Foundation, New Zealand) In 4th New Zealand Energy Conference sponsored by University of Auckland, Auckland, New Zealand May 1979 V. 1. pp. 166-173.

Identifies major barriers to the commercialisation of small-scale energy technologies in New Zealand, including wind power. Established technologies are the primary barrier to market penetration of new technology. Lack of financing and competitive price supports represent additional obstacles. (5 refs.)

5.1(2) FINANCIAL PROBLEMS FACING THE MANU-FACTURERS OF SMALL WIND ENERGY CONVERSION SYSTEMS. FINAL REPORT.
Bolle, T.G. (United Technologies Research Centre, Connecticut, USA) US Dept. of Energy Report no: DOE/DP/03533-T2 November 1979 69 pp.

Financial barriers faced by the manufacturers of small wind energy conversion systems are assessed. Problems are aggravated by high expectations of accelerated SWECS industry growth and lack of investment capital. Issues related to SWECS commercialisation, such as inflation, taxation, regulation, and federal R & D procurement policies are analysed.

5.1(3) LOAD MATCHING EFFECTS ON WIND ENERGY CONVERSION PERFORMANCE.
Dixon, J.C. (Open University, U.K.) In proc. International Conference on Future Energy Concepts, London, U.K. 30 January-1 February 1979. London, U.K., Institution of Electrical Engineers 1979 pp. 418-421. (IEE Conf. Publ. no: 171)

The manufacturers' claimed performance of typical water lifting windpumps indicates an overall efficiency of about 5%. The cost-effectiveness of the whole system is evidently related to the efficiency. Investigates the causes of this low efficiency and suggests areas for development. In particular the low efficiency is largely the result of poor load matching of the typical reciprocating pump. Improvement of this load matching offers much more potential for improvement in performance than aerodynamic refinement of the rotor in the opinion of the author.

5.1(4) ORGANISATIONAL INTERFACE AND FINANCIAL BARRIERS TO THE COMMERCIAL DEVELOPMENT OF COMMUNITY ENERGY SYSTEMS.
Schladale, R. and Ritschard, R. (LBL) LBL Report no: LBL-11188 December 1979 118 pp.

Four community energy technologies were investigated - combustion of municipal solid waste, small wind power, industrial co-generation, and photovoltaic electricity. Interface issues include purchase price structuring, fuel availability, and environmental impact. Lack of financing, markets and insufficient investment capital are significant financial barriers.

5.1(5) THE EVALUATION OF INNOVATIVE WIND ENERGY CONCEPTS.
South, P. and Jacobs, E.W. (Solar Energy Research Institute, Golden, CO, USA) In SERI 2nd Wind Energy Innovative Systems Conference, Colorado Springs, USA 3-5 December 1980. Golden, USA, Solar Energy Research Institute 1980 Session 1 pp. 11-32. Solar Energy Research Institute Report no: SERI/CP-635-938.

Describes the process being developed at SERI to efficiently, consistently, and objectively evaluate innovative wind energy concepts. Costing methods are being devised whereby developmental, pre-production, and production costs can be assessed for innovative wind energy systems in their conceptual and developmental stages.

5.1(6) MARKETS FOR WIND ENERGY SYSTEMS - WHEN, WHERE AND AT WHAT PRICE.
Johanson, E.E. and others. (JEF Sci. Corp., Wilmington, Massachusetts, USA) In Collection of Technical Papers - AIAA/SERI Wind Energy Conference, Boulder, USA 9-11 April 1980. New York, NY, AIAA - CP803 1980 Paper 80-0613 pp. 55-60.

The authors synthesise the available economic and marketing data on wind systems into a state-of-the-art picture of the 1980 market place. The resource, the machines, the loads, the rate structures, the utility mix of equipment and the potential users' purchase decision criteria are considered, and a market emergence picture is presented indicating where WECS penetration should begin and under what conditions. (11 refs.)

5.1 (7) STUDY OF PRODUCT LIABILITY INSURANCE ISSUES RELATED TO SMALL WIND ENERGY CONVERSION SYSTEMS.
Bass, L. and others. (Hans W. Winholds Co., Cupertino, CA, USA) US Dept. of Energy Report no: RFP-3178-80/21 November 1980 155 pp.

This study was sponsored by the Department of Energy to analyse the products' safety and liability status of the SWECS industry and to propose strategies for addressing their needs. Eigth strategies were identified which would reduce the manufacturers' risk, redistribute the economic burden of products liability, and redefine the concept of acceptable risk. (See also paper in ISES-AS/Et Al 1980 Annual Conference, Phoenix 2-6 June V. 3.2. pp. 1522-1525 by J. Noun.)

5.1(8) A SCREENING METHOD FOR WIND ENERGY CONVERSION SYSTEMS.
McConnell, R.D. (Solar Energy Research Institute, Golden, CO, USA) Paper presented at International Solar Energy Society American Section/ Et Al 1980 Annual Conference, Phoenix, USA 2-6 June 1980 V. 3.2 pp. 1507-1511.

Both value indicators and simplified cost estimating techniques are used. The value indicators are selected ratios of engineering parameters involving energy, mass, area, and power. The method is applied to the analysis of a tracked-vehicle airfoil concept. (12 refs.)

5.1(9) WIND SYSTEM APPLICATIONS AND COMMERCIALISATION.
Eldridge, F.R. (Mitre Corp., Virginia, USA) Paper presented at International Solar Energy Society American Section/Et Al 1980 Annual Conference, Phoenix 2-6 June 1980 V. 3.2 pp. 1442-1451.

Reviews typical wind system applications producing electrical, thermal or mechanical power. Available types of wind machines are described; including vertical axis, horizontal axis and other designs. The commercialisation of such systems should reduce their initial and operating costs. (7 refs.)

5.1(10) A COST EFFECTIVE STUDY OF NATURAL ENERGY POWER SUPPLIES.
Komesaroff, M.B. Paper presented at Institute of Engineers Australia, Engineering in 1981 Conference, Canberra 23-27 March 1981 pp. 43-48.

Discusses economic advantages associated with solar cell and wind generator power supplies. Discounted cash flow models are employed to determine least cost alternatives for several sites in SouthEastern Australia and results show that wind generators are more cost effective than solar panels for loads in excess of one amp. (11 refs.)

5.1(11) COST ESTIMATES FOR ADVANCED/INNOVATIVE WIND ENERGY CONVERSION SYSTEMS.
Jacobs, E.W. (Solar Energy Research Institute, Golden, CO, USA) Solar Energy Research Institute Report nos: SERI/TP-211-1409 CONF-811205-1 September 1981 5 pp. (AIWA Terrestrial Energy Systems Conference, Colorado Springs, CO, USA 1 December 1981) US Dept. of Energy Report no: DE 82 001210.

Details the strengths and weaknesses of three computerised costing methods being adapted by the Solar Energy Research Institute for use with wind energy systems to provide cost estimates for the Advanced/Innovative Wind Energy Concepts programme. These are the Solar Array Manufacturing Industry Costing Standards, the RCA Programmed Review of Information for Costing and Evaluation, and the Freiman Analysis of Systems Technique models. Some preliminary costing results from the ongoing validation efforts at SERI are outlined.

5.1(12) ECONOMIC EVALUATION.
Hoffman, L. (Regensburg University, W. Germany) In Implementing Agreement for Programme of Research & Development on Wind Energy Conversion Systems 1981 Meeting of Experts of Annex III & IIIa: Integration of Wind Power into National Electricity Supply Systems, Regensburg, W. Germany 29-30 January 1981. Julich, W. Germany, Kernforschungsanlage Julich GmbH April 1981 Session A Paper A3 pp. 28-37. (Juel/Spez-108)

The economic importance of wind energy conversion plants is considered under three headings: supply potential, competitiveness, and social desirability.

5.1(13) ECONOMIC FEASIBILITY OF WIND ENERGY FOR THE US VIRGIN ISLANDS.
Ball, D.E. and others. (Southern Solar Energy Centre, Atlanta, GA, USA) Southern Solar Energy Centre Report nos: SSEC/TP-31261/2 June/September 1981 26 pp. and 110 pp. US Dept. of Energy Report nos: DE 82 009584 & DE 82 009476.

Presents a preliminary assessment of the economic feasibility of wind energy conversion systems for the Virgin Islands. Current and near-term economics of available WECS for residential, commercial and wind farm applications were analysed and the sensitivity of WECS economic worth to a variety of factors investigated. Preliminary conclusions highlighted the favourable economics regarding larger, commercial and wind farm scale machines.

5.1(14) ECONOMICS OF SELECTED WECS DISPERSED APPLICATIONS.
Krawiec, S. (Solar Energy Research Institute, Golden, CO, USA) J. Energy, 5 (2) March-April 1981 pp. 72-78.

Paper analyses the cost of electricity generated by selected wind-energy systems in both applications; the impact of major economic factors on the cost performance index; and the breakeven cost of wind systems able to compete economically with conventional power sources in dispersed applications. Life-cycle cost and breakeven period are two major measures of economics. (2 refs.)

5.1(15) A LIFE CYCLE COSTING MODEL FOR SMALL
WIND ENERGY CONVERSION SYSTEMS.
Laulainen, L. (Mid-American Solar Energy Complex,
Minnesota, USA) Paper presented at 2nd Inter-
national Solar Energy Society American Section/
Et Al Wind Power Energy Alternative for MidWest
Conference, Minnesota, USA 3-4 April 1981
pp. 41-44.

Presents a life-cycle costing model that can
be applied to determine the costs and benefits
associated with an investment in a specific
small wind energy conversion system. Total
installed cost, operating expenses, maintenance
cost, and insurance cost are examined in
relation to the revenue that can be generated.

5.1 (16) PRODUCT LIABILITY INSURANCE AND SMALL
WIND ENERGY CONVERSION SYSTEMS.
Bass, L. (Kairos Co., Cupertino, CA, USA) J.
Product Liability, 4 (3-4) 1981 pp. 265-273.

A number of wind energy manufacturers have
experienced difficulty in obtaining products
liability insurance at reasonable costs. Such
insurance is available for all but the smallest
and newest manufacturers, but can be quite
expensive. Costs of products liability
insurance varies considerably, depending on
both the manufacturer's performance and the
insurer's perception of exposure.

5.1(17) PROSPECTS FOR FOREIGN APPLICATIONS OF
WIND-ENERGY SYSTEMS. PRELIMINARY REPORT IN
RESPONSE TO PUBLIC LAW 96-345.
Department of Energy, Washington, DC. Assistant
Secretary for Conservation and Renewable Energy.
Washington, DC, USA, US Dept. of Energy Report
no: DOE/NBM-1005 4 November 1981 18 pp.

Identifies potential foreign applications and
specific systems which would most closely match
the applications requirements from a list of
representative US wind energy systems. The
export potential was determined by analysing
a country's applications requirements, cost
of alternative energy, financial condition,
interest in the development of renewable energy
technologies, and level of indigenous
competition. A summary of results is presented.

5.1(18) RISK MANAGEMENT FOR ALTERNATE TECH-
NOLOGIES.
King, W.T. (Johnson & Higgins of California)
Paper presented at American Nuclear Society
Alternative Energy Sources for Electrical
Power Conference, MA 4-7 October 1981 5 pp.

Reviews fundamental aspects of external
capital markets and internal controls in
alternative energy risk management. These
technologies - wind power and solar energy -
will require the full normal insurance
coverage on a large scale for workers, the
public, and for assets and cash flows. These
technologies require capital attraction
based on high technology risk investment.
(2 refs.)

5.1(19) SMALL WIND SYSTEMS APPLICATIONS
ANALYSIS: A REPORT ON A STUDY.
Liske, C. and Johanson, E.E. (Rockwell Inter-
national; JBF Sci. Corp., USA) In proc. 5th
Biennial Wind Energy Conference & Workshop,
Washington, DC, USA 5-7 October 1981. I.E. Vas
(ed.) V. 2. Palo Alto, USA, Solar Energy
Research Institute 1981 Session 3A
pp. 499-513. Report nos: SERI/CP-635-1340
CONF-811043.

Discusses a detailed market study of small
wind energy conversion systems for five
selected applications conducted for the US
Department of Energy. The study was intended
to aid manufacturers and government planners
in finance allocation and to aid planning of
research and development programmes as an
estimate of the potential market for wind
power systems under 100 kW.

5.1(20) WIND ENERGY ON ITS WAY TO COMMERCIAL-
ISATION.
Selzer, H. In proc. International Colloquium on
Wind Energy, Brighton, U.K. 27-28 August 1981.
Cranfield, U.K., BHRA Fluid Engng. 1981
Session 3 pp. 152-160.

Reviews European efforts for large Wind
Energy Converters before discussing the next
steps when these WEC will be used in larger
quantities. The commercial aspects will gain
great importance, but the direct maintenance
costs have to be minimised before the WEC
become economically viable.

5.1(21) WIND ENERGY: SOME COMMENTS ON THE
ECONOMICS.
Musgrove, P.J. (Reading University, U.K.)
In proc. International Colloquium on Wind
Energy, Brighton, U.K. 27-28 August 1981.
Cranfield, U.K., BHRA Fluid Engng 1981 Session 2
pp. 119-126.

Although present generation wind turbines are
only marginally economic in the USA they are
shown to have very attractive economics in the
U.K. Reasons for this discrepancy are considered
and factors affecting the optimum size of machine
for U.K. applications are assessed.

5.1(22) ECONOMICS OF SCALE IN MASS PRODUCTION OF
SMALL WIND ENERGY CONVERSION SYSTEMS.
Mattila, J.R. (Central Solar Power Research
Corp., USA) Intl. J. Energy Syst., 2 (2) 1982
pp. 82-86.

A general approach to engineering economic cost
analysis of manufacturing small wind energy conversion systems in mass production is discussed.
The Solar Array Manufacturing Industry Costing
Standard (SAMICS) methodology developed by the
Jet Propulsion Laboratory is modified for
analysis of small wind energy conversion systems.
Manufacturing process analysis techniques are
applied in sequence along with SAMICS methodology
to reach manufacturing cost estimates. Preliminary manufacturing cost estimates on five
alternative annual production volumes (1,000,
5,000, 10,000, 25,000 and 50,000) for a US
Department of Energy prototype 15 kW horizonta
axis, downwind turbine are reported. (From
author's abstract.)

5.1(23) FOREIGN APPLICATIONS AND EXPORT
POTENTIAL FOR WIND-ENERGY SYSTEMS.
Griffith, S.K. and others. (Planning Research
Corp., McLean, VA; DHR Inc., Washington, DC)
Solar Energy Research Institute Report no:
SER/STR-211-1827 December 1982 307 pp. US Dept.
of Energy Report no: DE 83 009715.

Identifies foreign countries in which wind
applications appear to be competitive with
conventional energy sources and in which the
barriers to US exports do not appear to be
excessive. Data on wind resources, imports,
import restrictions, licensing requirements,
exchange controls, dependence on energy imports
and balance-of-payments are presented for 184
countries and territories. Data on 27 wind
applications were also collected, a catalogue
of wind plants suitable for export was developed
and each application was evaluated as a function
of wind speed, projected wind machine cost and
appropriate discount rate.

5.1(24) IMPLEMENTING AGREEMENT FOR CO-OPERATION
IN THE DEVELOPMENT OF LARGE SCALE WIND ENERGY
CONVERSION SYSTEMS.
Kernforschungsanlage Juelich GmbH, F.R. Germany,
Projektleitung Energieforschung. Report no:
Juel-Spez-147 CONF-8111156 March 1982 104 pp.
(Meeting of Experts on Costings for Wind Turbines,
Copenhagen, Denmark 18 November 1981.) US Dept.
of Energy Report no: DE 83 750640. (See also
Juel-Spez-100.)

The Meeting covered cost data for wind turbines in different countries and methods of
calculating wind energy cost. Papers from
Denmark, Germany, Sweden and the United States
were presented and discussed.

5.1(25) PRELIMINARY ESTIMATION OF THE ECONOMICS
OF WIND POWER IN PENNSYLVANIA.
Hoch, L.J. and Kennet, D.M. In proc. 1982 Wind
and Solar Energy Technology Conference, Kansas
City, USA 5-7 April 1982. Columbia, USA,
Missouri-Columbia University 1982 pp. 184-195.

Monitored data regarding wind turbine performance and wind availability was used to
assess the economic feasibility of Small Wind
Energy Conversion Systems under specific wind
conditions at sites in Pennsylvania.

5.1(26) RENTABILITY ANALYSIS OF SMALL-SCALE
WIND ENERGY SYSTEMS.
Wijnant, W. (Rijksuniv. Vent., Antwerpen,
Belgium) Ingenieursblad, 51 (9) 1 September 1982
pp. 208-210.

Presents a computer programme for determining
the rentability of small wind turbines. The
principal relevant factors involved in wind-
energy conversion systems are discussed with
some emphasis on the economic viability of
such systems. (6 refs.)

5.1(27) WIND ENERGY.
Carpentier, M. (Dept. of Energy, Mines &
Resources, Renewable Energy Div., Ottawa,
Canada) J. Can. Pet. Technol., 22 (6)
November-December 1983 pp. 40-42.

Wind energy is now considered marginally
economical in Canada. There is a good wind
regime and if the production of electricity
is very expensive, wind energy is now
economical. Wind energy efforts have been
mainly concentrated on the Darrieus turbine.

5.1(28) COMMERCIAL APPLICATIONS OF WIND POWER.
Vosburgh, P.N. New York, U.S.A., Van Nostrand
Reinhold Co. Inc. 1983 292 pp. (ISBN 0-422-
29036-5)

Discusses wind energy conversion systems
capable of generating substantial quantities of
electricity, arbitrarily defined as WECS which
utilise one or more wind turbines with generating capacity of 20 kW or more. Limited to
commercially promising systems which generate
utility power needed by most homes, farms, and
businesses.

5.1(29) ESTIMATION OF WIND SYSTEM PERFORMANCE
AND ECONOMICS USING UNIT AVAILABILITY CRITERIA.
Unione, A.J. and others. (Sci. Appl. Inc.,
U.S.A.) In Alternative Energy Sources III,
V. 4 - Indirect Solar/Geothermal Energy, proc.
3rd Miami Intl. Conference on Alternative
Energy Sources, Miami Beach, U.S.A.
15-17 December 1980. T. Nejat Veziroglu (ed.)
pp. 69-80. Washington, USA, Hemisphere Publ.
Corp. 1983 (ISBN 0-89104-336-0)

Describes a methodology for evaluating the
impact of unit availability on the overall
performance and costs of wind energy conversion systems. The WECS is modelled as a
collection of wind turbines, possibly with
backup diesel generators or batteries, tied
to a time varying demand and a variable wind
resource. Each wind turbine is modelled as a
series of subsystems. The reliability of a
subsystem is given as a probability distribution
of the time to failure for the subsystem.

5.1(30) ECONOMIC ANALYSIS OF THE HURRICANE
WINDMILL DESIGN - CASE STUDY.
Bolie, V.W. (Univ. of New Mexico, Albuquerque,
NM, USA) Eng. Econ. 30 (1) Autumn 1984
pp. 73-82.

An economic analysis is presented for a
vertical-axis windmill of sufficient strength
to operate continuously in gale-force winds.
The results are illustrated by computerised
plots of energy cost vs. wind speed for several
different interest rates. The corresponding
points for several possible installation sites
are superimposed for comparison.

5.2 Electric Power Generation Using WECS

5.2.1 TECHNICAL ASPECTS OF ELECTRIC POWER GENERATION USING WECS.

5.2.1(1) NEW GENERATION SCHEME FOR LARGE WIND ENERGY CONVERSION SYSTEMS.
Yadavalli, S.R. and Jayadev, T.S. (University of Wisconsin, Milwaukee, USA) In 11th Intersociety Energy Conversion Engineering Conference, State Line, Nevada, USA 12-17 September 1976. New York, NY, AIChE 1976 V. 2 (SAE Paper 769303) pp. 1761-1765.
The scheme utilises the principle of an induction generator which supplies power to the grid from both stator and rotor of an induction machine. Theoretical evaluation of the performance of this system is presented and a comparison is made with conventional induction generators in terms of overall efficiency and annual energy collection. It is concluded that the proposed generation scheme could provide an economic and efficient method to convert wind energy to electrical energy. (5 refs.)

5.2.1(2) IMPLICATIONS OF LARGE SCALE INTRODUCTION OF POWER FROM LARGE WIND ENERGY CONVERSION SYSTEMS INTO THE EXISTING ELECTRIC POWER SUPPLY SYSTEM IN THE NETHERLANDS.
Bontius, G.H., Manders, A.H.E. and Stoop, Th. (N.V. Kema, Netherlands) In 2nd International Symposium on Wind Energy Systems, paper presented at Amsterdma, Netherland 3-6 October 1978. Cranfield, U.K., BHRA Fluid Engng. 1978 pp. G3. 27-G3. 38.
Compared to conventional production units, a considerable amount of rated power must be installed in order to produce a given amount of energy because of the low utilisation factor of the wind-turbine. Even in a densely populated country with a complex electricity grid, the existing medium voltage distribution grid (1-20 kV) will not be sufficient to absorb the power from arrays or parks of wind turbines and it can be shown that only transmission networks and substations with voltages from 50 kV upwards can be used. General criteria are discussed for the conversion of the mechanical power into electric power with suitable characteristics and some economic considerations are given. (5 refs.)

5.2.1(3) CONTROL STRATEGY FOR A VARIABLE-SPEED WIND ENERGY CONVERSION SYSTEM.
Jacob, A., Veillette, D. and Rajagopalan, V. Washington, DC., NASA-TM-75512 Nov. 1979 10 pp.
Describes a control system for a variable speed wind energy conversion system is described. A self-excited asynchronous cage generator is used together with an array of thyristor converters. (See also paper by these authors in PESC Record. IEEE Electronic Power Spec.Conference, Syracuse, NY. 13-15 June 1978 pp. 69-75.)

5.2.1(4) WIND ENERGY CONVERSION SYSTEM WITH ELECTROMAGNETIC STABILISER.
Kant, M., Berna, M. and Vidoni, E. (Univ. de Technol. de Compiegne, France) Proc. Inst. of Electrical Engineers (London), 126 (11) November 1979 pp. 1201-1203.
Describes the development of a novel medium-power (<50 kW) system composed of a constant-pitch rigid rotor and a double-stage generator, uses electromagnetic braking to assure speed stabilisation.

5.2.1(5) ALTERNATOR DESIGN FOR DIRECT COUPLING TO REMOTE WIND ENERGY SYSTEMS.
Menzies, R.W., Mathur, R.M. and Bullock, W.R. (Manitoba University; Bristol Aerospace Ltd.) In proc. 3rd International Symposium on Wind Energy Systems, Lyngby, Denmark 26-29 August 1980. Cranfield, U.K., BHRA Fluid Engng. 1980 Paper F1 pp. 269-278.
The mechanical power available from the shaft of the wind turbine is not suitable for direct use by conventional alternators or generators as the shaft speed is very low, especially as the power rating increases, and the shaft power varies as the cube of the shaft speed. In the past, this problem has been solved by the use of gears and several alternators and/or generators to match the wide power range of the turbine which has led to reduced efficiency and reliability of the overall system. Examines the desirable performance specifications of direct-drive alternators suitable for meeting diverse load requirements and compares adaptations of several conventional generators/alternators with several novel designs.

5.2.1(6) ANALYSIS OF SMALL NONCONVENTIONAL ELECTRIC POWER SYSTEMS FOR REOMTE SITE APPLICATIONS.
Boehman, L.I. and others. (Dayton University, and other bodies) In IECEC '80, Energy to the 21st Century proc. 15th Intersociety Energy Conversion Engng. Conference, Seattle, USA 18-22 August 1980 V. 1. New York, USA, American Institute Aeronaut. & Astronaut. 1980 no: 809165 pp. 828-834.
Considers electric power systems with energy conversion by wind, solar and hybrid wind solar configurations and energy storage in flywheels, hydrogen, batteries and thermal devices. Relative performance, cost, availability and reliability are compared for the conceptual systems. A modular configuration with two 8 kW wind energy converters and sealed lead acid batteries is analysed in detail for a remote site military application in northern Alaska. The system analysed can provide 5 kW on a continuous basis with 5.6 metres per second average wind velocity and have 12 hours of reserve capacity stored in the battery energy storage system.

5.2.1(7) CHARACTERISTICS OF ELECTRO-GAS-DYNAMIC WIND ENERGY DEVICES.
De Mey, G. (Ghent State University, Belgium) Energy Conversion & Management, 20 (3) 1980 pp. 201-202.

Fresents a mathematical model to describe the electro-gas-dynamic (EGD) generator for converting wind energy directly into electricity. The basic operating concept is that the wind carries charged particles and hence bears an electric current. Characteristics of the EGD generator are very similar to those of a photovoltaic solar cell. The efficiency of the unit is improved if the mobility of the ions carried by the wind is low. (2 refs.)

5.2.1(8) DEVELOPMENT. MANUFACTURE AND TESTING OF A 52 M DIA/265 KW WIND ENERGY CONVERTER.
Spittler, W. and others. Statusber. Windenerg. (Semin.), Hamburg, W. Germany 24-26 June 1980 Dusseldorf, Germany, VDI 1980 pp. 47-76. (In German)
Describes the control system of the wind energy converter rotor. Control of the speed and power is effected by means of a micro-computer controlling the two rotor blades which can be adjusted in their longitudinal axis. The structure of the control arrangement, control circuit diagram, and control programme flow chart are given. The second part of the report describes the structural layout of the rotor blades of Voith wind turbine of the 52/265 wind power converter. The third part deals with the structural dynamics of the Voith converter. Part IV briefly describes the assembly of the rotor.

5.2.1(9) DOUBLE OUTPUT INDUCTION GENERATOR SCHEME FOR WIND ENERGY CONVERSION.
Bolton, H.R., Lam, W.C. and Freris, L.L. (Imperial College Sci. Technol., U.K.) In proc. 2nd British Wind Energy Association Wind Energy Workshop, Cranfield, U.K. 17 April 1980. London, U.K., Multi-Sci. Publishing Co. Ltd. 1980 pp. 30-40.
The mains-connected, three phase induction machine continues to be one of the best generator options for wind energy schemes over almost the entire power range. However, the standard machine can only operate efficiently over a relatively narrow speed range and this has a major impact on the design and operating mode of the turbine. The results of a steady-state simulation of a system comprising a three bladed, horizontal axis turbine, step-up transmission, slip-ring induction generator, mains and heater are analysed and the influence of step-up ratio and heater resistance on system performance are indicated. The scheme's advantages and disadvantages along with other generation options are discussed.

5.2.1(10) INTEGRATION PROBLEMS WITH LARGE WIND ENERGY CONVERSION SYSTEMS.
Thomas, R.J. and Thorp, J.S. (Cornell University, Ithaca, NY, USA) In proc. 18th Annual Allerton Conference Community Control Computing, Monticello, Ill. 18-10 October 1980. Sponsored by University of Illinois, Dept. of Electr. Engng. and Coord. Sci. Lab., Urbana-Champaign pp. 206-213.
Problems associated with integrating clusters of large wind-tunnel generators into a utility

grid system are considered. The use of a series-connected ac/dc/ac interface, specifically designed for intermittent sources of energy such as wind-turbine generators, is suggested as a means for circumventing the economic and technical problems associated with protection and interconnection schemes. (4 refs.)

5.2.1(11) ADAPTING PERMANENT MAGNET ALTERNATORS TO WIND SYSTEMS.
Gillette, W.D.(Zephyr Wind Dynamo Co., ME, USA) Alternative Sources of Energy, (50) July-August 1981 pp. 19-21.
Discusses the advantages and disadvantages of permanent magnet alternators for wind energy systems and examines efficiency, inductance, regulation and tuning considerations. Magnet design modifications can make these devices suitable for application to wind systems.

5.2.1(12) CONTROL POLICIES FOR WIND-ENERGY CONVERSION SYSTEMS.
Buehring, I.K. and Freris, L.L. (Imperial College of Sci. & Technol., London, U.K.) IEE Proc. Part C, 128 (5) September 1981 pp. 253-261.
The wind-rotor/generator dynamics are investigated for a number of control systems, and it is shown that the system response is a function of wind speed. Because of this relationship, control strategies based on static optimum matching premises are unlikely to be optimal under continuously changing conditions. For a given recorded windspeed sample, the energy delivered was measured for a number of control strategies. Results indicated that for the wind sample used and aerogenerator simulated, sophisticated control policies do not necessarily result in maximum energy yield. (11 refs.)

5.2.1(13) ELECTROENGINEERING ASPECTS OF WIND ENERGY CONVERSION.
Van Leuven, J. (Rijksuniv. Cent. Antwerpen, Belgium) Ingenieursblad, 50 (11) November 1981 pp. 307-310. (In Dutch)
A survey of the currently available systems for conversion of wind power into electric power having constant voltage and frequency. The advantages and drawbacks of the individual systems are examined. (10 refs.)

5.2.1(14) THE GEMINI SYNCHRONOUS INVERTER.
Werking, T.L. (Windworks Inc., Wisconsin, USA) Paper presented at 2nd International Solar Energy Society American Section/Et Al Wind Power Energy Alternatives for MidWest Conference, Minnesota 3-4 April 1981 pp. 25-33.
Describes the Gemini Synchronous Inverter a line-commutated, line-feeding inverter capable of converting dc power from a variable voltage source into ac power at standard line voltages and frequencies. The Gemini Synchronous Inverter allows safe, economical interfacing of wind turbines, solar photovoltaic arrays, solar thermal electric systems, small hydroelectric installations, and certain

industrial processes with conventional ac electrical power utility grids. The synchronous inversion method of operation is described, and the quality of the inverted power is assessed.

5.2.1(15) INDEPENDENT POWER SYSTEMS AND INVERTERS.
Paul, T.D.(Best Energy Systems for Tomorrow, Wisconsin, USA) Paper presented at 2nd International Solar Energy Society American Section/ Et Al Wind Power Energy Alternatives for MidWest Conference, Minnesota 3-4 April 1981 pp. 11-23.

Criteria for designing an independent power system for a specific application and selecting the proper static power inverter are reviewed. The inverter must be sized to handle peak power loads, leaving room for expansion. Advantages and drawbacks of six types of power inverters are discussed.

5.2.1(16) INTERFACING A GEMINI SYNCHRONOUS (LINE-COMMUTATED) INVERTER WITH A LOCAL UTILITY AND A MINI-GRID: TEST RESULTS.
Atomics International Div., Golden, CO, Wind Energy Research Centre. US Dept. of Energy Report no: RFP-3449 October 1981 52 pp. (DE 83 007198)

Presents data generated by interfacing a single-phase synchronous inverter to a local utility and to a minigrid which simulated a 25 percent utility penetration. These inverters are used by some wind systems manufacturers for interfacing the direct current or rectified alternating current output of wind-turbine generators directly to grid ac power lines.

5.2.1(17) MULTISPEED ELECTRICAL GENERATOR APPLICATION TO WIND TURBINES.
Andersen, T.S. and Kirschbaum, H.S. (Westinghouse Electric Corp., Pittsburgh, USA)
J. Energy, 5 (3) May-June 1981 pp. 172-177.

Studies cost-effective methods to achieve increased energy conversion from wind turbine generators. The pole amplitude modulated multispeed induction generator, is demonstrated to achieve an adequate energy capture without the introduction of adverse costs or inefficiencies associated with other approaches. (5 refs.)

5.2.1(18) PROGRESS ON A PRACTICAL WIND ENERGY SYSTEM WITH AN INDUCTION GENERATOR, OVER-SYNCHRONOUS ELECTRONIC CASCADE AND MAXIMAL POWER CONTROL.
Van Wyk, J.D. and others. (Rand Afrikaans Univ.) In proc. International Colloquium on Wind Energy, Brighton, U.K. 27-28 August 1981. Cranfield, U.K., BHRA Fluid Engng. 1981 Session 3 pp. 185-190.

Describes the research and development of a system for conversion of wind energy into electrical energy and feed back into the supply system. The system uses a fixed pitch wind turbine and operates under variable speed conditions using an induction generator. An oversynchronous electronic rotor cascade is used, enabling operation over a two to one speed range. Progress in constructing a laboratory system and a wind tower system is reported, system aspects including maximal power control are discussed and future work outlined. (10 refs.) (See also paper by authors in Sun 2, ISES Silver Jubilee Congress, Atlanta, GA May 1979 V. 3. pp. 2291-2295. Oxford, U.K., Pergamon Press 1979.)

5.2.1(19) AN UPDATE OF THE ELECTROFLUID DYNAMICS WIND DRIVEN GENERATOR.
Minardi, J.E. and Lawson, M.Q. (Dayton University) In proc. 5th Biennial Wind Energy Conference & Workshop, Washington, DC, USA 5-7 October 1981 I.E. Vas (ed.) V. 1. Palo Alto, USA, Solar Energy Research Institute 1981 Session 1D pp. 433-444. Report nos: SERI/CP-635-1340 CONF-811043.

Describes a research programme for the development of Electrofluid Dynamic (EFD) wind driven generators. In such generators, the wind blows through arrays of electrodes; transports charged particles against an electrical potential gradient and thereby generates electrical power directly without moving parts. This offers a simpler, less expensive system, free of frontal area and velocity limitations of conventional rotating wind energy systems. For the EFD wind driven generator there are no limits on size; therefore, economics of scale can be realised. (Author's abstract)

5.2.1(20) WIND-ENERGY-CONVERSION SYSTEMS.
Van Leuven, J. (Rijksuniv. Cent. Antwerpen, Belgium) Electronics & Power, 27 (10) October 1981 pp. 742-743.

Presents a short review of the different systems used to convert the output of wind driven generators into electric power of constant frequency. Wind-energy-conversion systems belonging to the constant-speed category, as well as those employing commutator machines, are of doubtful value owing to the low-conversion efficiency and the well-known troubles associated with their use. When synchronisation with an existing utility grid is possible, the double-output induction generator has limitations on the speed range, decreasing the conversion efficiency, but a dc link is possible, employing cycloconvertors and the self-excited squirrel-cage machine with pulse frquency modulated inverter.

5.2.1(21) WIND-ENERGY RECOVERY BY A STATIC SCHERBIUS INDUCTION GENERATOR.
Smith, G.A. and Nigim, K.A. (University of Leicester, U.K.) IEE Proc. Part C, 128 (6) November 1981 pp. 317-324.

Describes a technique for controlling a doubly fed induction generator driven by a windmill, or other form of variable-speed prime-mover, to provide power generation into the national grid. The secondary circuit of the generator is supplied at a variable frequency from a current source inverter which for test purposes is rated to allow energy recovery, from a simulated windmill, from maximum speed to standstill. To overcome the stability problems, a novel signal generator, which is locked in phase with the rotor emf, controls the secondary power to provide operation over a wide range of subsynchronous and super synchronous speeds. (9 refs.)

5.2.1(22) AN AUTONOMOUS WIND ENERGY CONVERTER USING A SELF EXCITED INDUCTION GENERATOR FOR HEATING PURPOSES.
Milner, I.P. and Watson, D.B. (Canterbury University) Wind Engng., 6 (1) 1982 pp. 19-23.

Describes the design philosophy and construction for an autonomous wind energy conversion system using a self excited induction generator to provide heat for a plant propagation house. Emphasis is given to the method of optimisation of generator output to obtain maximum possible energy extraction from the wind.

5.2.1(23) HOW TO DESIGN A REMOTE POWER SYSTEM.
Paul, T. (Best Energy Systems, WI, USA) Solar Age, 7 (10) October 1982 pp. 34-39.

Presents guidelines for effectively and efficiently designing remote (including wind) systems. The system output must be planned first, and the system needed to supply output is next considered. Daily usage and peak power requirements must be calculated. Storage needs, as well as a choice between ac and dc appliances, must be determined. Minimum system voltages and power system size are also discussed.

5.2.1(24) THE INTEGRATION OF SMALL WIND TURBINES WITH DIESEL ENGINES AND BATTERY STORAGE.
Slack, G., Lipman, N.H. and Musgrove, P.J. (Reading University, U.K.) In proc. 4th British Wind Energy Association Wind Energy Conference, Cranfield, U.K. 24-26 March 1982 P.J. Musgrove (ed.). Cranfield, U.K., BHRA Fluid Engng. 1982 pp. 119-128.

Describes a wind turbine with battery storage and diesel back-up for supplying electricity to small consumers remote from a grid system. Optimisation of such a system may best be achieved by means of a computer simulation which examines the effect of type and size of windmill and batteries and the system control logic used.

5.2.1(25) INTERFACING WIND ENERGY CONVERSION EQUIPMENT WITH UTILITY SYSTEMS.
Meier, R.C. and Macklis, S.L. (Gen. Electr. Co.) Energy, 7 (1) January 1982 pp. 15-29.

Government and industry is increasingly aware of the intermittent power generation characteristics of solar systems, in contrast to most utility generation equipment. Paper explores the interface issues for wind energy systems, a solar technology likely to achieve early commercialisation. Government- and industry-sponsored assessments of the impact of wind energy devices on industry operations, controls, and protective subsystems are described.

5.2.1(26) SLIP ENERGY RECOVERY TECHNIQUES FOR WIND ENERGY GENERATORS.
Smith, G.A. and Nigim, K.A.(Leicester University) In proc. 4th British Wind Energy Association Wind Energy Conference, Cranfield, U.K. 24-26 March 1982 P.J. Musgrove (ed.). Cranfield, U.K., BHRA Fluid Engng. 1982 pp. 283-290.

Describes techniques that may be used for control of slip energy in induction generators used for variable speed constant frequency operation, the static Kramer system only generates at supersynchronous speeds and operating limits can restrict the range of on-line energy recovery, but static Scherbius system using either a cycloconverter or a current source inverter provides generation at both sub- and supersynchronous speeds.

5.2.1(27) AN ADAPTIVE ROTOR RESISTANCE CONTROLLER FOR WIND-DRIVEN SLIP-RING INDUCTION GENERATOR.
Velayudhan, C. and others. (Dept. of Electrical & Electronic Engng., University of Western Australia) In proc. 1983 International Electrical, Electronics Conference, Toronto, Ontario, Canada 26-28 September 1983 V. 1. pp. 114-117. IEEE New York, USA. IEEE 1983 2 vols. 689 pp.

An electronic technique is described for controlling a system-connected induction generator with wind turbine drive to allow the turbine to deliver its maximum available power at all rational speeds in the usable wind speed range. This strategy allows the turbine to operate continuously at optimum tip speed/wind speed ratios.

5.2.1(28) ANALYSIS OF THE ISOLATED INDUCTION GENERATOR.
Ouazene, L. and McPherson, G. (Inst. Nationale d'Electricite et d'Electronique Boumerdas, Algiers, Algeria) IEEE Trans. Power Appar. Syst., PAS-102 (8) August 1983 pp. 2793-2798.

The ability of induction generators to convert mechanical power over a wide range of rotor speeds has given rise to the possible contribution of wind energy to provide fuel displacement. An arrangement using a capacitor-excited induction generator and a variable speed drive to supply resistive load is described and a method for determining the output voltage and frequency in the steady-state is presented. A comparison is made of the predicted behaviour of this isolated system, with the experimental results.

5.2.1(29) THEORETICAL STUDY OF AN AUTONOMOUS SYSTEM COMBINING A PHOTOVOLTAIC GENERATOR AND WIND MACHINE UNDER REAL LOCAL DATA.
Samarakou, N.T. and others. (Div. of Electronics, University of Athens, Athens, Greece)

In proc. of MELECON 83 Mediterranean Electro-
technical Conference, Athens, Greece 24-26 May
1983 pp. D11.08/1-2 V. 2. IEEE. New York, USA
1983 2 vols 524 pp. and 580 pp.

An appropriate model is given for the computation and optimisation of the size of the photovoltaic generator and the wind machine, using
as parameters real local data and the cost per
watt peak of the photovoltaic generator and the
wind machine. The method of determining the
size of the storage system is based on the
analysis of the daily balance of energy and the
constraint of ensuring the autonomy of the
system. (A) (4 refs.)

5.2.1(30) 20-KW WIND ENERGY CONVERSION SYSTEM
(WECS) AT THE MARINE CORPS AIR STATION,
KANEOHE, HAWAII.
Pal, D. (Naval Civil Engineering Lab., Port
Hueneme, CA) Report for September 1978-December
1981. Report no: NCEL-TN-1655 January 1983
69 pp. (AD-A128761/4)

The wind turbine generator chosen for the
evaluation was a horizontal-axis-propeller-
downwind rotor driving a three-phase, self-
excited alternator through a step-up gear box.
The alternator is fed into the base power
distribution system through a three-phase, line-
commutated-synchronous inverter using SCRs.
The site has moderate wind conditions with an
annual average windspeed of 12 to 14 mph, and
the WECS turbine has a relatively high (29 mph)
rated windspeed.

5.2.1(31) WIND ENERGY: SOME NOTES ON ITS
COLLECTION, STORAGE AND APPLICATION.
Stobart, A.F. Energy World, (103) May 1983
pp. 4-6.

A review of the current utilisation methods
of wind for generation of electrical power.
It was pointed out that a wind energy collector
of 6 metres in diameter gave similar results
with one of 10 metres diameter. So a wind
energy collector is not, in principle, a good
source of shaft horsepower for driving
conventional electricity generators.
Batteries are only an economic storage system
for very small wind energy systems. An
approach for the future is the generation
of hydrogen, using wind power to electrolyse
water.

5.2.1(32) CYCLOCONVERTOR-EXCITED DIVIDED-
WINDING DOUBLY-FED MACHINE AS A WIND-POWER
CONVERTOR.
Holmes, P.G. and Elsonbaty, N.A. (University
of Leicester, Dept. of Electrical Engineering,
U.K.) IEE Proc. Part. B, 131 (2) March 1984
pp. 61-69.

Describes a slip-ring induction machine operating as a doubly-fed generator above and below
synchronous speed. A prime mover simulates a
wind turbine by producing a torque increasing
in proportion to the square of the speed which
is balanced in steady state operation by the
double-fed generator. The stator is arranged
in two electrically separate, magnetically
coupled layers connected to a cycloconvertor

operating with continuous circulating current.
This reduces the harmonics injected into the
supply and prevents line-to-line short circuits
under shock conditions. The secondary frequency
is locked to the difference between actual
speed and synchronous speed.

5.2.1(33) RELIABILITY MODELLING OF LARGE WIND
FARMS AND ASSOCIATED ELECTRIC UTILITY INTERFACE
SYSTEMS.
Xifan, W. and others. (Xian Jiaotong Univ.,
Dept. of Electrical Engineering, Xian, China)
IEEE Trans. Power Appar. Syst. PAS-103 (3)
March 1984 pp. 569-575.

Concerned with modelling the reliability
characteristics of large electric utility
application wind turbine generators together
with their associated utility interface equipment. An algorithm is developed and applied
to a wind farm with a special ac/dc/ac interface currently under design. The effects of
various wind turbine/interface forced outage
rates on the expected annual energy output of
the farm is examined.

5.2.1(34) WIND-DRIVEN PERMANENT MAGNET
ALTERNATORS HAVING APPRECIABLE INDUCTANCE:
SOME ASPECTS OF STEADY-STATE BEHAVIOUR AND
CONTROL.
de Paor, A.N. (University College, Dublin, Dept.
of Electrical Engineering, Dublin, Ireland)
Applied Energy, 16 (3) 1984 pp. 163-174.

The Author's earlier analysis is extended
by the inclusion of leakage inductance in the
alternator. The load resistance for optimal
steady-state power transfer now becomes a
double-valued function of mill angular velocity,
and beyond that it increases linearly. In
the latter range the inductive reactance prevents the mill from delivering the maximum
power of which it is aerodynamically capable.
Consideration of a simple scheme in which the
load resistance is fixed reveals the possibility
of hysteretic jumps in the curve of angular
velocity versus windspeed.

5.2.1(35) COMMERCIAL WIND ENERGY CONVERSION
SYSTEMS (WECS) MONITORING AND UTILITY IMPACTS
STUDY.
Finch, T. and Stafford, R.W. New York, USA,
Bronx-Frontier Dev. Corp. September 1981
167 pp. (NYSERDA-82-12) (PB83-173047)

Performance of a 25 kW wind turbine operating
at an industrial site in the Bronx was monitored
and evaluated, concentrating on the operation
of the equipment necessary for safe and reliable
utility interconnection. The turbine was
tested for a number of different generator/
power conditioning configurations and a number
of problems were encountered, including a
lightning strike and turbine overspeeds. The
system was interconnected to Consolidated
Edison's power grid to slow backfeed of excess
power and the quality of the backfeed was then
measured.

5.2.1(36) CONSTRUCTION, COMMISSIONING AND OPERATION OF THE 300 KW WIND TURBINE AT CARMARTHEN BAY.
Young, T. and others. (James Howden & Co. Ltd. and C.E.G.B.) In Wind Energy Conversion, proc. 5th BWEA Wind Energy Conference, Reading, U.K. 23-25 March 1983. P. Musgrove (ed.) pp. 296-302. Cambridge, U.K., Cambridge Univ. Press 1984 375 pp. (Isbn: 0-521-26250-X)
Discusses construction and operation of the 300 kW Carmarthen Bay wind turbine, the MPS-200, a three bladed upwind machine with synchronous generator. Describes CEGB safety rules and safety features of the design.

5.2.1(37) DISTURBANCES IN ELECTRICITY SUPPLY NETWORKS CAUSED BY A WIND ENERGY SYSTEM EQUIPPED WITH A STATIC CONVERTER.
Looijesteijn, C.J. In Wind Energy Conversion, proc. 5th BWEA Wind Energy Conference, Reading, U.K. 23-25 March 1983. P. Musgrove (ed.) pp. 303-317. Cambridge, U.K., Cambridge Univ. Press 1984 375 pp. (Isbn: 0-521-26250-X)
Describes measurements taken at the Netherlands Energy Research Foundation (ECN) Petter, on the 25 m horizontal axis wind turbine and DC generator/static DC to AC converter. Reasons for selecting this conversion system include: variable rotation speed of the rotor shaft, constant and high power factor for varying wind speed, and reduced torque fluctuations.

5.2.1(38) MEASURED INTERCONNECTED BEHAVIOUR OF WIND TURBINE INVERTERS.
Park, G.L. and Miller, J.M. (Michigan State Univ., East Lansing, Michigan, USA) IEEE Trans. Power Appar. Syst., PAS-103 (10) October 1984 pp. 3074-3079.
Distribution engineers are concerned about harmonic currents injected in the distribution network by switching inverters and load controls. When an inverter is driven by a random power source such as a windmill, the effects may be exacerbated due to power output variations due to the source as well as the cycle by cycle distortion due to switching semi-conductors in the inverter. During a performance measurement programme on nearly 20 consumer-owned wind turbines in Michigan, the harmonic and voltage behaviour of several single-phase line-commutated inverters supplied with variable-speed wind-turbine generators were recorded.

5.2.1(39) NEW AUTOMATIC GENERATION CONTROLLER FOR A WIND-DRIVEN SLIP-RING INDUCTION GENERATOR.
Velayudhan, C. and Bundell, J.H. (Univ. of Western Australia, Dept. of Electrical & Electronic Engineering, Nedlands, Aust.) IEEE Proc. 72 (9) September 1984 pp. 1226-1229.
Presents a new scheme for controlling the stator power output of a system-connected slip-ring induction generator by electronic variation of its rotor resistance. The application of this scheme in overcoming major operational difficulties experienced by wind turbine generators operating in parallel with a small power grid is recommended.

5.2.1(40) OPERATIONAL EXPERIENCE, MP-200 WIND TURBINE DESIGNS.
Rose, M.B. (WTG Energy Syst. Inc.) In Intl. Wind Energy Symposium 5th Annual Energy-Sources Technology Conference, New Orleans, USA 7-10 March 1982 pp. 23-24. New York, American Soc. Mechanical Engineers 1982.
Describes developments of the MP-200 wind turbine based on the Gedser machine. The mechanical system consists of a three bladed steel rotor positioned upwind of the tower. Two commercial units are in operation at Wreck Cove, Nova Scotia and at the Whiskey Run site on the Pacific Coast. Design improvements on the prototype include increased tip area and tips designed as proportional torque control devices. The MP-200 also includes an asynchronous generator.

5.2.1(41) OVERVIEW OF WIND TURBINE TECHNOLOGY FOR UTILITY APPLICATIONS.
Goodman, F.R. (Electr. Power Research Inst.) In Intl. Wind Energy Symposium 5th Annual Energy-Sources Technology Conference, New Orleans, USA 7-10 March 1982 pp. 299-302. New York, American Soc. Mechanical Engrs. 1982.
Discusses the technological status and prospects of wind turbines for electric utility applications. A summary of utility activities is given and major projects are described in more detail. The wind power programme of the Electric Power Research Institute is discussed.

5.2.1(42) RURAL ELECTRIC UTILITY CONCERNS FOR THE EFFECTIVE USE OF WIND POWER.
Prichett, W. (Nat. Rural Electr. Coop. Assoc.) In Intl. Wind Energy Symposium 5th Annual Energy-Sources Technology Conference, New Orleans, USA 7-10 March 1982 pp. 61-90. New York, American Soc. Mechanical Engineers 1982.
Presents a list of known interconnected small power producers and co-generation on U.S. rural power lines in 1982. Most are small wind 1 kW - 50 kW generators. Lists rural electric utilities wind energy research and demonstration projects, giving brief details of the scope and progress of each project. Notes that major problems are the effective use of wind power for utility power supply purpose and interconnection of private power producers' wind machines.

5.2.1(43) SMALL WIND-ENERGY CONVERSION SYSTEMS.
Twidell, J. and Grainger, B. Electronics and Power, 30 (4) April 1984 pp. 285-289.
The system includes generator, controller, distribution lines, energy store and end use. Estimates of the number of machines up to 100 kW in use worldwide are given. The mechanical requirements are discussed, and recent research on electric generators and controls is described. Recent development programmes are reported particularly those in Denmark and the U.S.A.

5.2.1(44) WIND ELECTRIC CONVERSION AND
UTILISATION.
Ramakumar, R. (Oklahoma State Univ.,
Stillwater, OK., USA) In Solar Electric
Systems 1984 edited by G. Warfield pp. 15-36.
Washington, USA, Hemisphere Publ. Corp. 1984
(SBN: 0 89116 327 1)
   Energy storage and reconversion options and
other utilisation aspects are discussed,
followed by a simplified approach to assess
the economics of WECS. Research and development needs, and some of the key technical,
operational, and economic problems involved
in the entry of WECS into the energy systems
of the future are also examined. (48 refs.)

5.2.1(45) WIND POWER PARKS: 1983 SURVEY.
Palo Alto, USA, Electr. Power Research
Institute August 1984 176 pp. (Report no:
EPRI/AP-3578)
   Summarises results of an EPRI survey on
wind power parks. 85 wind parks provided
information on installed generation, expansion
plans and total electric energy generated.
Aggregate installed generation was 87 mW,
and most of the parks are in California, where
utilities are able to take on the wind
generated electricity. 30 turbine manufacturers
were represented in the survey. However, the
need for demonstrations of reliability of wind
turbine operation is apparent.

5.2.1(46) WIND TURBINE CLUSTER MODEL.
Chan, S.M. and others. (Systems Control Inc.,
CA., USA) IEEE Trans. Power Apparatus & Systems,
103 (7) July 1984 pp. 1692-1698.
   The Cluster Model analyses the effects of
wind turbulence and tower shadow on voltage
fluctuations at the substation bus and neigh-
bouring nodes. External network disturbances
such as faults and line switching can also be
simulated. (14 refs.)

5.2.1.1 ENERGY STORAGE MECHANISMS FOR WECS
ELECTRIC POWER GENERATION.

5.2.1.1(1) WIND ENERGY SYSTEM UTILISING HIGH
PRESSURE ELECTROLYSIS AS A STORAGE MECHANISM.
Allison, H.J. (Oklahoma State University,
School of Electr. Engng., Stillwater, USA)
In 1st World Hydrogen Energy Conference, Miami
Beach, Florida 1-3 March 1976. Coral Gables,
Florida, USA, University of Miami 1976 V. 2
Session 2B pp. 3-14.
   The technique of energy storage which utilises
electrolysis cells to disassociate water into
its component gases, then stores the evolved
hydrogen as a high pressure gas, liquid, or
hydride, has been widely discussed for several
years. Describes the system, presents performance parameters for those components of the
system which have reached the prototype stage,
and discusses the basic problems and economics
which must be satisfied before such a system
can become practical on a large scale. (7 refs.)

5.2.1.1(2) CONTROL AND DYNAMIC ANALYSIS OF A
WIND ENERGY CONVERSION AND STORAGE SYSTEM
OPERATING AT CONSTANT VELOCITY RATIO.
Simkovits, H.R. and Kassakian, J.G. (MIT,
Electr. Power Syst. Engng. Lab., Cambridge,
Massachusetts) Energy Dev. (New York), 3 IEEE
Power Eng. Society Paper (77CH1215-3-PWR).
New York, NY., IEEE 1977 pp. 48-55.
   Lead-acid batteries are used for energy
storage and consideration is given to the number
of battery sections required to produce
efficient operation of the windmill. A charge
control algorithm is developed and the system
energy extraction efficiency calculated. System
dynamics caused by both windspeed transients
and battery switching are investigated.

5.2.1.1(3) ANALYSIS OF REMOTE SITE ENERGY
STORAGE AND GENERATION SYSTEMS.
Crisp, J.N. and others. (University of Dayton,
School of Engineering, USA) NTIS Report
AD-A074 869/9 July 1979 148 pp.
   Presents the results of an investigation of
energy storage systems and alternative energy
sources for remote site applications. Energy
storage systems and converters studied included
hydrogen storage, thermal storage, batteries,
flywheels and wind turbines. Describes the
design and expected operating performance of an
energy system composed of separate 8 kW wind
turbine modules and a lead-acid battery
storage unit that would operate with an existing power grid system at Bar Main, Alaska.

5.2.1.1(4) APPLICATION OF WIND ENERGY TO A
SYSTEM WITH AN INHERENT ENERGY STORAGE MEDIUM.
Kobylarz, T. and Al-Shehri, A. (University of
Pet. & Miner., Dhahran, Saudi Arabia) In proc.
14th Intersociety Energy Conversion Engng.
Conference, Boston, Massachusetts 5-10 August
1979. Washington, DC, American Chem. Society
1979 pp. 312-318.
   Economics has severely limited the exploitation of wind energy to a few special situations.
Thus a tremendous energy source is wasted. One
of the most costly aspects of a wind energy
system is that of energy storage. It is suggested
that wind energy be utilised by systems having an
inherent storage medium and one such system is
described. (18 refs.)

5.2.1.1(5) HIGH TEMPERATURE STORAGE FOR A
WIND ENERGY SYSTEM.
Ramshaw, R. and Bowman, D. (University of
Waterloo, Ontario, Canada) In Alternative
Energy Sources 2, proc. 2nd International
Conference V. 1. Solar Energy, Miami Beach,
Florida, USA 10-13 December 1979. Washington,
DC, USA, Hemisphere Publ. Corp. 1981
pp. 453-458.
   A wind energy system design was developed for
supplying supplementary energy to a space heating load. Availability of wind energy and
load demand do not always coincide, so storage
is necessary. High temperature high energy
density storage is well suited when the energy
is transferred by electric generation. This
storage technology has been developed for the

design of off-peak heat-storage furnaces and may be applied to the wind energy system. (2 refs.)

5.2.1.1(6) RESIDENTIAL FLYWHEEL WITH WIND TURBINE SUPPLY.
Place, T.W. (Airesearch Manufacturing Co. of California, Torrance, USA) Paper presented at Department of Energy Mechanical & Magnetic Energy Storage Contractors 1979 Review Symposium, Washington, DC August 1979 pp. 287-293.
  Describes a flywheel system that stores energy from a wind turbine source and converts the energy to a 60 HZ, 220 V output for residential use. A cost-benefit analysis of using this flywheel system for a 1500 sq.ft. residence and a maximum wind turbine power level of kW was made. The system and associated components meet the objectives of low acquisition cost, safety and reliability, and high efficiency. (4 refs.)

5.2.1.1(7) ENERGY STRATEGIES TOWARD A SOLAR FUTURE (ENERGY STORAGE AND TRANSMISSION).
Union of Concerned Scientists Report 1980 pp. 201-255.
  Because of the intermittent nature of solar and wind energy, storage will be an essential feature of a renewable energy economy. Thermal, thermochemical, electrochemical, battery, and mechanical energy storage systems are discussed. Technologies evaluated include compressed air, pumped storage, flywheels, and superconducting magnets. (116 refs.)

5.2.1.1(8) STORING WIND ENERGY AS A LIQUID FUEL.
Miller, D. (Haverford College, USA) In proc. 2nd British Wind Energy Association Wind Energy Workshop, Cranfield, U.K. 17 April 1980. London, U.K., Multi-Sci. Publishing Co. Ltd. 1980 pp. 143-150.
  Formic acid is synthesised in an electrolytic reactor using inexpensive catalysts. The vertical axis wind turbine has high solidity and fixed, straight blades. Its low aspect ratio contributes to a power coefficient of 0.43. Power is transmitted hydraulically, and the generator is a modified Faraday disk. The turbine operates at maximum power over the full range of wind speeds, because the area of the reactor electrodes is varied.

5.2.1.1(9) ADVANCED BATTERIES FOR ENERGY STORAGE.
Jensen, J. and Tofield, B.C. (Odense University, AERE Harwell) In proc. International Conference on Energy Storage, Brighton, U.K. 29 April-1 May 1981 V. 1 Cranfield, U.K., BHRA Fluid Engng. 1981 Paper L2 pp. 205-216.
  Rechargeable electric batteries offer considerable promise in energy storage systems, in particular for direct load levelling within the electrical utility network, for indirect load levelling and fuel substitution using electric vehicles, and for storage of renewable energy inputs, particularly solar energy. A

review of these applications is given with emphasis on applications in Europe of electric vehicles and solar energy storage. Battery performance specification are discussed and compared with current engineering developments. New data on battery storage in combination with photovoltaic and wind energy systems in Northern Europe are given.

5.2.1.1(10) ANALYSIS OF BATTERY STORAGE IN WIND-ENERGY SYSTEMS FOR COMMERCIAL BUILDINGS.
Caskey, D.L., Broehl, J. and Skelton, J. (Sandia National Labs., Albuquerque, NM,; Battelle Columbus Labs., OH, USA) US Department of Energy Report no: SAND-81-7171 30 September 1981 52 pp. (DE 82 010875)
  Using the SOLSTOR programme, life-cycle energy cost and performance measures were calculated for different wind turbine and storage capacity levels. The analyses focused on Dodge City (average wind speed of 5.8 m/s) and Washington, DC (wind speed 2.9 m/s). Levelised system costs are computed for warehouse and office applications. To assess the sensitivity of the system performance measures and cost, two series of sensitivity tests were performed. The first determined the increase in system cost for an increase of storage capacity, and the second examined the effect of doubling the battery cost for the office building application.

5.2.1.1(11) ENERGY STORAGE AND THE ENERGY SYSTEMS.
Kurti, N. (University of Oxford, U.K.) Paper presented at CEC Thermal Energy Storage Symposium, Ispra, Italy 1-5 June 1981 pp. 3-20.
  Energy storage systems form an important part of energy strategies as storage enhances the use of renewable energy sources and permits the more efficient use of high capital cost conversion equipment. Energy storage applications for wind, solar, geothermal, tidal, and other forms of renewable energy are considered. (3 refs.)

5.2.1.1(12) RECHARGEABLE ALKALINE ZINC/FERRI-CYANIDE HYBRID REDOX BATTERY.
Adams, G.B., Hollandsworth, R.P. and Littauer, E.L. (Lockheed Palo Alto Research Labs., California, USA) Paper presented at 16th ASME Intersociety Energy Conversion Engineering Conference, Atlanta, GA 9-14 August 1981 V. 1 pp. 812-816.
  Describes a zinc/ferricyanide battery which can be used for utility load levelling and solar photovoltaic/wind applications. One advantage of such a battery system is its high level of electrochemical performance. Mean prices for such a system are estimated. (13 refs.)

5.2.1.1(13) SMALL SODIUM-SULFUR BATTERY FOR SOLAR AND WIND ENERGY SYSTEMS.
Haskins, H.J. and Domaszewicz, A.G. (Ford Aerosp. & Commun. Corp., Newport Beach, California, USA) In proc. 16th Intersociety

Energy Conversion Conference V. 1 Atlanta, GA, USA 9-14 August 1981. New York, NY, USA, ASME 1981 pp. 836-840.

Presents a conceptual design of a 1 MWh sodium-sulphur storage battery. The battery is intended for use in small (15 kW), stand-alone solar or wind electrical power systems. The design incorporates approximately 1400 sodium-sulphur cells of a new, high energy capacity configuration. (3 refs.)

5.2.1.1(14) STATIONARY FLYWHEEL ENERGY STORAGE SYSTEMS.
Gilhaus, A. and others. (Maschinenfabrik Augsburg-Nuremberg Aktiengesellschaft, Munich) NTIS Report PB82-105006 1981 161 pp.
(In German)

Study of industrial applications for stationary flywheel energy storage systems. Their economic value for the consumer and their effect on the power supply grid are investigated. Large-scale application provided savings in energy costs that were small compared to investment costs and the systems may prove to be profitable in settings where energy may be lost rather than stored. (Numerous refs.)

5.2.1.1(15) ENERGY STORAGE OPTIONS FOR HARNESSING WIND ENERGY.
Ramakumar, R. (Oklahoma State University, Stillwater, USA) New York, USA, American Society Mechanical Engineers 7-10 March 1982 8 pp.
(ASME Paper no: 82-Pet-9)

Presents a brief survey of the technical aspects of the various energy storage and reconversion options proposed for use with wind energy conversion systems. Energy can be stored in mechanical, chemical, thermal, or electrical form and the schemes discussed include hydrofilming, compressed air, flywheels, hydrogen, secondary batteries, and thermal storage.

5.2.1.1(16) MATHEMATICAL PROGRAMMING MODELS FOR THE ECONOMIC DESIGN AND ASSESSMENT OF WIND ENERGY CONVERSION SYSTEMS.
Reipert, K.A. (Granville Corp., Washington, DC) Wind Engng., 7 (1) 1983 pp. 43-50.

System reliability is one of the important determinants of the economic and technical feasibility of wind energy conversion systems. The inclusion of storage facilities into these systems greatly increases their reliability. Recognising this, two inter-period mathematical programming models are formulated which optimise the design of a wind energy conversion system, locating wind turbines in a number of distinct arrays, sizing a storage facility, providing rules for the operation of the storage facility, and ensuring that a pre-specified level of demand is met. In both models, the storage facility is modelled via linear decision rules.

5.2.1.1(17) STORAGE OPTIONS FOR HARNESSING WIND ENERGY.
Ramakumar, R. (Oklahoma State University, Stillwater, OK, USA) Mech. Engng., (USA) 105 (11) November 1983 pp. 74-83.

Schemes discussed include hydrofilming, compressed air, flywheels, hydrogen, secondary batteries and thermal storage. Presents a review of various energy storage and reconversion options that are possible or have been proposed for use with wind energy conversion systems. (30 refs.)

5.2.1.1(18) WIND-ENERGY BATTERY-STORAGE PROJECT. PHASE I. FINAL REPORT.
Neill, D.R. and Curtis, G.D. (Sandia National Labs., Albuquerque, NM) Washington, DC, USA, US Dept. of Energy Report no: SAND-82-7211 January 1983 100 pp. (DE 83 009389)

Describes a programme involving data collection, development of a computer model, and use of the data in the model to simulate use of the WEBS in a specific locality. The data collection involved hourly wind samples, load data, and performance and cost information on equipment. The model enables the user to call these data from files, configure the system, and obtain performance measures and operational statistics. Comparative economic analyses on probable configurations indicate that an array of wind machines would be cost-effective but that battery storage would be economically marginal unless it can interact usefully with the utility. Batteries do however contribute to better system operation.

5.2.1.1(19) ECONOMICS OF SMALL MULTI-SOURCE POWER SYSTEMS.
Vermij, L. and Meijer, B.J. (Holec Nederland B.V., Utrecht, Netherlands) In 4th International Conference on Energy Options. The Role of Alternatives in the World Energy Scene, London, U.K. 3 6 April 1984 pp. 17-20. London, U.K., IEE 1984 421 pp. (SBN 0-85296-290-8)

A useful application of solar or wind energy often requires a transformation of an energy source to a power source by applying some form of energy storage. In many cases such a system requires a small energy storage to absorb fast changes in supply or demand. The major criterion for judging the economics of the different types of systems is the price per kWh to be paid by the end-user. The authors discuss smaller systems (with power ratings up to a few hundred kW), suited for rural electrification in areas that lack public supply.

5.2.2 COSTS AND ECONOMICS OF WECS POWER GENERATION FOR ELECTRIC UTILITIES.

5.2.2(1) ECONOMIC POTENTIAL FOR WIND ENERGY CONVERSION.
Dubey, M. and Coty, U. (Lockheed-California Co., Burbank) In proc. Greater Los Angeles Area Energy Symposium, California 3 April 1975 pp. 112-121. Publ. by West Period Co. (Los Angeles Council of Eng. and Sci., Proc. Ser. V. 1) North Hollywood, California 1975.

Aims to prove that wind-energy conversion is economically competitive with conventional systems, compatible with the user's applications, and acceptable to the public. An approach is suggested which may succeed in defining the potential market and thus portend the birth of a new industry. Topics discussed are the availability of wind as an energy source, some wind-energy conversion system concepts, potential applications of wind power, and land-use considerations. (6 refs.)

5.2.2(2) IMPLEMENTATION ISSUES OF WIND ENERGY. Coty, U. and Vaughan, L. (Lockheed-California Co., Burbank) In New Options in Energy Technol., AIAA/EEI/IEEE Conference, San Francisco, California 2-4 August 1977. New York, NY, USA, AIAA 1977 pp. 97-105.

First is determined the effect of initial production quantity on the selling price of a two megawatt wind turbine generator. To this is added the effect of loan interest rates, taxes, and other annual operating expenses. The cost of electrical energy generated is determined and compared to the cost of conventional fuel for private utilities and to the cost of buying energy wholesale for public utilities. From this comparison, the initial production parameters of wind turbine generators is determined which would bring cost of wind energy down to a competitive level. (3 refs.)

5.2.2(3) WIND ENERGY CONVERSION SYSTEMS (WECS) FOR CENTRAL STATION AND DISPERSED POWER APPLICATIONS. Kornreich, T.R. and Tompkins, D.M. (JBF Science Corp., Arlington, VA, USA) In Alternative Energy Sources, Miami International Conference, International Compendium, Miami Beach, Florida 5-7 December 1977. Washington, DC, Hemisphere Publ. Corp. 1978 V. 4 pp. 1865-1886.

Gives an analysis of the current economic status of large wind energy systems for power station use. Regional wind energy assessments with a view toward establishing a common basis for comparing their results and establishing a reasonable estimate of the current state of knowledge concerning large WECS economics. The study data are normalised to the same basis and estimates of capital cost and energy cost are provided. The economics of several dispersed power applications of small WECS are also considered. The applications considered may be distinguished by the provisions made for WECS energy storage and the source of energy against which the WECS is competing. Wind energy systems with battery storage are analysed for rural and remote residences. The WECS with synchronous inverter application for linking a rural residence to the utility network when the WECS output exceeds the residential energy demand, is also analysed. Unconstrained demand applications using WECS without provisions for energy storage are also investigated. (17 refs.)

5.2.2(4) RELIABILITY PLANNING IN DISTRIBUTED ELECTRIC ENERGY SYSTEMS. Kahn, E. NTIS Report LBL-7877 October 1978 57 pp.

Discusses the reliability planning requirements in utility power systems based on conventional power technologies with controllable output and in utility systems based on intermittent resources, such as wind and solar energy. Significant differences in the reliability planning requirements and costs for the various systems are identified. (38 refs.)

5.2.2(5) ASSESSMENT OF WIND ENERGY SYSTEMS IN A UTILITY FRAMEWORK. Macklis, S.L. and Oplinger, J.L. (GE, Valley Forge, Philadelphia, USA) In proc. 14th Intersociety Energy Conversion Engineering Conference, Boston, Massachusetts 5-10 August 1979. Washington, DC, USA, American Chemical Society 1979 pp. 319-324.

Describes the work done under two contracts sponsored by the Electric Power Research Institute. Both studies were concerned with the technical and economic aspects of wind energy use in a utility grid. The first project, "Requirements Assessment of Wind Energy Systems", concentrated on the methodology for assessing the benefits and possible barriers to wind energy application without concern for specific siting or distribution of the WES in the grid. The second project, "Assessment of Distributed Wind Power Systems", is concerned with the impact of dispersed versus centralised wind power facilities and the comparison of small versus large machines. (4 refs.)

5.2.2(6) A FEASIBILITY STUDY OF WINDPOWER FOR THE NEW ENGLAND AREA. Martin, P. and others. (SRI International, California) NTIS Report AD-A076 614, October 1979 244 pp.

Examines the applicability of large-scale windpower energy conversion systems as an alternative source of energy in the New England area in general, and in the U.S. Navy Portsmouth Shipyard at Kittery, ME. Parametric economic analyses led to the conclusion that WECS electricity can be at economic parity with electricity from conventional generators, provided that the annual utilisation factor for the WECS is at least 50%. Such utilisation appears to be achievable at heights of 620-1240 m. Thus the environmental impact of a WECS towers and other equipment was also investigated.

5.2.2(7) REQUIREMENTS ASSESSMENT OF WIND POWER PLANTS IN ELECTRIC UTILITY SYSTEMS. FINAL REPORT AND APPENDICES. Marsh, W.D. (General Electric Co., NY, USA) Report no: EPRI-ER-978 V. 1-3 January 1979 339 pp., 98 pp. (Electric Power Res. Inst.)

A requirements assessment and preliminary impact and penetration analyses are included. Evaluations are based on the comparison of utility generation system costs with and without wind plants and the energy requirements that could be replaced by the operation of the wind plants. Conventional utility loss of local probability and production simulation methods were used in conjunction with a wind turbine generator performance model.

5.2.2(8) UTILITY ROLES IN IMPLEMENTING COMMUNITY RENEWABLE ENERGY SYSTEMS.
Page, A.C. and Mitsock, M.J. (Mitre Corp. MA Inc., USA) Paper presented at US Department of Energy/Solar Energy Research Institute Community Energy Self-Reliance Conference, Boulder, Colorado 20-21 August 1979 pp. 424-433.
   Summary of issues and options that have been identified regarding utilities' roles in implementing community-scale renewable energy systems. Utility buy-back power of small-system power producers is discussed and direct utility participation in promoting distributed energy systems considered. Sale of back-up power to the independent system is reviewed.

5.2.2(9) WIND ENERGY CENTRAL STATION POWER GENERATION IN THE SOUTHWEST AND SOUTHEAST.
Guild, D.H. (Stone & Webster Engng. Corp., USA) In proc. Workshop on Economic and Operational Requirements
Large Scale Wind Systems, Monterey, USA 28-30 March 1979. Palo Alto, USA; Electr. Power Research Institute 1979 pp. 158-170.
(Special Report no: ER-1110-SR and US DOE Conference Report no: 790352.)
   Emphasis is on the integration of wind power conversion systems into conventional power stations on the electricity grids. Costs of integration and operational economics are considered.

5.2.2(10) WIND ENERGY FROM A UTILITY PLANNING PERSPECTIVE.
Butler, N.G. (Bonneville Power Adm., Portland, Oregon) In proc. of Conference and Workshop on Wind Energy Characteristics and Wind Energy Siting, Portland, Oregon 19-21 June 1979. Richland, Washington, Pacific NorthWest Lab. Technical Report no: PNL 3214 1979 pp. 305-310.
   Describes one method a utility could use to assess the potential for wind generation. The primary objective would be to extract the maximum amount of energy per year at the lowest cost, possibly 2-3 cents per kW.

5.2.2(11) WIND ENERGY SYSTEMS. APPLICATION TO REGIONAL UTILITIES.
JBF Scientific Corp., Wilmington, MA. US Dept. of Energy Report no: DOE/ET/20063-T1 (Exec.Sum.) June 1979 40 pp.
   This study developed a generic planning process that utilities can use to determine the feasibility of WECS as part of their future mix of equipment. While this is primarily an economic process, other questions dealing with WECS availability, capacity credit, operating reserve, performance of arrays, etc. are dealt with. (See also V. 1 of this Report DOE/ET/ 20063-T1 (V.1) for an economic impact assessment of WECS.)

5.2.2(12) THE ECONOMICS OF ELECTRICITY FROM COAL, NUCLEAR AND WIND ENERGY.
Baker, A. and Prior, M. IEA Coal Research 1980 64 pp. (Report no: IEA EAS H1/78)

Seeks to illuminate the relative economics, and the part played by methodological conventions in determining the economics of electricity produced from coal, nuclear and wind generating plants. The report considers single plants, and does not seek to give guidance on the choice of plant within an electricity system, though the principles of electricity system analysis are discussed.

5.2.2(13) ESTIMATING SIZES AND OUTPUTS FROM WIND ENERGY SYSTEMS.
van Leersum, J. (Commonwealth Scientific & Industrial Research Organisation) Inst. Engrs. Aust. Electr. Engng. Trans., EE 16 (3) September 1980 pp. 120-127.
   Gives a simple method for evaluating the annual energy output from wind turbo generators connected into an existing electrical grid. A model for determining the relationship between load fraction supplied by wind and storage size is developed for wind energy conversion systems containing energy stores and auxiliary power supplies.

5.2.2(14) INTEGRATION OF WIND POWER INTO AUSTRALIAN ELECTRICITY GRIDS WITHOUT STORAGE: A COMPUTER SIMULATION.
Diesendorf, M. and Martin, B.(Commonwealth Scientific & Industrial Research Organisation; Australian National University, Canberra) Wind Engng., 4 (4) 1980 pp. 211-226.
   A computer simulation is performed of the operation of two conventional fuel-based electricity grids in South and Western Australia with zero storage and added hypothetical wind power capacity. It is found that wind energy contributions of 20 percent (Western Australia) and 30 percent (South Australia) of the annual grid energy output can be achieved before the losses of unutilised wind energy increase to 20 percent of wind energy generation.

5.2.2(15) A REGIONAL WIND-HYDRO ELECTRICITY SUPPLY SYSTEM.
Sorensen, B. (Copenhagen University) In proc. 3rd International Symposium on Wind Energy Systems, Lyngby, Denmark 26-29 August 1980. Cranfield, U.K., BHRA Fluid Engng. 1980 Paper L1 pp. 533-543.
   The combination of a wind-based electricity generating system with a hydro-based system is investigated by use of a simulation model. The hydro-power system is an idealised version of the Norwegian system, and the wind-power system corresponds to generating all the electric power needed in Denmark by wind. A transmission capability between the two countries is assumed to allow a power transfer up to 2.5 times the maximum capacity of the present interconnections. The model has been used to simulate the performance of such a system, using wind and hydro data for the respective countries. The combined system is as dependable as the hydro system alone, and the combination thus offers a way of large-scale electricity supply by use of wind energy. The extra trans-

mission losses associated with the power exchange between the wind and the hydro system amounts to 8.9% of the wind energy production.

5.2.2(16) A SIMULATION MODEL OF AN ELECTRICITY GENERATING SYSTEM INCORPORATING WIND TURBINE PLANT.

Whittle, G.E. and others. (Reading University; Rutherford Lab., U.K.) In proc. 3rd International Symposium on Wind Energy Systems, Lyngby, Denmark 26-29 August 1980. Cranfield, U.K., BHRA Fluid Engng. 1980 Paper L2 pp. 545-554.

Describes a simulation model of an electricity generating system incorporating wind turbine plant appropriate to the Central Electricity Generating Board grid system. Wind plant ratings of up to 25 GW - 60% of maximum system demand are considered. The effect of the fluctuating wind on the output of the plant is identified as well as the amount of wind energy lost due to short term excess of wind power. The use of steam plant spinning reserve and gas turbine power to compensate for uncertainties in wind power availability is studied.

5.2.2(17) SYSTEM ECONOMIC THEORY FOR WECS.

Rockingham, A.P. (C.E.G.B.) In proc. 2nd British Wind Energy Association Wind Energy Workshop, Cranfield, U.K. 17 April 1980. London, U.K., Multi-Sci. Publishing Co. Ltd. 1980 pp. 109-117.

When wind energy conversion systems are used as electricity producers for an integrated electrical utility system, their value is determined not only by their own output characteristics, but also by the characteristics of other plant in the system and the shape of the system load curve. Outlines a cohesive framework for the analysis of the value of WECS if they were to be used as part of a large electricity grid such as that in the U.K.

5.2.2(18) WIND ENERGY - A UTILITY PERSPECTIVE.

Fung, T.K., Scheffler, R.L. and Stolpe, J. (South California Edison Co., Rosemead) In IEEE Power Engng. Society Summer Meeting, Minneapolis, Minnesota 13-18 July 1980. Publ. by IEEE Power Engng. Society, Piscataway, NJ 1980 Paper no: 80-SM-564-5 7 pp.

Of all the renewable energy systems, wind turbine generators are likely to make the earliest significant cost-effective contribution to the electrical utility grid. Presents some of the issues facing the utility industry as they relate to implementation of an effective wind energy programme. In addition, the Southern California Edison Company's wind energy programme is presented, which is designed to provide pertinent answers to the technical, economic and environmental issues concerning WTG installations and their commercial viability as a future generation resource. (6 refs.)

5.2.2(19) WIND-ENERGY-CONVERSION SYSTEM ANALYSIS.

Smith, L. (Southern Solar Energy Centre, Atlanta, GA) Southern Solar Energy Centre Report no: SSEC/SP-31164 October 1980 6 pp. (US Dept. of Energy Report no: DE 82 009092.)

Presents a life cycle cost analysis of a 2 kW wind energy conversion machine for Charleston, South Carolina. The total cost of the machine minus tax credits is analysed and includes operation and maintenance. A total cost per year is finally produced and a summation of total costs is presented.

5.2.2(20) WIND ENERGY SYSTEMS WITH BATTERY STORAGE AND DIESEL BACK-UP FOR ISOLATED COMMUNITIES.

Slack, G. and others. (Reading University, U.K.) In proc. 2nd British Wind Energy Association Wind Energy Workshop, Cranfield, U.K. 17 April 1980. London, U.K., Multi-Sci. Publishing Co. Ltd. 1980 pp. 134-142.

Asserts that a wind turbine with battery storage and diesel generator back-up may be regarded as the standard wind energy system for supplying electricity to remote, low level consumers. Describes a computer model of such a system used to investigate the effect of windmill size/battery capacity mix, under various economic conditions, using a year's hourly mean wind speed data for a site in the Scilly Isles.

5.2.2(21) ASSESSMENT OF TECHNICAL POTENTIAL: PROBLEMS, METHODS AND RESULTS.

Obermair, G. (Regensburg University) In Implementing Agreement for Programme of Research & Development on Wind Energy Conversion Systems, 1981 Meeting of Experts of Annex III & IIIa: Integration of Wind Power into National Electricity Supply Systems, Regensburg, Fed. Rep. Germany 29-30 January 1981. Julich, Fed. Rep. Germany, Kernforschungsanlage Julich GmbH April 1981 Session A Paper A2 pp. 17-27 (Jul/Spez-108).

Summarises technical and economic problems related to integrating wind energy into electricity grids. Natural potential is reduced to 'Technical Potential', which is transformed to an 'Economic Potential', using a technique involving competitiveness with other resources, comparative social costs and socio-economic preferences, and the actual cost and existing use of land for wind sites.

5.2.2(22) THE CAPACITY CREDIT OF WIND POWER: A THEORETICAL ANALYSIS.

Haslett, J. and Diesendorf, M. (Trinity College, Dublin; Commonwealth Scientific & Industrial Research Organisation, Australia) Solar Energy-26 (5) 1981 pp. 391-401.

A probabilistic model of capacity credit is constructed for wind power in an electricity grid, based on the assumptions that electricity supply, demand and output of a system of wind generators are normally distributed. Two concepts of wind power capacity credit, the equivalent conventional capacity, and the equivalent firm capacity are analysed and loss of load probabilities are also computed. (14 refs.).

5.2.2(23) ECONOMIC ANALYSIS OF SMALL WIND-ENERGY-CONVERSION SYSTEMS FOR TEN MASEC SITES.
Mid-American Solar Energy Complex, Minneapolis, MN. Washington, DC, USA, US Department of Energy. Report no: MASEC-R-81-071, W-101-4 September 1981 304 pp. (DOE Report no: DE 82 003976)
The economic feasibility of Small Wind Energy Conversion Systems was evaluated for ten sites in the region served by the Mid-American Solar Energy Complex. A computer-based life cycle costing programme was utilised in the economic evaluations. The wind energy model is an interactive computer tool that may be used to perform economic analyses for many different scenarios, and includes a package of computer routines that create an interactive computer environment. The programme deals with information specific to investing in a wind turbine and uses inform-ation on electrical rates, turbine performance, power demand, and wind data to form inter-mediate quantities which are fed to a life cycle costing programme.

5.2.2(24) THE ECONOMIC VALUE OF WIND POWER IN ELECTRICITY GRIDS.
Diesendorf, M., Martin, B. and Carlin, J. In proc. International Colloquium on Wind Energy, Brighton, U.K. 27-28 August 1981. Cranfield, U.K., BHRA 1981 pp. 127-132.
Wind power capacity tends to substitute for conventional base load capacity with the same annual average energy production. For quite a wide range of cost parameters, the dollar value of the net capital saving by wind power is comparable in magnitude with the dollar value of the fuel saving.

5.2.2(25) ECONOMICS OF WIND ENERGY FOR UTILITIES.
McCabe, T.F. and Goldenblatt, M. (JBF Sci. Corp.) In proc. Workshop on Large Horizontal-Axis Wind Turbines, Cleveland, USA 28-30 July 1981. R.W. Thresher (ed.) Washington, DC, USA, NASA 1982 pp. 783-801. (NASA Conference Public-cation 2230; DOE Publication nos: CONF-810752 & SERI/CP-635-1273.)
Presents preliminary results from a study to establish the economic value of central station wind energy to certain electricity utilities. The results for the various utilities are com-pared specifically in terms of wind resource, mix of conventional generation sources, and specific financial parameters including pro-jected fuel costs.

5.2.2(26) ECONOMICS OF DAWT WIND ENERGY SYSTEMS.
Foreman, K.M. (Grumman Aerosp. Corp.) In proc. 5th Biennial Wind Energy Conference & Workshop, Washington, DC, USA 5-7 October 1981. I.E. Vas (ed.) V. 1. Palo Alto, USA, Solar

Energy Research Institute 1981 Session 1D pp. 469-485. (Report nos: SERI/CP-635-1340 CONF-811043)
A diffuser augmented wind turbine (DAWT) preliminary design was investigated to assess the economic viability of its electrical energy generation. Unit costs were estimated for three output ratings and for three different construc-tion approaches. A limited production run of 100 to 500 units is considered for factory-built subassemblies and on-site final assembly and erection. Regional production centres are assumed within about 350 km of installation.

5.2.2(27) EFFECTIVE OUTPUT AND AVAILABILITY OF WIND TURBINES FOR HOUSEHOLD LOADS.
Feron, P. and Lysen, E.H. (Eindhoven University of Technology) Wind Engng., 5 (4) 1981 pp. 194-206.
Presents a theoretical model of a small-scale wind energy system producing electricity for a group of households connected to the grid. By means of a Weibull distribution of the wind velocity, a linear output model of the wind turbine and a measured load distribution function of a group of Dutch households, the average values of the electricity surplus and electricity deficit are calculated for the wind regime in Den Helder. The effective energy output and the power availability were useful descriptive parameters of the system.

5.2.2(28) THE IMPACT OF DECENTRALISED RENEWABLE ENERGY SYSTEMS ON ELECTRIC UTILITY COMPANIES.
D'Aquanni, R.T. and Oullette, D.L. (Chas. T. Main Inc., Boston, Massachusetts, USA) Paper presented at International Solar Energy Society-American Section/Et Al 1981 Annual Conference, Philadelphia 26-30 May 1981 V. 2 pp. 1310-1314.
Discusses the impact of residential adoption of decentralised solar technologies on electric utilities. Wind energy conversion systems are briefly considered and homeowner investment, tax credits, operating costs, and base fuel costs are examined. The energy displacement potential of solar space heating and wind energy systems are demonstrated with two case studies. (3 refs.)

5.2.2(29) INCORPORATION AND IMPACT OF A WIND ENERGY CONVERSION SYSTEM IN GENERATION EXPANSION PLANNING.
Schenk, K.F. and Chan, S. (University of Ottawa, Ontario, Canada) IEEE Power Engng. Society Summer Meeting Conference Paper, Portland, Oregon, USA 26-31 July 1981. Piscataway, NJ, USA, IEEE 1981 Paper no: 81-SM-419-1 9 pp.
WASP, a state-of-the-art computer model and the IEEE Reliability Test System are used in the evaluations. The optimal plans satisfy given levels of reliability, in terms of loss of load probability and capacity reserve margin, and minimise the present worth of capital and oper-ating costs minus a salvage value at the end of the current planning phase. The analyses were carried out with 75 mW and 150 mW wind energy conversion systems as expansion candi-dates and evaluating their impact on optimal

mix of capacity for displacement, fuel savings and capital and operating costs over the 1981-2010 planning period. As the results show, the wind energy conversion systems have a significant effect on optimal mix of generation and construction timing of conventional units. (7 refs.) (A)

5.2.2(30) MARKET EXPERIENCE WITH SMALL WIND ENERGY CONVERSION SYSTEMS (SWECS) USING CONJOINT ANALYSIS.
Heiko, L.K. and Nainis, W.S. (Renewable Energy Co.; Arthur D. Little Inc., USA)
In proc. International Colloquium on Wind Energy, Brighton, U.K. 27-28 August 1981. Cranfield, U.K., BHRA Fluid Engng. 1981 Session 4 pp. 213-218.
Personal interviews were performed to determine residential demographics, energy use, and sites; and to characterise the owners' wind machines. Interviews included ranking alternative SWECS using conjoint techniques. Useful results were obtained for orienting R & D programmes and for understanding SWECS interconnections to electric power grids.

5.2.2(31) WHO'LL WIN THE WIND RACE?
Swift-Hook, D.W. (British Wind Energy Association) Consulting Engineer, 45 (8) August 1981 pp. 10-11, 13, 15.
Discusses the prospects for electricity generation from wind energy.

5.2.2(32) WIND ENERGY.
Riaz, M. (University of Minnesota, USA)
Paper presented at Mid-American Solar Energy Complex/Et Al Utility Planning Symposium, Minneapolis 27-28 April 1981 pp. 158-177.
Reviews small wind energy electric conversion systems below the 100 kW capacity. Potential benefits to both customers and utilities are surveyed and wind characterisation techniques, generation models, economic models, and interface issues are discussed.

5.2.2(33) WIND ENERGY CONVERSION.
Oppedahl, C.J. and Tarduno, M.E. (Harvard University, USA) Harvard Env. Law Review, 5 (2) 1981 pp. 431-450.
The energy resource potential of wind and the operation of conversion systems are discussed. Major technical, economic, and environmental considerations associated with this technology are surveyed. Small systems for individual homes, medium-sized systems to provide power to neighbourhoods and collectives, and large-scale wind projects capable of providing energy for electric grids are considered. (References)

5.2.2(34) WIND POWER MAKES A BID FOR COMMERCIAL STATUS.
Chemical Week, 128 (19) 13 May 1981 pp. 76-78.
Many U.S. electric utilities are participating in the commercialisation of wind energy systems, an indication of the commitment by major power

suppliers to develop such new technology. By 2000, DOE expects wind-turbine generators to supply roughly 4% of U.S. power requirements. Only a fraction of potential wind energy is likely to be accessible, nevertheless, many industries are proceeding to develop and refine wind energy system components.

5.2.2(35) WIND TURBINE RESPONSE AND SYSTEM INTEGRATION.
Bossanyi, E.A. and others. (Reading University, U.K.) Presented at 3rd International Conference on Future Energy Concepts, London 27-30 January 1981 pp. 296-303. London, Institution of Electrical Engineers 1981.
Discusses the integration of wind energy systems into existing electricity grids. System simulation models depicting integration with solid fuel generating stations are described. Wind turbine responses to system integration are examined. Variability of output and wind turbulence spectrum factors are considered. (15 refs.)

5.2.2(36) AN ASSESSMENT OF WIND CHARACTERISTICS AND WIND ENERGY CONVERSION SYSTEMS FOR ELECTRIC UTILITIES.
DeWinkel, C.C. (Wisconsin University-Madison Inst. for Environmental Studies. Environmental Research Lab.-Duluth, MN) NTIS Report no: PB82-258971 June 1982 175 pp.
Evaluation of wind speed data from 12 airport sites in Wisconsin, Minnesota, Iowa and Illinois, and from five Coast Guard stations along Lakes Superior and Michigan, indicates annual average wind speeds of 4.5 to 6 m/s and wind power densities of 100 to 200 W/sq.m. at 7 m. height. The economic analysis of wind energy conversion systems applied to the Dairyland Power Cooperative (DPC) system indicates that it can be economically attractive for the DPC to install WECS in the 1980s. This analysis does not include benefits due to the potential replacement of conventional capacity by the WECS. A preliminary study of WECS in combination with directly controlled water heaters shows that this combined system may delay conventional generating capacity additions longer than will controlled heaters only.

5.2.2(37) ANALYSIS METHOD FOR NON-SCHEDULABLE GENERATION IN ELECTRIC SYSTEMS.
Moretti, P.M. and Jones, B.W. (Oklahoma State University and Kansas State University) Solar Energy, 28 (6) 1982 pp. 499-508.
Presents various methods that can be used to quantify the economic costs and benefits of modifications to an electric energy system, such as solar or wind power. A load-duration methodology to assess the economic impact of such modifications is developed, treating unschedulable generation as a negative load while obtaining a modified load duration curve. The benefits per unit of an individual alternative-energy device are greatest for small penetrations. (17 refs.)

5.2.2(38) ECONOMIC ANALYSIS OF SMALL WIND-ENERGY CONVERSION SYSTEMS.
Haack, B.N. (Ball State University, Muncie, Indiana, USA) Appl. Energy, 11 (1) May 1982 pp. 51-60.

The financial costs of obtaining electricity from small wind-energy conversion systems are calculated and compared with the cost of electricity from traditional utility grids. A 3 kW-rated wind electric system for residential use is examined. (12 refs.)

5.2.2(39) A FREQUENCY AND DURATION METHOD FOR THE EVALUATION OF WIND INTEGRATION.
Janssen, A.J. (Netherlands Energy Research Foundation, Petten) Wind Engng., 6 (1) 1982 pp. 37-58.

Supply reliability can be expressed in various reliability indices, e.g. loss-of-load probability and frequency of loss-of-load events. Calculation results show that the different indices are affected differently by wind power integration. Thermal capacity savings due to wind power are therefore somewhat dependent on the adopted reliability criterion. Furthermore, there is no unambiguous relation between wind energy production and the type of fuel that is saved. It is shown that even for high wind power penetration levels flexibility exists in thermal plant operating strategy to achieve savings in accordance with the prevailing fuel mix consumption policy. (A)

5.2.2(40) REGULATORY BOUNDARIES AND THE DEVELOPMENT OF ALTERNATIVE ENERGY SOURCES.
Kahn, E. and Merritt, M. California Energy Commission Report November 1982 167 pp.

The possible roles that regulated utilities might play are discussed, and economic motives for substituting alternative systems for oil-based technologies are cited. State regulation of utilities, utility internal diversification, holding companies, and subsidiaries are examined. Analyses risk and stabilisation policies for decentralised electricity generation for wind turbines, small hydro projects, geothermal power plants, and co-generation. Rate equalisation, debt guarantees, federal tax policies, and the value of utility conservation programmes are also reviewed.

5.2.2(41) A CASE STUDY OF WIND ENERGY CONVERSION SYSTEMS IN AN ELECTRIC UTILITY SYSTEM.
Jong, M.T. and Thomann, G.C. Electric Power Syst. Res., 6 (2) 1983 pp. 117-127.

The potential of wind-energy conversion systems (WECS) in a regional electric-utility system was investigated. The Kansas Gas and Electric Co. (KG & E) with 1800 mW generating capacity, served as the case study subject. The general method utilised to conduct a case study of this nature are outlined and are applied to investigate the effects of WECS on the specific electric-utility system selected for this study. Five years of con-

current wind/load data were used to simulate 15 years of KG & E system operation both without and with WECS. The KG & E system-simulation computer programmes were used to model hour-by-hour operation and to generate fuel requirements, production costs and other reports. The 'loss-of-load profitability' (LOLP) index was also calculated for the system both without and with WECS; from this the effective capacity of WECS was determined. Economic analysis was performed to assess the value of WECS in the KG & E system; break-even WECS costs were calculated by considering WECS as 'fuel-saver', plus extra tax credit and capacity credit.

5.2.2(42) ECONOMIC EVALUATION OF WIND ENERGY APPLICATIONS FOR REMOTE LOCATION POWER SUPPLY.
Bandopadhayay, P.C. (Commonwealth Scientific and Industrial Research Organisation, Div. of Energy Technology, Highett, Victoria, Australia) Wind Engng. 7 (2) 1983 pp. 67-68.

Proposes an economic evaluation methodology based on the Present Value Analysis. Graphical representations of the system performance and the economic requirements are used to examine the economic implications of using Wind Energy Conversion Systems to supply electrical power to an isolated user. The method can also be used to determine the optimum sizes of the wind energy convertor and the electrical storage size of the simple WECS. (4 refs.)

5.2.2(43) ECONOMICS OF LARGE SCALE WIND POWER IN THE U.K.: A MODEL OF AN OPTIMALLY MIXED CEGB ELECTRICITY GRID.
Martin, B. and Diesendorf, M. (Australian National University, Faculty of Science, Canberra) Australia Energy Policy, 11 (3) September 1983 pp. 259-266.

Presents a simple but powerful numerical generation planning model that has been constructed for grids containing wind farms and three classes of thermal power station, but no storage. Electricity demand and available power are specified by empirically based probability distribution functions and the plant mix which minimises the total annual costs of the generating system is determined. Using the model, the breakeven costs of wind energy in a model British CEGB grid, containing coal, nuclear, oil and wind driven power plant, are evaluated under various conditions.

5.2.2(44) SMALL INVESTORS SELLING WIND POWER TO UTILITIES.
Turner, W. New York Times, 14 February 1983 pp. 14.

Looks at the economic, financial and power aspects of several wind generators sited 50 miles East of San Francisco. A discussion of the rules under which electricity from these privately owned generators is sold to utility companies - such as Pacific Gas & Electric Co. and Southern California Edison Co. - and the amounts of electricity generated is included.

5.2.2(45) WIND ENERGY - A PROGRESS REPORT.
Bedford, L.A.W. Coal & Energy Q., (37) 1983
pp. 16-26.
Wind energy could contribute significantly
to energy supplies and, although much research and work is being done in the U.K. and
overseas, its eventual use on a large scale
depends on the electricity supply industry
being satisfied that performance and costs are
reliable and acceptable. Reviews recent
developments in wind energy and concludes that
substantial progress had been made towards
demonstrating wind energy as a viable alternative source of energy.

5.2.2(46) APPLICATION EXAMPLES FOR WIND-TURBINE
SITING GUIDELINES.
Wegley, H.L. and Pennell, W.T. (Palo Alto, USA,
Electr. Power Research Inst.) Electric Power
Res. Inst. Report no: EPRI/AP 2906 March 1983
119 pp. (See also EPRI/AP 2795)
Includes labour and time estimates for each
stage and some economic assumptions. Examines
the sensitivity of wind turbine economics to
two types of performance models. Evaluates
the guidelines and notes the major cost
uncertainties observed in the trial application.

5.2.2(47) THE COST OF ELECTRICITY FROM WIND
TURBINES - THE THROUGH LIFE COST CONCEPT.
Riddell, J.C. (J.C. Riddell & Assocs.) In Wind
Energy Conversion, proc. 5th BWEA Wind Energy
Conference, Reading, U.K. 23-25 March 1983.
P. Musgrove (ed.) pp. 54-61. Cambridge, U.K.,
Cambridge Univ. Press 1984 375 pp.
Looks at costs from an engineering standpoint
and shows how the power produced between over-
hauls can be used both as a design objective
and as an effective basis for evaluating unit
costs of electricity generated. In this way,
the sales price for such equipment may be shown.
The method of investigation is also used to
suggest the necessary conditions where the integration of a wind turbine and diesel electric
generator will be cost effective.

5.2.2(48) ECONOMIC ASSESSMENT OF WIND TURBINE
IN AN ELECTRICITY SUPPLY SYSTEM.
Talbot, J.R.W. and Taylor R.H. (CEGB, London,
England) In 4th International Conference on
Energy Options. The Role of Alternatives in the
World Energy Scene, London, England 3-6 April
1984 pp. 155-162. London, England, Instn.
Electr. Engrs. 1984 421 pp. (SBN 0-85296-290-8)
Presents the results of a study which uses
data on wind speeds at different types of site
and data on the operation of several large
machines currently operating or being designed
to estimate for a range of situations, the
target costs for wind turbines which will need
to be achieved before wind energy can make a
substantial contribution to electricity
generation in England and Wales. (21 refs.)

5.2.2(49) THE ECONOMICS OF EXISTING WIND
TURBINES IN THE SIZE RANGE 10 TO 100 METRES
DIAMETER.
Musgrove, P.J. (Reading University, U.K.)
In Wind Energy Conversion, proc. 5th BWEA
Wind Energy Conference, Reading, U.K.
23-25 March 1983. P. Musgrove (ed.) pp. 34-45.
Cambridge, U.K., Cambridge Univ. Press 1984
375 pp.
Present wind turbine costs are reviewed and
it is concluded that wind turbines in the size
range 15 to 30 metres diameter offer the most
attractive economics, at least in the immediate
future. The cost of electricity is approximately
2.2 pence/kWh, competitive with electricity from
coal, though long wind turbine life and low
operating and maintenance costs have to be
demonstrated. For private users in rural areas
small wind turbines operated in parallel with
the grid currently offer payback periods of 6 to
9 years, exclusive of any subsidies or tax
incentives.

5.2.2(50) EFFECTS OF CLUSTERS ON THE ELECTRIC
POWER FROM WINDFARMS.
Rahman, S. and Chowdbury, B.H. (Dept. of
Electrical Engineering, Virginia Polytech.
Inst., Blacksburg, VA., USA) IEEE Trans. Power
Apparatus & Systems, 103 (8) August 1984
pp. 2158-2166.
A methodology to analyse the effects of
clusters on the capacity factors of WTGs in a
windfarm is presented. The atmospheric
boundary layer effect is used to calculate
the reduction in the wind velocity due to
turbulence caused by WTGs (upwind) in a cluster.
Using a Weibull distribution for the wind speed
the capacity factor for the WTG in each row
of the cluster is then calculated. (21 refs.)

5.2.2(51) SMALL CHEAP WINDMILLS OUTDO HIGH
TECH.
Broad, J. New York Times, 14 August 1984 p. 7.
A new efficient windmill design resembling
an eggbeater has appeared on hillsides across
the USA. A U.S. Sandia National Lab.'s eggbeater prototype generates 100 kW of electricity
at peak power.

5.2.2(52) WIND VARIABILITY EFFECTS ON THE
RELIABILITY AND ECONOMICS OF ELECTRICITY
GENERATED FOR INDUSTRIAL APPLICATIONS.
Thabit, S.S. and Stark, J. (Manchester Univ.
Inst. Sci. & Technol., and Southampton Univ.,
U.K.) Wind Engng., 8 (1) 1984 pp. 36-49.
For industrial consumers of electricity,
local siting of wind energy conversion systems
is shown to exhibit an appropriate match with
maximum demand tariff chargeable for the
centrally supplied electricity. A levelised
life cycle costing methodology was applied to
evaluate the cost of kWh generated or supplied.
The introduction of a wind integrated electricity supply system is shown to reduce the
electricity bill of the industrial consumer.

5.2.2.1 VALUE ANALYSIS OF WECS FOR UTILITY GRIDS.

5.2.2.1(1) A PROBABILISTIC SIMULATION MODEL FOR THE CALCULATION OF THE VALUE OF WIND ENERGY TO ELECTRIC UTILITIES.
In proc. 1st British Wind Energy Association Wind Energy Workshop, Cranfield, U.K. April 1979. London, U.K., Multi-Science Publishing Co. Ltd. 1979 pp. 147-156.
The optimal design of windmills for the production of electricity for utility grids will depend on the type of conventional plant present, the type of loads and utility supplies, and the wind sites available. An example is presented which shows the effect of design changes on the value of a windmill to a given electric utility. The model used to value the electricity produced is described.

5.2.2.1(2) A COMPARISON OF STUDIES OF WECS ECONOMICS FOR UTILITY APPLICATIONS.
Taylor, R.H. and Rockingham, A.P. (C.E.G.B.)
In proc. 2nd British Wind Energy Association Wind Energy Workshop, Cranfield, U.K. 17 April 1980. London, U.K., Multi-Science Publishing Co. Ltd. 1980 pp. 118-126.
Several studies world-wide have now been completed which analyse the economic value of wind energy conversion systems when incorporated in electric utility grids. Reviews the models which have been used and compares the main conclusions. Results of general applicability suggest that for most utilities, the largest part of the value results from fuel saving, but that capacity credit may be significant in some cases.

5.2.2.1(3) A METHOD FOR ANALYSING THE VALUE AND AVOIDED COSTS ASSOCIATED WITH DISPERSED TECHNOLOGIES.
Davitian, H. (Entek Research, NY, USA) Paper presented at Mid-American Solar Energy Complex/ Et Al Utility Planning Symposium, Minneapolis 27-28 April 1981 pp. 108-128.
A method of electric utility value and marginal cost analysis is described. Technologies considered are cogeneration, wind energy, and solar energy.

5.2.2.1(4) VALUE ANALYSIS OF WIND ENERGY SYSTEMS TO ELECTRIC UTILITIES.
Percival, D. and Harper, J. Solar Energy Research Institute, Golden, CO.; Department of Energy, Washington, DC. In Annual Systems Simulation, Economic Analysis/Solar Heating and Cooling Operational Results Conference, Reno, NV, USA 27 April 1981. Solar Energy Research Institute Report no: SERI/TP-732-1064 & Conference Report no: CONF-810405-10 January 1981 9 pp.
The analysis is performed by computer models that interface with most conventional utility planning models. Weather data are converted to wind turbine output powers, which are used to modify the utility load representation. Execution of the utility planning models with both the original and modified load represent-

ation yields the gross and marginal value of the added wind energy systems. This value is then compared with cost estimates to determine if for economic reasons the wind energy system should be integrated into future generation plans.

5.2.2.1(5) THE VALUE OF WIND TURBINES TO LARGE ELECTRICITY UTILITIES.
Rockingham, A.P. and Taylor, R.H. (C.E.G.B.)
Paper presented at 3rd International Conference on Future Energy Concepts, London 27-30 January 1981 pp. 348-352. London, Institution of Electrical Engineers 1981.
The economics of wind power will be influenced by future fuel prices, plant type existing in an integrated system and alternative investment choices. A preliminary analysis of how each of these factors affects the economics of wind energy conversion is described. The sensitivity of this economic analysis for conditions in the U.K. is examined. (10 refs.)

5.2.2.1(6) WIND SYSTEM VALUE ANALYSIS FOR ELECTRIC UTILITIES: A COMPARISON OF FOUR METHODS.
Harper, J. and others. (Solar Energy Research Institute; Florida University) In proc. 5th Biennial Wind Energy Conference & Workshop, Washington, DC, USA 5-7 October 1981. I.E. Vas (ed.) V. 2 Palo Alto, USA, Solar Energy Research Institue 1981 Session 3A pp. 485-497. Solar Energy Research Institute Report no: SERI/CP-635-1340 CONF-811043.
Compares methods developed for conducting economic assessments of wind energy conversion systems in utility applications, under Department of Energy sponsorship. The purpose of this comparative analysis is to identify any discrepancies among the methods and corresponding changes that might be required to improve their accuracy.

5.2.2.1(7) ECONOMIC OPTIMISATION OF WIND ENERGY CONVERSION SYSTEMS FOR ISOLATED USERS.
Bandopadhayay, P.C. (Commonwealth Scientific & Industrial Research Organisation, Australia) Inst. Engrs. Aust. Mech. Engng. Trans., ME7 (1) April 1982 pp. 30-37.
The system comprises a wind driven electrical generator, electrical storage, control and power conditioning equipment, and an auxiliary power plant. The performance of the system at a given location is investigated by discrete time simulations. The requirements for the performance of such a system to be economically advantageous in complementing a conventional internal combustion engine driven power plant are obtained by Present Value Analysis. A method is demonstrated where the system performance and economic requirements optimise the size of the wind energy converter and the storage capacity of the system for a given location.

5.2.2.1(8) ELECTRIC-UTILITY VALUE ANALYSIS FOR WIND-ENERGY-CONVERSION SYSTEMS.
Harper, J.R. (Solar Energy Research Institute,

Golden, CO) Solar Energy Research Institute Report no: SERI/TR-98336-1A April 1982 153 pp. (DE 82 014249)

Demonstrates an analytical methodology for assessing the value of WECS installed in an electric utility generation system. Two case studies were conducted: one at Southern California Edison, San Gorgonio, CA, and the other at Consumers Power Company, Ludington, MI. The study years for these analyses ranged from 1980 to 1995, and levels of WECS penetration spanned up to 10%. Cost-of-production values and capacity values were obtained for all combinations of utility, study year, and WECS penetration levels.

5.2.2.1(9) ELECTRIC UTILITY VALUE ANALYSIS METHODOLOGY FOR WIND ENERGY CONVERSION SYSTEMS.
Bush, L.R., Cretcher, C.K. and Davey, T.H. (Solar Energy Research Institute, Golden, CO; Aerospace Corp., El Segundo, CA. Energy and Resources Div.; Department of Energy, Washington, DC) Solar Energy Research Institute Report no: SERI/TR-98336-1 September 1981 31 pp.

Summarises a methodology developed for a study of the value of augmenting conventional electric energy generation with wind energy conversion systems. The pertinent measures of value include both the displacement of conventional fuels measured in volumetric and economic terms, and the potential for capacity credit, realised either through the sale of capacity or deferred starting of new installations.

5.2.2.1(10) ELECTRIC-UTILITY VALUE DETERMINATION FOR WIND ENERGY. VOLUME 1: A METHODOLOGY.
Percival, D. and Harper, J. (Solar Energy Research Institute, Golden, CO; Department of Energy, Washington, DC) Solar Energy Research Institute Report no: SERI/TR-732-604-V.1 & 2 January 1982 94 pp. & 121 pp. (See also SERI/TR-98336-2 September 1981 35 pp. for original report on methodology.)

Describes a method electric utilities can use to determine the value of wind energy systems. It is performed by a package of computer models available from SERI that can be used with most utility planning models. The final output of these models gives a financial value ($/kW) of the wind energy system under consideration in the specific utility system. The value determination method is described and detailed discussion on each computer programme available from SERI is given. The report is in two volumes, the second being a users' guide to the computer programmes.

5.2.2.1(11) WIND SYSTEM VALUE ANALYSIS FOR ELECTRIC UTILITIES: A COMPARISON OF FOUR METHODS.
Harper, J.R. and others. (Solar Energy Research Institute; Florida University) ASME Trans. J. Solar Energy Engng., 104 (2) May 1982 pp. 70-76.

Reports on the only known effort that used more than a single methodology for the value analysis of WECS to a specific utility. Presents and compares the WECS utility value analysis methodologies of Aerospace Corp., JBF Scientific Corp. and the Solar Energy Research Institute (SERI). Results of the application of these three methodologies were found for two large utilities. The amounts a utility can pay for a wind turbine over its lifetime and still breakeven economically were found to be from 1600 dollars to 2400 dollars per kW of wind capacity in 1980 dollars. The reasons for variation in the results are discussed.

5.2.2.1(12) WIND-ENERGY SYSTEMS FOR ELECTRIC UTILITIES: A SYNTHESIS OF VALUE STUDIES.
Hoock, S and Flaim, T. (Solar Energy Research Institute, Golden, CO) Solar Energy Research Institute Report no: SERI/TP-214-1976 CONF-830622-23 May 1983 11 pp. Also in American Solar Energy Society meeting, Minneapolis, MN, USA 1 June 1983.

The results from five major studies that assess the value of wind-energy-conversion systems to 14 electric utilities have been normalised by a set of standard economic assumptions to facilitate comparisons across studies. The results indicate that WECS breakeven value is highly dependent on wind resource, utility generation mix; assumed WECS penetration, and year of WECS installation. The studies also show that WECS increase system reliability in many cases and thus can displace some capacity. This capacity displacement - measured in terms of effective load-carrying capability (ELCC) - declines with increasing WECS penetration. Sensitivity cases were examined to determine the effects of changes in utility financial parameters, fuel cost projections, WECS operation and maintenance costs, and the generation mix over-time. The results of the normalisation indicate that the value of WECS to utilities ranges from $1500 to $4000 per kW, depending on the factors listed above.

5.2.2.1(13) ELECTRIC-UTILITY VALUE DETERMINATION FOR WIND ENERGY. VOLUME II. A USER'S GUIDE.
Percia, D. and Harper, J. Golden, USA, Solar Energy Res. Inst. January 1982 121 pp. Report no: SERI/TR-732-604/V.2.

A method is described for determining the value of wind energy systems to electric utilities. It is performed by a package of SERI computer models that can be used with most utility planning models. The final output of these models gives a financial value of the wind energy system under consideration in the specific utility system.

5.2.2.1(14) WIND TURBINE VALUE ANALYSIS FOR ELECTRIC UTILITIES.
Dub, W. Regensburg, Fed. Rep. Germany, Regensburg Univ. June 1982 20 pp.

A methodology for the analysis of the value of large scale wind turbines to electric utilities and its application to utility, socio-economic and meteorological data in two case studies is described. The value analysis included different levels of wind penetration. The utility planning procedures in current use and the unique problems of the integration of wind power into the power system are discussed.

5.2.2.1(15) ANALYSIS OF WIND ENERGY SYSTEMS FOR SELECTED ELECTRIC UTILITIES. A FINAL SUB-CONTRACT REPORT.
McCabe, T.F. (JBF Scientific Corp., Wilmington, MA., USA) Solar Energy Res. Inst. Report no: SERI/STR-211-2379 July 1984 332 pp.
Nine electric utilities were analysed taking into account different wind turbine installation years to determine the effects of changes in generation mix and how capacity displacement would alter fuel savings. The value results are compared on wind resource and WECS performance, current and projected utility characteristics, and utility financial parameters relative to carrying investments in production facilities.

5.2.2.1(16) ASSESSMENT OF DISTRIBUTED WIND-POWER SYSTEMS. FINAL REPORT.
Kaupang, B.M. (Gen. Electric Co., Schenectady, USA) Electric Power Research Inst. Report no: EPRI/AP-2882 February 1983 354 pp.
Discusses a method for distributed wind power system evaluation which is based on conventional utility planning techniques. Compares total utility system cost with and without wind power plants in terms of wind power plant value and cost. Value is measured by the worth of displaced energy and capacity of conventional power plants, transmission and distribution equipment deferrals and loss savings. Value is dominated by the generation of energy and capacity value.

5.2.3 ENERGY MANAGEMENT AND OPERATION OF INTEGRATED SYSTEMS.

5.2.3(1) ENERGY ANALYSIS OF WIND ENERGY CON-VERSION SYSTEM FOR FUEL DISPLACEMENT.
Devine, D.W.Jr. (Oak Ridge Assoc. University, Tennessee, USA) In Alternative Energy Sources, Miami International Conference International Compendium, Miami Beach, Florida 5-7 December 1977. Washington, DC, Hemisphere Publ. Corp. 1978 V. 4 pp. 1901-1924.
An input/output approach is employed to estimate the energy embodied in a horizontal-axis wind electric generating station used to displace fossil fuel in an electric utility system. Five ratios comparing delivered electrical energy to the energy requirement of the wind machine are displayed. Results indicate that the system considered could be a large net producer of energy and

should displace a quantity of fossil energy equivalent to that embodied in the machine in considerably less than one year. (12 refs.)

5.2.3(2) USING WIND ENERGY FOR PEAK ELECTRICAL LOAD LEVELLING.
Soderholm, L.H. (USDA, Sci. & Educ. Adm., Ames, Iowa, USA) Paper for ASAE Summer Meeting, Utah State University, Logan, USA 27-30 June 1978. St. Joseph, Michigan, ASAE 1978 Paper 78-3040 12 pp.
Examines the possibilities and value of using wind energy for levelling peak electrical loads on rural electric power systems for both power suppliers and users. A preliminary analysis of experimental and simulation data for the Grumman "Windstream 25" wind system indicates that wind energy can make a substantial contribution to load levelling particularly for rural heating loads.

5.2.3(3) LEWIS RESEARCH CENTRE STUDIES OF MULTIPLE LARGE WIND TURBINE GENERATORS ON A UTILITY NETWORK.
Gilbert, L.J. and Triezenberg, D.M. (NASA Lewis Research Centre; Purdue University, USA) In proc. Workshop on Economic and Operational Requirements and Status of Large Scale Wind Systems, Monterey, USA 28-30 March 1979. Palo Alto, USA, Electr. Power Research Institute 1979 pp. 388-402. (Special Report no: ER-1110-SR & DOE Conference Report no: 790352.)
The wind energy project office of NASA Lewis Research Centre has undertaken and studied the anticipated performance of a wind turbine generator farm on an electric utility network. Preliminary results on an in-house simulation of two Mod-2 systems tied to an infinite bus indicate favourable system performance.

5.2.3(4) WIND ENERGY INTEGRATION STUDY.
Bonneville Power Administration, Portland, Oregon. Washington, DC, USA, US Department of Energy August 1980 77 pp. (US Dept. of Energy Report no: DOE/BP/10552-17)
The objectives of the study were to investigate the feasibility of integrating a simulated 3000 megawatt wind energy conversion network into the Pacific NorthWest hydro-thermal generation system. To identify those significant effects which require planning consideration and further study prior to extensive development of wind energy, the following areas for preliminary analysis were identified: seasonal power planning; surplus energy; energy reserve planning; peak reserves; and hourly planning.

5.2.3(5) ANALYSIS OF SMALL, NONCONVENTIONAL ELECTRIC POWER SYSTEMS FOR REMOTE SITE APPLICATIONS.
Boehman, L.I. and others. (Dayton University; Kansas University; San Jose University, USA) In IECEC '80 'Energy to the 21st Century' proc. 15th Intersociety Energy Conversion Engineering Conference, Seattle, USA 18-22 August 1980 V. 1.

New York, USA, American Institute Aeronaut. &
Astronaut. 1980 Paper no: 809165 pp. 828-834.
Considers electric power system with energy
conversion by wind, solar and hybrid wind/solar
configurations and energy storage in flywheels,
hydrogen, batteries and thermal devices.
Relative performance, cost, availability, and
reliability are compared for the conceptual
systems and a modular configuration with two
8 kW wind energy converters and sealed lead
acid batteries is analysed. The system can
provide 5 kW on a continuous basis with 5.6
metres per sec. average wind velocity and have
12 hours of reserve capacity stored in the
battery.

5.2.3(6) HYBRID ENERGY SYSTEM WIND AND WATER.
Fordham, J.W. and others. (University of
Nevada, USA) NTIS Report PB80-171218 February
1980 78 pp.
A prototype wind-hydro system is designed for
a small area Nevada site. It includes a reservoir, hydroelectric generation plant, generator,
and wind turbine. Cost-benefit analysis of
six preliminary designs for a wind-hydro system
located on Slide Mountain in Nevada is given.

5.2.3(7) LARGE WIND TURBINES: A UTILITY OPTION
FOR THE GENERATION OF ELECTRICITY.
Robbins, W.H., Thomas, R.L. and Baldwin, D.H.
NASA Lewis Research Centre, Cleveland, USA
1980 18 pp. (NASA-TM-81502)
An overview of the NASA activities with regard
to Wind Energy Conversion Systems which are
being developed. Emphasis is placed on the
application of large wind turbines for generation of electricity by utility systems for the
grid.

5.2.3(8) COMMERCIAL WIND ENERGY CONVERSION
SYSTEMS (WECS) MONITORING AND UTILITY IMPACTS
STUDY.
Finch, T. and Stafford, R.W. (Bronx Frontier
Development Corp., NY) New York State Energy
Research and Development Authority, Albany.
Final Report September 1981 167 pp.
The performance of a 25 kW wind turbine operating at an industrial site in the Bronx was
monitored and evaluated. The turbine was
tested for a number of different generator/
power conditioning configurations and a number
of problems were encountered, including a
lightning strike and turbine overspeeds. The
system was interconnected to Consolidated
Edison's power grid to allow a quantity assessment of backfeed of excess power.

5.2.3(9) EFFECTIVE OUTPUT AND AVAILABILITY
OF WIND TURBINES FOR HOUSEHOLD LOADS.
Feron, P. and Lysen, E.H. (Eindhoven
University of Technology, Netherlands)
Wind Engineering, 5 (4) 1981 pp. 194-206.
Presents a theoretical model of a small-
scale wind energy system producing electricity
for a group of households connected to the
grid. By means of a Weibull distribution of
the wind velocity, a linear output model of the

turbine and a measured load distribution function of a group of Dutch households, the
average values of the electricity surplus
and electricity deficit are calculated for
the wind regime in Den Helder. (21 refs.)

5.2.3(10) BIVALENT WIND - DIESEL POWER STATION.
Fritzsche, A. and Muller, W. In proc. International Colloquium on Wind Energy, Brighton,
U.K. 27-28 August 1981. Cranfield, U.K., BHRA
Fluid Engng. 1981 Session 3 pp. 161-166.
An economic fuel saving power supply for
remote rural areas can be achieved by combining a diesel generator and a wind energy
converter. Problems of operation modes and
system requirements are discussed.

5.2.3(11) NET ENERGY ANALYSIS OF SMALL WIND
ENERGY CONVERSION SYSTEMS.
Haack, B.N. (Ball State University, Muncie,
Indiana, USA) Appl. Energy, 9 (3) November 1981
pp. 193-200.
Net energy is the amount of energy remaining
for consumer use after deducting the energy
required to construct and maintain the electricity generating system. A 3 kW rated wind
electric system for residential use is examined.
The amount of energy obtained from this system
is estimated by using a computer-operated simulation model which incorporates wind speeds,
residential electricity demands, and parameters
from the generator, inverter, and storage components. (15 refs.)

5.2.3(12) CASE STUDY EVALUATING THE POTENTIAL
FOR SMALL WIND ENERGY CONVERSION SYSTEMS (SWECS)
AS AN INTEGRAL PART OF THE GENERATING MIX FOR
A REGIONAL UTILITY.
Jong, M.T. and Thomann, G.C. (Wichita State Univ.,
Kansas, USA) In proc. IEEE International Conf-
erence on Electrical Energy 1981, Oklahoma City,
Oklahoma, USA 13-15 April 1981. Piscataway, NJ,
USA, IEEE (Cat no: 81CH1666-7) pp. 121-129.
A condensed version of the final report to
evaluate the potential for SWECS as an integral
part of the generating mix for the Kansas Gas
& Electric Co. system. (19 refs.)

5.2.3(13) SUBSTITUTION OF DIESEL POWER PLANTS
BY SOLAR AND WIND ELECTRICITY GENERATORS A
CASE STUDY FOR A TROPICAL ISLAND.
Minder, R. and Gilby, D. (Electrowatt Engineering Services, Switzerland; Electrowatt
Engineering Services, U.K.) Natural Resources
Forum, 5 (3) July 1981 pp. 271-276.
Electricity supply systems on many small
islands and in isolated regions of developing
countries are based on power generation by
diesel engines. The availability of solar
and wind energy on a Caribbean island were
examined to investigate the feasibility of
displacing diesels with renewable energy
systems. A combination of a 100 kW solar
thermal power plants with ten 40 kW wind
generators is envisioned.

5.2.3(14) WIND ENERGY AND ELECTRIC UTILITIES -
A PERSPECTIVE.
Ramakumar, R. (Oklahoma State University, Stillwater, USA) In proc. IEEE International
Conference on Electrical Energy 1981 Oklahoma
City, Oklahoma, USA 13-15 April 1981. Piscataway,
NJ, USA, IEEE (Cat no: 81CH1666-7) pp. 112-120.
 Utilisation of wind energy on a meaningful
scale will require the operation of arrays of
large 100 kW wind electric conversion systems
in parallel with conventional electric utility
grids. This paper examines some of the key
technical, operational, and economic issues
involved in their successful integration in
the power systems of the future. (20 refs.)

5.2.3(15) WIND AS A MAJOR ENERGY SOURCE.
Sorensen, B. (Roskilde University, Denmark)
Mazingira, 5 (1) 1981 pp. 34-43.
 Assesses potential contributions of wind
energy to electricity supplies. Wind
resource assessment methodologies, hybrid wind
energy systems (particularly wind-hydroelectric
power technology) prospects for realisation of
such technology and associated environmental
impacts are examined.

5.2.3(16) LARGE-SCALE USE OF WIND ENERGY IN
THE NETHERLANDS ELECTRICITY SUPPLY.
van Hoek, G.A.L. and van Nielen, N.S. Energiespectrum, 5 (7-8) July-August 1981 pp. 207-212.
(In Dutch)
 Further research is necessary to overcome
the problems of large-scale full-time power
generation from wind as a viable alternative
to fossil fuels. Most of the wind generator
investment should be offset against the fuel
costs of thermal generators e.g. a wind
capacity proportion of 15% of the thermal
capacity.

5.2.3(17) FEASIBILITY OF AN UNDERGROUND PUMPED
HYDRO STORAGE IN THE NETHERLANDS.
Lohuizen, H.P.S. and Halberg, N. (Delft Univ-
ersity of Technology, The Netherlands) In
AIAA/EPRI International Conference on Under-
ground Pumped Hydro and Compressed Air Energy
Storage, Collection Technical Papers, San
Francisco, USA 20-22 September 1982. New York,
USA, American Institute Aeronaut. & Astronaut.
1982 Paper no: 82-1665 pp. 140-155.
 Examines the possibility of using wind energy
in underground pumped hydro storage in the
Netherlands.

5.2.3(18) A GUIDE FOR THE ASSESSMENT OF TECHNO-
LOGIES FOR GENERATING ELECTRICITY.
US Dept. of Energy Report no: EIA-0344 June 1982
387 pp.
 Reviews various electricity generating technologies, both commercialised and still in the
demonstration state, including wind energy conversion systems. Background information,
development and use, and a detailed description
of the electricity producing process are
provided for each type of electric generating
plant. (Numerous references)

5.2.3(19) IMPLEMENTING AGREEMENT FOR A PRO-
GRAMME OF RESEARCH AND DEVELOPMENT ON WIND
ENERGY CONVERSION SYSTEMS.
(Kernforschungsanlage Juelich GmbH, Fed. Rep.
Germany, Projektleitung Energieforschung)
Meeting of Experts on Integration of Wind
Power into National Electricity Supply Systems,
Juelich, Fed. Rep. Germany 8 March 1982.
Report no: Juel-Spéz-155 CONF-8203124
June 1982 248 pp.
 All papers deal with problems of inte-
gration of wind power into national elec-
tricity supply systems. (Some of the papers
appear in other relevant sections of this
bibliography.)

5.2.3(20) WIND ENERGY PROGRESS AT PG AND E.
Everett, L.H. and Steeley, W.J. (Pacific Gas
& Electr. Co., USA) In proc. IECEC '82, 17th
Intersociety Energy Conversion Engineering
Conference, Los Angeles, USA 8-12 August 1982
V. 4. New York, USA, Institute of Electrical &
Electronic Engineers 1982 Paper no: 829348
pp. 2102-2105.
 A complete year of meteorological data has
confirmed that winds of commercial quality
exist over hundreds of acres of hilly grazing
land near Bay Area. This led to the siting
and construction of PG and E's own 2.5 mW
Boeing wind turbine (Model 2560) in Solano
County. An extensive two-year monitoring programme is planned with particular emphasis
on the machine's performance, mechanical
reliability, and environmental impacts.
Identification of promising sites has resulted
in privately financed development of wind
farms whose power will be sold to PG and E.

5.2.3(21) UNITED STATES ELECTRIC UTILITY
ACTIVITIES IN WIND POWER.
Goodman, F.R. and Vachon, W.A. (Electr. Power
Research Institute; Arthur D. Little Inc., USA)
In papers presented at 4th International
Symposium on Wind Energy Systems, Stockholm,
Sweden 21-24 September 1982 V. 2. Cranfield,
U.K., BHRA Fluid Engng. 1982 Session M Paper M4
pp. 305-321.
 The Electric Power Research Institute (EPRI)
has evolved a wind power programme aimed at the
development of planning tools; the evaluation
of state-of-the-art technology, and the trans-
fer of technical information related to wind
power. An overview is provided of this EPRI
programme as well as its relationship to the
Federal, private and utility wind programmes.

5.2.3(22) THE FEASIBILITY OF ELECTRIC POWER
GENERATION BY THE WIND ON THE UNIVERSITY OF
NEW ORLEANS CAMPUS.
Hilbert, L.B. and Janna, W.S. (New Orleans
University, USA) New York, USA, American Soc.
of Mechanical Engineers 7-10 March 1982.
8 pp. Paper no: 82-Pet-1.
 Presents a feasibility study for using wind
as an alternative or supplementary energy sources
for the campus. The University is located on
the southern shore of Lake Pontchartrain, and
the results of the study could be considered

valid for the entire lakeshore area of New Orleans. The paper covers wind characteristics study, brief introduction to wind energy systems; an energy and economic analysis of some wind energy systems; and, conclusions and recommendations.

5.2.3(23) WIND ENERGY RESEARCH PROGRAMME FOR NORTHERN STATES POWER COMPANY.
Kasmarik, M.D. and others.
(Northern States Power Co.; Environ. Consult. Inc.) In proc. 1982 Wind and Solar Energy Technology Conference, Kansas City, USA 5-7 April 1982. Columbia, USA, Missouri-Columbia University 1982 pp. 267-284.
Discusses Northern States Power Company's wind energy research programme objectives, the results from WECS siting studies and planned wind monitoring and analysis programmes.

5.2.3(24) THE 80 MEGAWATT WIND POWER PROJECT AT KAHUKU POINT, HAWAII.
Laessig, R.R. (Windfarms Ltd.) In proc. NASA Workshop on Large Horizontal-Axis Wind Turbines, Cleveland, USA 28-30 July 1981. R.W. Thresher (ed.) Washington, DC, USA, NASA 1982 pp. 689-709. (NASA Conference Publication 2230; US Dept. of Energy Report no: CONF-810752 SERI/CP-635-1273.)
Describes the two largest wind energy projects in the world, being developed by Windfarms Ltd, designed to produce 80 megawatts at Kahuku Point, Hawaii and 350 megawatts in Solano County, California. These projects will be the prototypes for the future large-scale wind power installations throughout the world. Details of turbines and siting are given.

5.2.3(25) WINDPOWER IN THE U.K. ELECTRICITY SUPPLY INDUSTRY.
Milborrow, D.J. Electron. & Power, 28 (10) October 1982 pp. 665-669.
Several pilot projects are underway in the U.K. to assess the feasibility of using wind energy as a primary energy source. The prospects for the use of wind energy in U.K. electricity supply are assessed.

5.2.3(26) MEDICINE BOW WIND PROJECT.
Nelson, L.L. (U.S. Bureau of Reclamation) ASME Trans. J. Solar Energy Engng., 104 (2) May 1982 pp. 77-83.
The Bureau conducted studies for a wind turbine field of 100 mW at a site near Medicine Bow, Wyoming one of the windiest areas in the United States. The wind turbine system would be interconnected to the existing Federal power grid through the substation at Medicine Bow. Power output from the wind turbines would thus be integrated with the existing hydroelectric system, which serves as the energy storage system. The report concludes that a 100 mW wind field at Medicine Bow has economic and financial feasibility and the Bureau's construction of the Medicine Bow

wind field could demonstrate to the industry the feasibility of wind energy.

5.2.3(27) OPERATING AND MAINTENANCE EXPERIENCE WITH A 6-KW WIND ENERGY CONVERSION SYSTEM AT NAVAL STATION, TREASURE ISLAND, CALIFORNIA.
Pal, D. (Naval Civil Engineering Lab., Port Hueneme, CA) Technical Note September 1979-June 1981 Report no: NCEL-TN-1641 June 1982 55 pp.
Describes in detail the experience gained and lessons learned from the 6-kW grid-integrated Wind Energy Conversion System demonstration at San Francisco Bay. The objective was to develop operating experience and maintenance information on the 6-kW WECS using a combination of a permanent magnet alternator with a line commutated synchronous inverter. On-site measurements conducted during the demonstration indicate that the WECS site has annual average wind-speeds of about 8 to 10 mph. The test results indicate a satisfactory performance of the WECS except for two failures involving arcing at the electrical terminals located on the yaw shaft.

5.2.3(28) CONNECTING RENEWABLE POWER SOURCES INTO THE SYSTEM.
Watzler, F.U. IEEE Spectrum, 19 (11) November 1982 pp. 42-45.
Distributed storage and generation sources, such as windmills, fuel cells, and photovoltaics, have many advantages for utilities and consumers. Interconnecting such systems to a power grid can present a number of technological challenges and equipment reliability and alternative power device protection must be ensured. Problems associated with decentralised energy system distribution and grid interconnection are surveyed.

5.2.3(29) UTILITY EXPERIENCE WITH TWO DEMONSTRATION WIND TURBINE GENERATORS.
Wehrey, M.C. (Southern California Edison Co., USA) In proc. NASA Workshop on Large Horizontal-Axis Wind Turbines, Cleveland, USA 28-30 July 1982. R.W. Thresher (ed.) Washington, DC, USA, NASA 1982 pp. 727-739. (NASA Conference Publication 2230; US Dept. of Energy Publication CONF 810752 SERI/CP-635-1273.)
Two demonstration WTGs have been installed and operated at Edison's Wind Energy Centre near Palm Springs, California: a 3 mW horizontal axis Bendix/Schachle WTG and 300 kW vertical axis Alcoa WTG. The performance of the WTGs is evaluated, their system operation and environmental impact assessed and the design criteria of future WTGs identified. Edison's experience with these two WTGs is summarised and the problems encountered with the operation of the two machines are discussed.

5.2.3(30) ELECTRICITY SUPPLY IN THE SOUTH EAST.
SCLERP/Central Electricity Generating Board Standing Conference on London and South East Regional Planning February 1983 12 pp. (SC 1780)

Covers electricity supply capacity in the S.E. region of the U.K., station decommissioning programme, electricity demand forecasts, transmission links and other developments such as combined heat and power and wind power.

5.2.3(31) WIND PARKS, AN ECONOMIC ENERGY ALTERNATIVE.
D'Aquanni, R.T. and Church, C.B. (Parkway Energy Products) Energy Economics Policy & Management, 2 (4) Spring 1983 pp. 4-18.
    Discusses the economic and technical aspects of using wind parks to provide substantial amounts of electricity to U.S. towns. Wind farms sites throughout the U.S. are evaluated, and the construction, operating, and maintenance costs of a wind park are calculated. Wind farms must be carefully sited and their developers must be prepared to wait for higher electricity prices to make wind power competitive.

5.2.3(32) SOURCE RELIABILITY IN A COMBINED WIND-SOLAR-HYDRO SYSTEM.
De Almeida, A. and others. (University de Coimbra, Portugal) IEEE Trans. Power Apparatus & Systems, 102 (6) June 1983 pp. 1515-20.
    Describes a combined wind-solar-hydro power plant system. A simple multivariable weather model is used to calculate wind speeds, solar radiation, and rainfall. Using the projected hydro power capacity in 2000 for Portugal different percentages of solar-wind power re calculated to gain insight into the reliability of the combined system. In addition, the optimum share of wind power in the system is examined in relation to the corresponding load loss probability. (10 refs.)

5.2.3(33) UTILITY OPERATING STRATEGY AND REQUIREMENTS FOR WIND POWER FORECASTS.
Dub, W. and Pape, H. (Regensburg University) J. Energy, 7 (3) May-June 1983 pp. 231-236.
    Examines the problem of obtaining wind power forecasts based on wind speed forecasts and studies the fitting of time series models to wind speed data for 3 hours and 6 hours. Discusses unit availability, shut down times, and economics of wind power.

5.2.3(34) DISPERSED STORAGE AND GENERATION IMPACTS ON ENERGY MANAGEMENT SYSTEMS.
Kirkham, H. and Klein, J.) NASA Jet Propulsion Lab., California, USA) IEEE Trans Power Apparatus & Systems, 102 (2) February 1983 pp. 339-344.
    Examines the integration of dispersed storage and generation systems into the total energy management system. Dispersed storage and generation systems are defined as sources of electricity connected to the utility distribution system. The impact these sources will have on automatic generation control, economic dispatch, voltage control, protection, stability, and system planning and design are discussed. Includes wind energy. (11 refs.)

5.2.3(35) THE WORLD'S FIRST MUNICIPAL WIND FARM.
Reinemer, V. Public Power, 41 (2) March-April 1983 pp. 10-13.
    A municipal wind energy farm is being operated by Livingston, MT. and wind energy is seen as a logical and economical source of revenue and electricity for the city. Its evolution is traced, and the design and performance of the six wind-electric generators are described.

5.2.3(36) THE OPERATING EXPERIENCES AND PERFORMANCE CHARACTERISTICS DURING THE FIRST YEAR OF OPERATION OF THE CROTCHED MT. NEW HAMPSHIRE WINDFARM.
Sumanski, D.P. (New England Power Service Co., Maine, USA) IEEE Trans. Power Apparatus & Systems, 102 (6) June 1983 pp. 1637-1641.
    The windfarm has an array of generators whose combined electrical output is fed directly into the utility power grid. The windfarm manufactured by a private company, allows examination of such operating parameters as energy and capacity displacement, reliability, metering and safety. (3 refs.)

5.2.3(37) MEASURED EFFECT OF WIND GENERATION ON THE FUEL CONSUMPTION OF AN ISOLATED DIESEL POWER SYSTEM.
Stiller, P.H., Scott, G.W. and Shaltens, R.K. (Westinghouse Electric Corp., Pittsburgh, PA, USA) In IEEE Trans. Power Apparatus & Systems, 102 (6) June 1983 pp. 1788-1792.
    The Block Island Power Company, on Block Island, Rhode Island, operates an isolated electric power system consisting of a 150 kW wind turbine, designated MOD-OA by the U.S. Department of Energy, operated in parallel with two diesel generators to serve an average winter load of 350 kW. Wind generation serves up to 60% of the system demand depending on wind speed and total system load. Diesel fuel consumption measurements are given for the diesel units operated in parallel with the wind turbine and again without the wind turbine and consumption data are used to calculate the amount of fuel displaced by wind energy. Results indicate that the wind turbine displaced 25,700 lbs. of the diesel fuel during the test period, representing a calculated reduction in fuel consumption of 6.7% while generating 11% of the total electrical energy.

5.2.3(38) WIND LOAD CORRELATION AND ESTIMATES OF THE CAPACITY CREDIT OF WIND POWER: AN EMPIRICAL INVESTIGATION.
Martin, B. and Carlin, J. Wind Engng. 7 (2) 1983 pp. 79-84.
    Most previous investigations of the capacity credit of wind power in electricity grid systems have paid little attention to the effect of correlation between wind power and electrical demand. Calculations made with several years of data from Western Australia indicate that empirical estimates of capacity credit may vary greatly with small changes in the joint distribution of wind and load. These changes are not reflected in the ordinary sample-correlation coefficient. It

is shown that capacity credit estimates are very sensitive to the availability of wind power at a few periods of high load. A simple summary measurement is described which gives an indication of the strength of wind load association in relation to capacity credit.

5.2.3(39) METHOD FOR DETERMINING HOW TO OPERATE AND CONTROL WIND TURBINE ARRAYS IN UTILITY SYSTEMS.
Javid, S.H. and others. (General Electric Co., Schenectady, NY. Electric Utility Systems Engineering Department, Washington, DC) U.S. Dept. of Energy Report no: CONF-840112-2 January 1984 8 pp. In IEEE Power Engineering Society Winter Meeting, Dallas, Texas, USA 29 January 1984. (DE 84 005817)
Describes a method for determining how arrays should be controlled and operated on the load frequency control time-scale. Initial considerations for setting wind turbine control requirements are followed by a description of open loop operation and of closed loop and feed forward wind turbine array control concepts. The impact of variations in array output on meeting minimum criteria are developed.

5.2.3(40) ANALYSIS OF THE EFFECTS OF INTE-GRATING WIND TURBINES INTO A CONVENTIONAL UTILITY CASE STUDY.
Goldenblatt, M.K. and others. (JBF Scientific Corp., MA., USA) NTIS Report no: DE83-011154 March 1983 69 pp.
Three aspects of integration into the utility generating system of the Los Angeles Department of Water & Power were investigated. The sensitivity of the economic impact of wind turbine generation to wind speed sampling frequency, wind turbine performance model, and wind speed forecasting accuracy were examined. Simulations estimated the output of the Boeing MOD-2 turbine. The utility's ability to forecast wind speeds accurately can increase the production cost savings from wind turbine generation by nearly 20% annually.

5.2.3(41) FEASIBILITY OF WIND TURBINE DIESEL HYBRID GENERATORS AT MCMURDO STATION, ANTARCTICA.
Scott, L.B. and others. Tucson, USA, Arizona Univ. March 1983 354 pp. (PB83-189860)
Gives a short description of past and present energy systems at McMurdo, and describes some projected energy systems within which wind turbine generators would function. Feasibility is then discussed in terms of certain variables and performance parameters characterising diesel electric and wind turbine generator systems.

5.2.3(42) WIND ENERGY APPLICATIONS ON ISOLATED DIESEL SYSTEMS.
Tennis, M. (JBF Scientific Corp., Wilmington, MA., USA) Research Report Can. Electr. Assoc. 139 267 September 1983 117 pp.

The economic feasibility of wind turbines on diesel systems was investigated and the sensitivity of the economic analysis to changes in uncertain inputs or assumptions was evaluated. The technical feasibility of the wind turbines was also examined with particular attention to wind turbine size selection and protection and to the impacts of wind turbine operation on diesel operations.

5.2.3(43) AN ANALYSIS OF THE INTEGRATION PROBLEM FOR WIND POWER INPUTS INTO SMALL GRID SYSTEMS.
Lipman, N.H. (Rutherford Appleton Lab., Didcot, U.K.) Wind Engng., 8 (1) 1984 pp. 9-18.
The problems are caused by the fluctuating outputs of the wind turbines. A wind turbine could be adequately meeting a given load one moment and failing seconds later. Therefore, computer modelling techniques are being used for integration analysis.

5.2.3(44) ANALYSIS OF WIND SPEED DATA RECORDED AT 14 WIDELY DISPERSED U.K. METEOROLOGICAL STATIONS.
Halliday, J.A. (Energy Res. Support Unit, Rutherford Appleton Lab., Didcot, U.K.) Wind Engng., 8 (1) 1984 pp. 50-73.
Examines the problems associated with integrating electricity generated by wind turbines and other renewable sources of energy into the U.K. National Grid. The importance of knowledge of the site and the various options for scaling wind data to a given height are reviewed. Results of the statistical analysis, the various frequency distributions available and the use of the Weibull distribution are examined. (33 refs.)

5.2.3(45) THE FAIR ISLE WIND POWER SYSTEM.
Stevenson, W.G. and Somerville, W.M. In Wind Energy Conversion, proc. 5th BWEA Wind Energy Conference, Reading, U.K. 23-25 March 1983. P. Musgrove (ed.) pp. 171-184. Cambridge, U.K., Cambridge Univ. Press 1984 375 pp. (Isbn: 0-521-26250-X)
Describes design and installation of a 55 kW wind turbine generator for Fair Isle. The 14 m diameter 3 bladed turbine system uses a step-up gear transmission and braking system, and the support tower, a lattice construction, is 15 m high. Outlines the safety features incorporated. A data logger records wind speed and direction, generator output, and output to dump load. Describes how the wind power distribution scheme is linked to the existing diesel generator system for four groups of load.

5.2.3(46) SMALL SCALE WIND/DIESEL SYSTEMS FOR ELECTRICITY GENERATION IN ISOLATED COMMUNITIES.
Infield, D.G. (Rutherford Appleton Lab., U.K.) In Wind Energy Conversion, proc. 5th BWEA Wind Energy Conference, Reading, U.K. 23-25 March 1983. P. Musgrove (ed.) pp. 151-162. Cambridge, U.K., Cambridge Univ. Press 1984 375 pp. (Isbn: 0-521-26250-X)
Discusses problems facing integrated wind power/diesel generator systems. Variability

of wind turbine outputs leads to a high number of stop/start cycles for the diesel generator, and low load conditions for the diesel generator means increased wear and maintenance. Considers possible energy storage options and develops a methodology to assess these and the simple pay-back period is calculated.

5.2.3(47) WIND TURBINE GENERATOR INTERACTION WITH CONVENTIONAL DIESEL GENERATORS ON BLOCK ISLAND, RHODE ISLAND. VOLUME 1 : EXECUTIVE SUMMARY. VOLUME 2 : DATA ANALYSIS.
Wilreker, V.F. and others. Westinghouse Electric Corp., Pittsburgh, PA., USA, Advanced Systems Technology Division. U.S. Dept. of Energy Report no: DOE/NASA-0354-1/2; NASA-CR-168318/9 February 1984 45 pp. and 140 pp. (DE84 015874)
The MOD-OA installation was the third of four experimental nominal 200 kW wind turbines connected to various utilities under the Federal Wind Energy Programme. The three-phases of the study analysis address: fuel displacement, dynamic interaction, and three modes of reactive power control.

5.3 Domestic Applications: Self-Sufficiency and Water/Space Heating

5.3(1) TECHNOLOGICAL SELF-SUFFICIENCY.
Clarke, R. Faber 1976 302 pp.
Reports on conversion of a hill farm and cottage into a commune and the learning and application of practical skills of house building, plastering and joinery. Methods of energy conservation through insulation, solar energy, and wind power, and attempts made to remain self sufficient in domestic aspects such as heat, water, plumbing, waste, compost, transport and food are described. Bibliography.

5.3(2) AMBIENT ENERGY.
Mcnelis, B. and Stevens, E. Building Design, 25 March 1977 pp. 16-19.
Looks at costs and performance of solar energy, wind generators, heat pumps, and water turbines.

5.3(3) URBAN HOMESTEADING WITH SUN AND WIND.
Buren, A.V. Architectural Design, 47 (4) April 1977 pp. 244-245.
Reports on the energy saving measures which form an integral part of a self-help tenant rehabilitation project in New York's lower east side. The measures include both energy conservation and the use of ambient sources of energy.

5.3(4) AMBIENT ENERGY AND BUILDING DESIGN.
Randall, J.E. Construction Press, 1978 166 pp.
Discusses use of alternative energy sources in building design. Includes wind energy, solar energy, use of heat pumps.

5.3(5) DESIGN OF A LOW-ENERGY HOUSE IN DENMARK HEATED BY A COMBINATION OF SOLAR AND WIND ENERGY.
Esbensen, T.V. and Strabo, F. In Sun; Mankind's Future Source of Energy: proc. of the International Solar Energy Society Congress, New Delhi, India 16-21 January 1978. New York, NY and Oxford, U.K., Pergamon Press 1978 V. 3 pp. 1454-1458.
Describes a low-energy house constructed in Skive, Jutland in Denmark. With energy conservation arrangements such as high-insulated construction, mobile insulation of the windows and heat recovery in the ventilating system, the heat requirement for space heating is calculated at 6000 kWh per year. The energy system consists of a flat-plate solar collector integrated into the roof construction, a windrotor and a water storage tank is provided with a water brake driven by the windrotor. This energy system supplies the house with 67% of the total heat requirement for space heating and hot water supply.

5.3(6) ECOLODGE - THE SELF-SUFFICIENT HOUSE.
Armstrong, G.H. House Builder, 37 (9) October 1978 pp. 414-415.
Describes with diagrams the construction of an autonomous house in Macclesfield, Cheshire, U.K. which draws its energy supplies from the sun, wind and rain.

5.3(7) LOW COST AERODYNAMIC HEATER REPRESENTING A FULLY MATCHED LOAD FOR WIND ENERGY SYSTEMS.
Lawson, M.O. (University of Dayton, Research Institute, Ohio, USA) In IEEE proc. National Aerospace Electronics Conference NAECON '78, Dayton, Ohio 16-18 May 1978. New York, USA, IEEE 1978 V. 2 pp. 874-880.
Early realisation of the broad use of wind energy for heating applications depends primarily on the total system cost. Major components of such a heating system are a wind turbine, an alternator, a tower, and a heat storage system. This order of listing may also represent the present order of costs, but the cost of the wind turbine can be reduced. Replacing the alternator with a mechanical heater, and in particular a heater acting on aerodynamic principles can provide a significant lowering of system cost.

5.3(8) SOFT TECH.
Baldwin, J. and Brand, S. Penguin Books, 1978 176 pp.
Collection of articles on alternative architecture and energy uses, including sections on tools, inventions, solar heating, wind power, transport, biofuels, building, etc.

5.3(9) WIND ENERGY-HEAT GENERATION.
Matzen, R. (Royal Vet. & Agric. University, Denmark) In 2nd International Symposium on Wind Energy Systems, Amsterdam, Netherlands 3-6 October 1978. Cranfield, U.K., BHRA Fluid Engng. 1978 pp. H2:17-H2.32.

A hot-water generator is an appropriate load for fixed pitch propeller wind-power plants as further special regulating systems are not required for normal service. This report deals with experiments on simple non-adjustable generators built into a 135-litre tank. An open impeller type for use in a heat-buffer tank or a highly insulated hot-water accumulator tank is also discussed. Different types of adjustable water-brakes are developed. Adjustment of power consumption is achieved by altering the stator-vanes' effective height by means of a disk. (5 refs.)

5.3(10) THE AUTARKIC HOUSE.
Littler, J.G.F. Electronics & Power, 25 (7) July 1979 pp. 489-493.

Describes a house that would be independent of all mains services by relying on energy conservation, and energy recovery, and the use of solar and wind power.

5.3(11) INVESTIGATION OF THE FEASIBILITY OF USING WIND POWER FOR SPACE HEATING IN COLDER CLIMATES.
Crcmack, D.E. (University of Massachusetts, USA) Rockwell International/DOE Report RFP-3060/67C25/3533/80-3 October 1979 110 pp.

A wind and solar powered space heating system has been operational in a fully automatic mode since September 1977 at the University. Performance data has been collected, and analysed for the heating system as a whole and for the wind turbine in particular. Results indicate that the wind furnace provided a significant amount of the heating load for the structure during November 1977 - April 1978. (24 refs.) See also paper by same authors in DOE Small Wind Turbine Systems R & D Requirements Conference, Boulder 27 February-1 March 1979 V. 1 pp. 159-165, covering design of the wind furnace.

5.3(12) LOW COST WIND ENERGY CONVERSION SYSTEM FOR HEATING OF DOMESTIC PREMISES.
Freris, L.L. and others. (Imperial College of Science & Technology, London, U.K.) In International Conference on Future Energy Concepts, London, U.K. 30 January-1 February 1979 pp. 282-285. London, Institution of Electrical Engineers 1979 IEE Conference Publ. no: 171.

Attempts to demonstrate that if a number of conditions are fulfilled and on the basis of certain assumptions, an economically viable case can be made for small scale wind energy exploitation, i.e. individual household applications in rural areas and the outer fringes of suburbia. (4 refs.)

5.3(13) PROPOSAL FOR GENERATING HEAT FROM WIND ENERGY FOR DOMESTIC AND OTHER USES.
White, P.W. In Energy for Industry, a Collection of Scientific and Engineering Papers concerned with Utilising Energy with Maximum Efficiency in Industry. New York and Oxford, U.K., Pergamon Press 1979 pp. 141-151.

The system has a heat pump with the compressor driven directly by a wind-powered turbine. Heat is absorbed in the evaporator of the heat pump by ambient convection and solar radiation. Mismatch between heat supply and demand is overcome by providing a thermal store using the latent heat of water which supports the wind-powered turbine and absorbs heat for transmission uniformly to the evaporator. The turbine is shrouded and driven at high speed due to the shape of the inlet duct. Advantages for this arrangement are low torque reaction, low noise and high aerodynamic efficiency.

5.3(14) WIND POWERED HOUSING PROJECT IN HULL.
Hodges, D.C. Sun at Work in Britain, (8) 1979 pp. 45-48.

Reports a project funded by Hull Corporation (U.K.) in which 32 houses were designed to reduce tenants' energy consumption by using high levels of insulation and energy inputs from natural sources such as sun and wind. The scheme has now developed into primarily a wind power project with back-up provided by coal burners.

5.3(15) A COMBINED SOLAR HEATING PLANT WITH SEASONAL HEAT STORAGE.
Margen, P. Stockholm, Sweden, Swedish Council Building Research 1980 28 pp. Report no: R90. (In Swedish)

A preliminary study of the feasibility of supplying space heating to a coastal residential area (e.g. 1000 dwellings) using an aerogenerator to drive a heat pump with the sea as its heat source. A storage capacity of only 15% of annual heat demand is thought necessary to maintain supply, compared with about 60% for a solar system. Combining the wind plant with solar collectors via a shared heat store is also considered.

5.3(16) OPERATIONS CHARACTERISTICS OF A UTILITY-FREE DWELLING IN KANSAS.
Pick, E.A. (Energy Productions, Kansas, USA) Paper presented at Oklahoma State University/Et Al Earth Sheltered Building Design Innovations Conference, Oklahoma City 18-19 April 1980 PIV-3 4 pp.

Describes an energy efficient, utility-free dwelling in Kansas. The earth sheltered home is supplied with electricity from a small wind energy conversion system. Solar features provide space heating and water heating needs.

5.3(17) THE SELF-SUFFICIENT HOUSE.
Vale, R. and Vale, B. Macmillan, 1980 214 pp.

Do it yourself handbook for those who wish to apply energy conservation measures in their own houses. Covers thermal insulation in floors, walls, ceilings and windows, windmills and solar collectors. Describes the use of trees and shrubs to create shelter belts and wind breaks and gives advice on water recycling and cooking techniques.

5.3(18) WIND ENERGY.
Energy Digest, 9 (5) October 1980 pp. 26-30.
   Wind energy siting studies have identified
the East coast of the U.K. as suitable for
establishing wind energy homes. Describes
a project which seeks to determine the
feasibility of using an aerogenerator to
provide domestic space and water heating.
The 32 houses to be supplied will be highly
insulated and contain heat recovery devices.
Design proposals are described.

5.3(19) A DESIGN PROCEDURE FOR WIND POWERED
HEATING SYSTEMS.
Manwell, J.F. and McGowan, J.G. (University
of Massachusetts, USA) Solar Energy 26 (5)
1981 pp. 437-445.
   Describes a generalised procedure for the
month-by-month prediction of performance of
wind powered residential heating systems.
The technique simulates residential heating
systems using conventional horizontal axis
wind turbines that dissipate their output
into water-based thermal storage systems.
The monthly heating energy fraction supplied
by a wind system can be determined with a
minimum of site, turbine, and system input
parameters. (14 refs.)

5.3(20) HEATING FROM WIND.
Stobart, A. Building Technology Management,
19 (7) July/August 1981 pp. 15-16.
   Discusses the potential of wind energy for
heating and describes the principal elements
for a heating system; a wind energy collector
(windmill), control system to compensate
for variations in wind speed, main heat store,
and water heater. Lists possible wind energy
applications, both domestic and commercial.

5.3(21) SMALL WIND ENERGY CONVERSION SYSTEMS
FOR RESIDENTIAL APPLICATION.
Shinyama, N.K. and Kuwanoe, C.A. (Maui Electr.
Co., Kahului, Hawaii) In IEEE Reg. 6 Conference
1981: Electrical Engineering Applications in
the Pacific, Honolulu, Hawaii 1-3 April 1981.
Piscataway, NJ, IEEE Cat. no: 81CH1664-2 1981
pp. 78-101.
   On the Island of Maui, Hawaii, 3 Enertech
1500 watts induction motor-generator windplants
have been installed and interconnected to Maui
Electric Company's distribution system. The
windplant is a horizontal axis, down wind
machine with a fixed propeller and mounted on
a 60 foot tower. As an induction motor/
generator, it requires external excitation from
the utility system to operate; therefore providing the safety feature of not being able to feed
back to the utility grid during outages. The
maximum output rating is 2100 watts in a 28 mph
wind and always produces an output of 115 volts,
60 Hz, synchronous with U.S. utility line
voltage.

5.3(22) WIND TO HEAT CONVERTER. FINAL TECHNICAL
REPORT.
Fisher, E.D. (Bellaire, USA) U.S. Dept. of Energy
Report no: DOE/R5/10120-2 6 March 1981 56 pp.

   The feasibility of a wind driven machine
which converts wind energy directly to heat
through the action of the impeller of a windmill driving a rotary hydraulic brake which
absorbs the power by converting the mechanical
energy into heat is demonstrated. A prototype
model was constructed and erected on a hilltop
site adjacent to a building connected to the
building's heating system and its performance
evaluated. The unit consists of a steel tower
nearly 50 feet high, and a two-bladed impellor
approximately 38 feet in diameter.

5.3(23) THE BEHAVIOUR OF A LONG TERM HEAT
STORAGE SYSTEM IN CONNECTION WITH A WIND ENERGY
CONVERTER (PHASE 1) FINAL REPORT MAY 1980.
Auer, F., Bley, H. and Mueller, M. (Battelle-
Inst. e.V., Frankfurt am Main, F.R. Germany)
Report no: BMFT-FB-T-82-057 April 1982 65 pp.
(In German with English Summary) NTIS Report
no: 82066754.
   Report on the technical and economical
feasibility of heating a museum on the
Wasserkuppe mountain in the Rhoen hills by
wind energy. It is possible to recover the
capital cost within a period of 15 to 19 years
if the service life of the wind energy converter and of the heat storage system is between
20 and 30 years. At a maximum thermal load of
120 kW, the minimum cost will be achieved with
a WEC power output of 120 kW and a two day
storage system. This combination permits 60%
of the heat demand to be met by wind energy
and the balance supplied by conventional energy
sources.

5.3(24) EVALUATION AND MODELLING OF A RESI-
DENTIAL WIND ENERGY CONVERSION SYSTEM.
Jong, M.T. (Wichita State University, USA)
In proc. 1982 Wind and Solar Energy Technology Conference, Kansas City, USA 5-7 April
1982. Columbia, USA, Missouri-Columbia University 1982 pp. 221-227.
   Presents a performance analysis of an
Enertech 1500 small wind energy conversion
system which was used to develop a model
for SWECS in a household.

5.3(25) HOME ENERGY CONSERVATION PASSIVE
SOLAR, WINDOWS, WOOD STOVES, WIND TURBINES,
APPLIANCES, AND HEAT PUMPS.
Biddle, S. and others. Renewable Energy News,
5 (7) October 1982 pp. 15-26.
   Discusses passive solar, wind energy, wood
fuel, and heat pump technologies which are
increasing their penetration in the residential
sector. Trends in residential use of wood
chips, wind turbines, and solar greenhouses
are reviewed.

5.3(26) OPERATION OF IOWA'S ONLY TOTALLY
ENERGY INDEPENDENT OFFICE BUILDING.
White, T. (Natural Energy Inc., Iowa, USA)
Paper presented at Iowa Energy Policy Council/
Iowa Solar Operational Results Conference,
Des Moines 21-22 June 1982 pp. 51-58.

Describes an office building in Davenport, Iowa that is energy self-sufficient by using an active and passive solar heating system, waste heat recirculator, and wind energy turbine. The 1800 sq.ft. structure costs $94,000, including its energy systems, making it economically competitive with similar commercial buildings in Davenport.

5.3(27) SPACE HEATING THROUGH ISENTROPIC COMPRESSION UTILISING WIND ENERGY.
Noon, R. (McGraw-Edison Co., Shawnee, Kansas, USA) ASME Trans., J. Solar Energy Engng., 104 (3) August 1982 pp. 204-207.

A preliminary examination of using wind power to provide hot air for space heating using the shaft power of a wind turbine to isentropically compress air. Excluding mechanical losses, 71 percent of the shaft power supplied to a single stage compressor can be converted to internal energy gain. Approximate economic parameters indicate that such wind space heating systems are more economic than active solar systems as long as the average heating season available wind energy per collection area is 20 percent higher than that for solar. (13 refs.)

5.3(28) TOTAL ENERGY WIND CONVERSION SYSTEM.
Bandopadhayay, P.C. (Commonwealth Scientific & Industrial Research Organisation, Highett, Victoria, Australia) Wind Engineering, 6 (2) 1982 pp. 85-94.

Proposes a system which supplies the air conditioning and the hot water demands as well as the electric load, and shows that such a system can improve the economic justification for the use of wind energy to supply electricity to an isolated consumer in a rural area.

5.3(29) HEATING WITH WIND ENERGY.
Kaier, U. and Czink, F. Bundesministerium fuer Forschung und Technologie, Bonn-Bad Godesburg, F.R. Germany) Report no: BMFT-FB-T-83-131 June 1983 317 pp. (In German)

In two parts: A guide for those interested in the practical use of wind energy for heating purposes including economic aspects and suggestions and recommendations as to possible problems, which might be encountered; A survey of usable wind energy resources for heating purposes at 10 locations in the country and at 3 locations along the off-coast region of the Federal Republic of Germany, and an international marketing analysis on wind energy converting systems including a selection of usable systems up to 20 kW.

5.3(30) MIXING RENEWABLES COMBINING WIND AND SOLAR ELECTRICITY.
Cock, S.P. (North Arkansas Community College, USA) Alternative Sources of Energy, (60) March-April 1983 pp. 20-22.

Examines factors to be considered when planning to install a combined solar-wind energy electricity system in a home. Operation of such a combined system in an Arkansas dwelling demonstrates that it is more practical compared with either solar or wind energy alone.

5.3(31) RATED ELECTRICAL POWER NEEDED IN A WIND TURBINE FOR A HOME.
Steinberg, T.A. and Reynolds, B.W. (New Mexico State University, Las Cruces, NM, USA) J. Energy, 7 (6) November-December 1983 pp. 745-746.

A rated power chart is developed here for sizing the rated power of a wind turbine needed to provide a specified annual amount of energy in a known wind regime. To utilise the chart the amount of electrical energy needed per year (in kilowatt hours) to be delivered by the wind turbine and the mean annual wind speed at the site where the wind turbine installation is proposed must be known.

5.3(32) AN ANALYTICAL STUDY OF A HYBRID WIND-PASSIVE SOLAR SYSTEM.
Bell, B.F. and McGowan, J.G. (Massachusetts University, USA) Solar Energy 32 (3) 1984 pp. 401-415.

Analytical performance and economic evaluation of a system for residential heat and electrical energy system in a simulated New England wind and weather environment, based on a model of a vertical axis wind turbine with a super insulated passive solar house. Results for heating, electrical supply and economic performance of the system based on site average wind speed, are given.

5.4 Agriculture Including Irrigation

5.4(1) ECONOMICS OF WIND ENERGY USE FOR IRRIGATION IN INDIA.
Tewari, S.K. (Nat. Aeronaut. Lab., USA) Science, 202 (4367) November 1978 pp. 481-486.

Shows that wind energy could be economically competitive for irrigation from open wells on small farms in rural and remote areas.

5.4(2) IRRIGATION PUMPING WITH WIND ENERGY.
Clark, R.N. and Schneider, A.D. (USDA, South-West Great Plains Research Centre, Bushland, Texas, USA) Paper for American Society of Agricultural Engineers Winter Meeting, Chicago, Illinois 18-20 December 1978. St. Joseph, Michigan, USA, ASAE 1978 Paper 78-2549 16 pp.

A 40-kW vertical axis, wind turbine was erected at the USDA SouthWestern Great Plains Research Centre, Bushland, Texas. Objectives were to assemble a complete wind-power pumping system, adapt or modify existing pumping equipment so that it could be effectively powered by a wind turbine and make economic analyses of wind pumping systems. The pumping system used both a wind turbine and an

electrical motor to power an existing deep
well irrigation pump.

5.4(3) TECHNICAL AND ECONOMIC FEASIBILITY OF
MAKING FERTILISER FROM WIND ENERGY, WATER AND
AIR.
Dubey, M. (Lockheed California Co., Burbank,
California, USA) In Sun; Mankind's Future
Source of Energy: proc. of International
Solar Energy Society Congress, New Delhi,
India 16-21 January 1978. New York, NY and
Oxford, U.K., Pergamon Press 1978 V. 3
pp. 1812-1821.
   Wind energy, air and water can be combined
to make anhydrous ammonia and ammonium nitrate,
two important nitrogenous fertilisers that are
usually synthesised using natural gas as the
primary feedstock. A study has shown that it
is technologically feasible to reduce the
scale of an ammonia processing plant to
produce a tiny fraction of the output rates
of full scale commercial plants. Such a
system can be adequately powered by a wind
turbine driving an electrolysis cell to
produce the required hydrogen feedstock.
(6 refs.)

5.4(4) APPLICATION OF WIND ENERGY TO GREAT
PLAINS IRRIGATION PUMPING. FINAL REPORT.
Hagen, L.J. and others. (Kansas Agricultural
Experiment Station, Manhattan, Kansas, USA)
Washington, DC, USA, U.S. Dept. of Energy
Report no: DOE/SEA-3707-20741/80/1 October
1979 160 pp.
   Looks at wind energy systems without energy
storage for irrigation in the Great Plains.
Major uses of irrigation energy were identified as pumping for surface distribution
systems, which could be supplied by variable
flow, and pumping for sprinkler systems using
constant flow. A computer programme was
developed to simulate operation of wind-
powered irrigation wells. Pumping by wind
turbine systems was simulated for 2 variable
and 2 constant flow operational modes in
which auxiliary motors were used in 3 of
the modes. Using the simulation programme,
the well yields and maximum pumping rates as
a function of drawdown in a typical well
are compared.

5.4(5) ENERGY FOR WORLD AGRICULTURE.
Stout, B.A. and others. (Michigan State University, USA) FAO Report 1979 286 pp.
   Examines current uses of commercial energy
in agriculture, such as inputs to fertilisers,
pesticides, irrigation and farm machinery.
Alternative energy sources, including wind
energy, are then evaluated. (232 refs.)

5.4(6) WIND ENERGY APPLICATIONS IN AGRICULTURE.
Kluter, H.H. and Soderholm, L.H. (Iowa State
University, Ames, Iowa, USA) In proc. of
Conference on Wind Energy Application in Agri-
culture, Ames, Iowa 15 May 1979. Washington,
DC, USA, U.S. Dept. of Energy Report no: CONF-
7905109 1979 325 pp.

5.4(7) WIND ENERGY APPLICATIONS IN AGRICULTURE:
EXECUTIVE SUMMARY. FINAL REPORT.
David, M.L. and others. (Development Planning
and Research Associates, Inc., Manhattan,
Kansas, USA) Washington, DC, USA, U.S. Dept. of
Energy Report no: DOE/SEA-1109-20401/79/2
August 1979 209 pp.
   Presents an assessment of the potential use
of wind turbine generator systems in U.S. agriculture. Identifies the agricultural WTG's
applications in terms of location, type and
size (complete farm and dedicated-use applications), the number of WTG's by wind machine
and generator size category; aggregate energy
conversion potential; and other technical and
economic WTG's performance data for particular
applications. It also describes the methodology, data and assumptions used for the
analysis. A major part of the study was the
development and use of a rigorous analytical
system to assess an application's wind power
generation and use potential.

5.4(8) WIND ENERGY POTENTIAL FOR AGRICULTURAL
APPLICATIONS IN SOUTH DAKOTA.
Verma, L.R., Lytle, W.F. and Hellickson, M.A.
(Louisiana State University, Baton Rouge,
Louisiana, USA) Paper for American Society of
Agricultural Engineers Winter Meeting, New
Orleans, Louisiana 11-14 December 1979.
St. Joseph, Michigan, USA, ASAE Paper 79-3509
1979 20 pp.
   Climatological data records were analysed for
Huron, South Dakota for 1965 to 1976 to
establish wind energy available in east central
South Dakota; intensity and variation on a
daily and monthly basis; relative potential
of wind energy as compared to other sources; combined potential of wind and solar energies in
east central South Dakota and potential of wind
energy for agricultural applications.

5.4(9) ECONOMIC POTENTIAL OF WIND ENERGY IN
CROP-DRYING AND PROCESS HEATING/COOLING APPLI-
CATIONS.
Harper, M.R. and Garling, W.S. (Aerosp. Corp.,
Germantown, Maryland, USA) In proc. Annual
Meeting of American Section International Solar
Energy Society 1980, Solar Jubilee 25 Years of
the Sun at Work V. 3.2, Phoenix, Arizona, USA
2-6 June 1980. Newark, Delaware, USA, ISES/AS
1980 pp. 1473-1477.
   Departments of Energy and Agriculture
currently are establishing the feasibility of
wind energy use in applications where the energy
can be used as available, or stored in a simple
form. These include production of hot water for
rural sanitation, heating and cooling of rural
structures and products, drying agricultural
products, and irrigation. (7 refs.)

5.4(10) ECONOMICS OF WIND ENERGY FOR IRRIGATION
PUMPING.
Lansford, R.R. and others. (SouthWest Research
and Development Co., Las Cruces) Washington,
DC, USA, U. . Dept. of Energy Report no:
DOE/SEA-7315-2074/81/2 14 July 1980 89 pp.

Addresses some of the economic questions associated with wind power as an energy source for irrigation under different situations in seven regions of the United States. The analysis will provide the United States Departments of Energy and Agriculture with target investment costs for wind turbines used for irrigation pumping; it also will provide policy makers with bases for adjusting taxes to make alternative sources of energy investments more attractive. The three types of wind powered irrigation systems evaluated for each region were: wind assisted combustion engines, wind assisted electric engines, with or without sale of surplus electricity, and stand alone reservoir systems with gravity flow reservoirs.

5.4(11) THE IMPACT OF VARIOUS ENERGY INNOVATIONS ON ENERGY CONSUMPTION AND NET INCOME FOR 48 SMALL FARMS: FINAL REPORT.
Centre for Rural Affairs Small Farm Energy Project Report, August 1980 60 pp. (USA)
The Small Farm Energy Project was initiated in 1977 to demonstrate the potential for small farmers to develop low-cost alternative energy systems that reduce production costs and increase net incomes. The project involves only appropriate use technologies that are cost-effective and relatively easy to manage and maintain, meet the constraints existing on small farms, and can be built at low cost, making use of locally available materials and common farm skills. Includes wind energy.

5.4(12) AN IMPROVED CONTROL STRATEGY FOR WIND-POWERED REFRIGERATED STORAGE OF APPLES.
Baldwin, J.D.C. and Vaughan, D.H. (Louisiana State University; Virginia Polytechnic Institute and State University) Paper presented at American Society of Agricultural Engineers National Energy Symposium, Kansas City 29 September-1 October 1980 pp. 496-500.
Describes a refrigerated apple storage facility, which includes a 10 kW electric wind generator, electrical battery storage, thermal storage, and auxiliary power. The facility, in the Virgina Polytechnic Institute's Horticultural Research Farm in Blacksburg, VA., was constructed to aid in the development of control strategies that enable wind powered systems to make the best use of available wind energy. Individual components are tested and in situ performance test are conducted. (8 refs.)

5.4(13) THE POTENTIAL FOR OPTIMISING THE ECONOMICS OF WIND ENERGY UTILISATION ON A DAIRY FARM THROUGH LOAD MANAGEMENT.
Kear, E.B. (Clarkson College, USA) In A Collection of Technical Papers from AIAA/SERI Wind Energy Conference, Boulder, USA 9-11 April 1980. New York, USA, American Institute Aeronaut. & Astronaut. 1980 Paper no: 80-0640 pp. 275-279.
Several different sized WECS were studied, along with various simple load management scenarios and potential storage concepts, to

determine their individual and cumulative effects on the percentage of the total energy generated by the WECS which would be used on the dairy farm.

5.4(14) WIND ASSISTED DEEP-WELL PUMPING.
Clark, R.N. (U.S. Dept. of Agriculture) Paper presented at American Society of Agricultural Engineers National Energy Symposium, Kansas City 29 September-1 October 1980 pp. 479-483.
Examines a wind energy project for irrigation pumping at the USDA Conservation and Production Research Laboratory in Bushland, Texas. A 37-kW vertical axis wind turbine connected to an existing irrigation pump is coupled by a combination gear drive between the pump and electric motor, thus reducing the load on the electric motor and saving energy. An overrunning clutch synchronises the two power sources, and results indicate that as much as 40% of the present energy consumption can be produced by wind power. (4 refs.)

5.4(15) A WIND DRIVEN WATER HEATING SYSTEM DEVELOPMENT AND FIELD TESTING.
Gunkel, W.W. and others. (Cornell University, USA) Paper presented at American Society of Agricultural Engineers National Energy Symposium, Kansas City 29 September-1 October 1980 pp. 257-261.
A USDA-SEA research project "Wind Energy Substitution at Dairy Milking Centres" studied the potential of a wind driven direct water heating system for dairy production. A commercial wind turbine, the Pinson cycloturbine, and an agitation-type water heater are used, and the power absorption characteristics of the agitator design are determined. The system was monitored for nine months, and data on available wind power density at the system site, monthly energy output, and operating time are included. The system is shown to have produced power during 50% of its operating time. (7 refs.)

5.4(16) WINDPOWER FOR AGRICULTURE.
Stobart, A. (Trimble Windmills) In proc. Conference on Energy for Rural & Island Communities, Inverness, U.K. 22-24 September 1980. J. Twidell (ed.) Oxford, U.K., Pergamon Press 1981 Topic D, Paper D2 pp. 137-143.
Describes the potential use of wind energy in agriculture. The most cost effective uses for wind power on a farm are heating and water pumping.

5.4(17) AGRICULTURAL APPLICATIONS OF SWECS.
Nelson, V. (West Texas State University, USA) In proc. 5th Biennial Wind Energy Conference & Workshop, Washington, DC, USA 5-7 October 1981 I.E. Vas (ed.) V. 1. Palo Alto, USA, Solar Energy Research Institute 1981 Session 1B pp. 227-235. Report nos: SERI/CP-635-1340 CONF-811043.
Discusses principal applications of wind energy for agriculture: farmstead power, mainly

electrical; building heating; irrigation pumping; product storage and processing; hot water
for residences and dairies; and associated
industries such as feedlots, fertilisers,
elevators, greenhouses, etc. Future applications
include small wind farms.

5.4(18) DECENTRALISED APPLICATIONS OF WIND
ENERGY.
Beurskens, H.J.M. Energiespectrum, 5 (7-8)
July/August 1981 pp. 182-188. (In Dutch)
    Discusses the extension of the Dutch National
Research Programme on Wind Energy to cover the
market, especially in the agricultural sector,
for medium-sized turbines of 7-25 m. rotor
diameter.

5.4(19) ECONOMICS OF WIND SYSTEMS FOR AGRI-
CULTURAL APPLICATIONS.
Soderholm, L.H. and Clark, R.N. (U.S. Dept. of
Agrculture, Ames, Iowa, USA) In proc. 5th
Biennial Wind Energy Conference and Workshop,
Washington, DC, USA 5-7 October 1981 I.E. Vas
(ed.) V. 2. Palo Alto, USA, Solar Energy Research Institute 1981 Session 3A pp. 535-547.
Solar Energy Research Institute Report nos:
SERI/CP-635-1340 CONF-811043.
    The economics of several major uses of wind
energy for agriculture, as determined from
U.S. Department of Agriculture and Department
of Energy studies, are given for irrigation
pumping, farmhouse and farm building heating,
crop drying, and food processing. Small wind
energy conversion systems are close to economic feasibility for many agricultural applications when used for other than limited
periods, are located in wind regimes of at
least 5-6 m/s and if presently available tax
incentives are used, based on data in 1980.

5.4(20) FEASIBILITY STUDY OF WIND-POWERED
SYSTEMS FOR DAIRY FARMS.
McGowan, J.G. and Wendelgass, P.F. (University
of Massachusetts, New York State Energy Office)
J. Energy, 5 (3) May-June 1981 pp. 164-171.
    Presents the results of an analytical per-
formance and economic evaluation of the potential
of wind-turbine-generator systems for supplying
electrical and thermal energy to dairy farms.
The model consists of four sections: dairy farm;
WTG system; wind system performance; and life
cycle economic models. Three different wind-
turbine systems are used to carry out a national
and regional evaluation. (17 refs.)

5.4(21) WIND ENERGY FOR IRRIGATION: WIND-ASSISTED
PUMPING FROM WELLS. FINAL REPORT.
Clark, R.N. and others. (Agricultural Research
Services, Bushland, Texas, Conservation and Production Lab.) Washington, DC, USA, U.S. Dept.
of Energy Report no: DOE/SEA-7315-20741/81/3
January 1981 77 pp.
    The project objectives were: to assemble a
complete wind-powered pumping system for irrigation wells in the Southern Great Plains; to
adapt or modify existing pumping equipment so
that it could effectively be powered by a

vertical-axis wind turbine; to develop a data
collection system for collecting, recording and
processing on-site wind data and mechanical,
electrical and hydraulic systems data; to
combine data from all systems in an overall
analysis that will permit an engineering
evaluation of the complete system; to develop
a dynamic mathematical model of the pumping
system which includes the most appropriate
model of the vertical-axis wind turbine; and
to make an economic analysis of the wind-
powered pumping system.

5.4(22) WIND ENERGY USE IN ON-FARM GRAIN
DRYING.
Garling, W.S. (Yellowstone Research Corp.,
Arlington, VA, USA) Intl. J. Ambient Energy,
2 (2) April 1981 pp. 73-78.
    On-farm grain dryers and their energy require-
ments are described, and the energy output and
cost of selected small wind systems are estimated. Life cycle cost analyses are performed
to determine economic feasibility. (10 refs.)

5.4(23) WIND TURBINES FOR IRRIGATION PUMPING.
Clark, R.N. and others. (Conservation & Production Research Lab., Texas, USA) J. Energy,
5 (2) March-April 1981 pp. 104-108.
    A vertical-axis and a horizontal-axis turbine
were tested in a mechanical wind-assist mode
and a horizontal-axis, electrical-output turbine was tested in an electrical-assist mode.
Data from all turbines indicated that power
produced was normally less than expected, but
was adequate for irrigation pumping. As much
as 40% of the present energy consumed in irrigation pumping can be generated by wind power.
(4 refs.)

5.4(24) AGRICULTURE WIND ENERGY APPLICATIONS
ANALYSIS.
Wagner, J.P. (Development Planning & Research
Associates Inc., Manhattan, Kansas, USA) Intl.
J. Ambient Energy 3 (4) October 1982
pp. 207-212.
    Presents simulation model results for a sample
agricultural wind energy application, depicting a hypothetical winter wheat and sorghum
farm. (11 refs.)

5.4(25) ECONOMICS OF WIND ENERGY FOR HEATING
RURAL STRUCTURES.
Soderholm, L.H. (Iowa State University, USA)
American Society of Agric. Engrs. Trans. (Gen.
Edn.), 25 (5) September-October 1982
pp. 1392-1395. (See also American Society of
Agricultural Engineers related papers and
abstracts from the ASAE National Energy
Symposium 29 September-1 October 1980, Kansas
City, USA. St. Joseph, Michigan, ASAE 1981
pp. 262-265.)
    Wind energy is very close to economic feasi-
bility to heat many rural buildings by means of
small wind energy conversion systems when they
are located in wind regimes of at least 5 to 6
m/s and when available tax incentives are used.
The building heating possibilities and pro-

cedures for determining systems output and economics are given.

5.4(26) FEASIBILITY OF WIND ELECTRIC PUMP FOR UTTAR PRADESH.
Nathan, G.K. and Srivastava, K.P. Indian J. Power & River Valley Development, 32 (7) July 1982 pp. 112-116.
Discusses the potential for wind power in the form of a wind electric pump for irrigation purposes in the United Provinces, India and notes various methods available for wind powered irrigation pumping.

5.4(27) METHODOLOGY FOR THE ECONOMIC EVALUATION AND OPTIMAL SIZING FOR WIND ENERGY HEATING OF FARM BUILDINGS.
Darvish, M. (Econ. Syst. Inc., USA) Engineering Econ., 27 (3) Spring 1982 pp. 197-216.
The methodology determines the cost of wind-powered heating systems which would break even with the cost of present heating methods. This enables a potential WECS user to determine the suitability and economic feasibility of using wind power heating for a range of future conventional energy costs. The associated market size of potential users of wind-powered heating systems is predicted as a function of conventional energy costs. (16 refs.)

5.4(28) SOLAR- AND WIND-POWERED IRRIGATION SYSTEMS.
Enochian, R.V. (Economic Research Service, DC, USA) NTIS Report PB82-177486 February 1982 34 pp.
Irrigation pumps may be powered by five different direct solar and wind energy systems but only two may produce unsubsidised energy for such pumps before the turn of the century. These utilise either concentrating collectors or wind turbines.

5.4(29) USE AND PERFORMANCE OF WIND ENERGY FOR BUILDING HEATING.
Soderholm, L.H. (Iowa State University, USA) Paper presented at Iowa Energy Policy Council/ Iowa Solar Operational Results Conference, Des Moines 21-22 June 1982 pp. 101-119.
Describes the use and performance of wind energy for farm structures. Wind energy in such a setting may first be converted into electrical energy, then into heat and thus stored in a water medium. Water storage, in combination with a heat pump, allows the farmer to store his wind energy and reduce energy requirements from conventional sources. (10 refs.)

5.4(30) WIND ENERGY IN AGRICULTURE.
Halliday, J.A. and Lipman, N.H. Wind Engineering, 6 (4) 1982 pp. 206-218.
Describes the evaluation of the wind energy potential of a site; different types of wind turbine; uses of wind power in agriculture; the integration of wind generated electricity into factors which can influence the growth of

the wind energy industry. (Also presented at the 2nd International Seminar on 'Energy Conservation and use of Renewable Energies in the Bio-Industries', Oxford, September 1982.)

5.4(31) DAIRY FARM WIND ENERGY SYSTEMS.
Abarikwu, O.I. and Meroney, R.N. (Colorado State University, USA) American Society Agricultural Engineers Trans. (Gen. Edn.), 26 (1) January-February 1983 pp. 255-259.
Presents a model for the selection of a wind energy system substituting wind generated electrical energy for utility electric energy to meet cooling and water heating energy requirements in dairy operations. The model determines sizes, installed costs and wind energy cost of the system for different herd sizes and wind speeds, with and without energy recovery systems. A system which combines conservation measures with a small wind machine is likely to prove economically attractive to the dairy farmer.

5.4(32) HARNESSING WIND IN WESTERN CANADA.
MacGregor, I. (Abax Energy Services Ltd., Calgary, Alberta, Canada) J. Can Pet. Technol., 22 (6) November-December 1983 pp. 43-45.
An Alberta-based manufacturing firm, Abax Energy Services Ltd., has developed a new type of large-scale wind turbine for water pumping. This unit, the largest direct water pumping wind turbine built to date, is now being tested in Calgary. The turbine has a substantial advantage in terms of capital cost versus outputs over other units currently being manufactured and therefore is competitive with conventional energy sources in areas of appropriate wind regimes.

5.4(33) WIND ENERGY FOR IRRIGATION PUMPING.
Rezachek, D. (Engng. Dept. Hawaiian Sugar Planters' Association, Aiea, HI, USA) In 6th Miami International Conference on Alternative Energy Sources, proc. of condensed papers, Miami Beach, Florida, USA 12-14 December 1983 pp. 473-474. Coral Gables, Florida, USA, University of Miami 1983 694 pp.
A significant proportion of the electrical energy used on Hawaiian sugar plantations is for irrigation pumping. Some electricity must be purchased during periods when electrical demand exceeds production or when plantations are not milling. Wind energy is one potential alternate energy source and Hawaii has some of the most favourable wind regimes in the world. For these reasons, a study was undertaken to assess the potential of wind energy for irrigation pumping and other applications.

5.4(34) DESIGN AND DEVELOPMENT OF A SMALL OUTPUT MULTIBLADE TYPE WINDMILL FOR PUMPING WATER FOR AGRICULTURAL PURPOSES.
Jagadish, B.S. and others. (Indian Inst. of Tech., Bombay, Dept. of Mechanical Engineering) Report no: NP-4901349 December 1983 101 pp. (DE84 901349)
Study is concerned with evolving a design suitable for operation with rotary pumps.

Wind data for Santa Cruz, Bombay is analysed and
designs for two horizontal axis, multiblade
type rotors suitable for operation in 8 to
30 kmph winds are evolved. Both have solidity
ratios considerably less than the conventional
design used for similar applications. The
design details and performance of the windmill
tested over a period of two years are presented
with cost estimates and economics of operation
to enable evaluation of the windmill with alternatives for water pumping applications.

5.4(35) PERFORMANCE AND ENERGY PRODUCTION OF
THE ENERTECH 44.
Vosper, F.C. and Clark, R.N. (ARS, Texas,
USA) In American Wind Energy Association Wind
Energy Expo National Conference, San Francisco
17-19 October 1983 pp. 294-306.
A wind turbine for agricultural applications
was installed in May 1982 at the USDA Research
Lab. in Bushland, Texas. The machine is intended
to provide electricity for irrigation pumping
and other agricultural loads. Over 7200 hours
of operation have been logged with 94% availability. The unit has produced over 75,000 kWh
of energy at an average windspeed of 13 mph at
a height of 10 m. (3 refs.)

5.4(36) GROUNDWATER PUMPING BY USING WIND
ENERGY.
Veltri, P. (Univ. di Calabria, Dipartimento
di Difesa del Suolo, Rende, Italy) Energ.
Elettr., 61 (6) June 1984 pp. 233-241.
(In Italian)
The possibility of pumping various groundwater volumes for some irrigation areas close
to Reggio Calabria by using wind electric
energy is explored. The wind speed data at
the meteorological station of Reggio Calabria,
which have been collected during 28 years, are
processed by making use of duration curves.
The rated wind speed of the WECS is chosen
by means of a simple criterion of economic
optimisation. Various plant configurations
are possible and energy storage problems
examined. (10 refs.)

5.4(37) IRRIGATION PUMPING WITH WIND ENERGY -
ELECTRICAL VS. MECHANICAL.
Clark, R.N. American Soc. Agric. Engrs. Trans.
(General Edition) 27 (2) March-April 1984
pp. 415-418.
System uses a vertical axis wind turbine to
generate electricity that is comparable with
utility grid power, and the pump is powered
by this electrical system using a conventional
electric motor. Another new system uses a
vertical axis wind turbine to produce mechanical
power or in combination with a diesel engine or
electric motor. The mechanical system is
claimed to provide more energy, whilst the
electrical system is more profitable.

5.5 Other Applications including Desalination Plants, Telecommunications and Wind Propulsion

5.5(1) HYDROGEN FUEL PRODUCTION BY WIND ENERGY
CONVERSION.
Ben-Dov, E., Naot, Y. and Rudman, P.S. (Israel
Electr. Corp., Haifa) In Alternative Energy
Sources, Miami International Conference International Compendium, Miami Beach, Florida, USA
5-7 December 1977. Washington, DC, Hemisphere Publ. Corp. 1978 V. 8 pp. 3563-3576.
Gives an analysis based on NASA experience
and capital cost estimates by General Electric
Co. and by Kaman Aerospace Corp. of wind energy
conversion systems for feeding an electric
utility grid. Considered are the cost of WEC
for feeding an electric utility grid as a
function of the mean wind speed; the use of WEC
for feeding an electric grid that has no storage
capacity, so that the WEC has a fuel saving
role only; and the use of WEC to produce
hydrogen by electrolysis of water as a substitute for gasoline.

5.5(2) APPLICATION OF SMALL WIND ENERGY CON-
VERSION SYSTEMS TO THE TELECOMMUNICATIONS
INDUSTRY.
Ferrell, G.C. and others. (Teknekron Research
Inc., Berkeley, California, USA) In proc.
INTELEC International Telecommunications
Energy Conference, Washington, DC
26-29 November 1979. Piscataway, NJ, IEEE
1979 pp. 438-444.
With regard to high reliability, recent
testing of wind-electric systems under harsh
atmospheric conditions, has proved and
documented a high level of necessary performance. Small wind systems are now capable
of remote, unattended, and near maintenance
free operation, performance characteristics
which are needed for applications in the telecommunications industry. (12 refs.)

5.5(3) PRODUCTION OF METHANE USING OFFSHORE
WIND ENERGY.
Young, R.B. and others. (AAI Corp., Maryland,
USA) NTIS Report PB80-129158 July 1979 63 pp.
Describes an investigation into the feasibility of converting wind energy to methane
gas. The process involves using offshore
winds to drive generators that supply electricity to electrolysis cells. Hydrogen
produced during the electrolysis of sea water
and carbon dioxide derived from underwater
carbonate deposits combine to form methane.
The type and availability of equipment to
implement such a process on a commercial
scale were also considered.

5.5(4) WIND AND SOLAR-POWERED REVERSE OSMOSIS
DESLINATION UNITS: DESCRIPTION OF TWO DEMON-
STRATION PROJECTS.
Petersen, G. and others. Desalination, 31 (1-3)
October 1979 pp. 501-509. From V. 2. proc. Inter-

national Congress on Desalination & Water Reuse, Nice 21-27 October 1979. (Also updated paper in proc. International Colloquium on Wind Energy, Brighton, U.K. 27-28 August 1981. BHRA Fluid Engng. Session 3 pp. 67-172.)

Two demonstration projects were under performance tests at the GKSS-research centre in cooperation with AEG-Telefunken and DIGAASES, Mexico using wind and solar power for the energy supply of reverse osmosis desalination units. Engineering design, site conditions and the operation mode of two RO-desalination plants with the GKSS plate module system supplied by a 6 kW wind energy converter and a 2.5 kW solar generator respectively are described.

5.5(5) WIND ENERGY CONVERSION AND HYDROGEN PRODUCTION: A FEASIBILITY STUDY FOR SOUTH AFRICA.
Harley, R.G. and Ben-Dov, E. (University of Natal, Durban, S. Africa) In International Conference on Future Energy Concepts, London, U.K. 30 January-1 February 1979. London, U.K., Institute of Electrical Engineers 1979 pp. 273-276. (IEE Conference Publ. no: 171)

Describes a feasibility study into the use of three different wind energy conversion schemes: WEC generates AC power and is synchronised to an electric utility grid which has energy storage capacity such as pumped water storage for example; similar to the first scheme, but the grid has no storage capacity; and WEC generated DC power is used in an electrolysis mode to generate hydrogen. (9 refs.)

5.5(6) WIND GENERATED ELECTRIC POWER FOR SANITATION SERVICES: A CASE STUDY.
Crawford, M.A. & Bergin, T.J. (Ellerbe Alaska, Fairbanks, Alaska Dept. of Env. Conservation, Juneau) Paper presented at 2nd Symposium of Environment Canada Utilities Delivery in Northern Regions, Alberta 19-21 March 1979 pp. 89-104.

Describes a wind energy system designed to provide electrical power for the water supply and sanitation delivery systems in the small community of Council, Alaska. Design and operation of the wind generator, electrical pumping and water storage systems are explained. Project experience to 1979 indicates that wind power systems were a viable alternative to conventional diesel generator power sources in remote villages.

5.5(7) ALTERNATIVE FUELS FOR MARITIME USE (POTENTIAL ALTERNATIVE FUELS FOR MARINE APPLICATIONS).
NAS Report, 1980 pp. 24-34. (USA)

Potential fuels evaluated include synfuels from coal, oil shale and tar sands; methanol/coal slurry; ethanol; gasoline/alcohol blends; hydrogen; ammonia; hydrazine; methane; nuclear power; coal wood; solar energy; OTEC; wave energy; and wind energy.

5.5(8) ALTERNATIVE FUELS FOR MARITIME USE (WIND-DRIVEN SHIP ALTERNATIVES).
NAS Report, 1980 pp. 125-147.

A resurgence of interest in wind-driven marine propulsion systems - in response to increasing petroleum fuel prices - is evident. Modern requirements for sailing ships are explained, and innovative ship designs are described.

5.5(9) WIND ENERGY FOR ELECTRIC VEHICLE RECHARGE.
Sammells, A.F. and Fejer, A.A. (Inst. of Gas Technol., Chicago, Illinois, USA) In proc. 15th Intersociety Energy Conversion Engineering Conference, Energy to the 21st Century, V. 1, Seattle, Washington, USA 18-22 August 1980. New York, USA, AIAA 1980 Paper 809166 pp. 835-839.

Wind energy is an attractive means of energy supply for electric vehicles intended for local traffic in suburban areas where individually owned windmills used for this purpose can be spaced at large enough distances from one another to avoid undesirable interference effects. Various aspects of systems of this type are examined leading to the conclusion that with the major components of the system already well developed, this source of energy could be utilised in a cost-effective manner in most parts of this country. (18 refs.)

5.5(10) WIND POWER FOR SHIPS - A GENERAL SURVEY.
Nance, C.T. (Medina Yacht Co. Ltd., U.K.) In proc. Symposium on Wind Propulsion of Commercial Ships, London, U.K. 4-6 November 1980. London, U.K., Royal Inst. Naval Archit. 1981 Paper no: 1 pp. 1-15.

Six principal wind-propulsion systems: square rig, fore and aft rig, aerofoils, magnus effect devices, wind turbines, and airborne sails (kites) - are summarised, and limitations considered. It is deduced that the central problem at the outset of the 1980s is how to match to each requirement the most suitable rig, ship size, wind/fuel engine power ratio, and service speed; and that the pressing need is for the acquisition and organisation of the necessary data to enable this to be done. Experience in the Government-funded wind turbine ship study is then drawn upon to suggest a rational methodology.

5.5(11) IN THE WAKE OF THE FLYING CLOUD.
Azarin, B. Science 81 2 (2) March 1981 pp. 80-85.

In an effort to find alternative methods of propulsion, various shipbuilders have been experimenting with the use of sails to supplement a conventional diesel engine. Sensors would feed data into the ship's computer, which will automatically trim the sails for optimal performance. Sailing ships ultimately could fulfill 50-75% of all ocean transport requirements.

5.5(12) INTEGRATED WIND AND SOLAR POWERED DESALINATION FACILITY.
Szostak, R.M. (Catalytic, Pennsylvania, USA) Paper presented at 16th ASME Intersociety

Energy Conversion Engineering Conference, Atlanta, GA 9-14 August 1981 V. 2 pp. 1825-1831.
Describes a solar power generation system combined with a reverse osmosis membrane filtration system. Power is generated by a wind and solar energy conversion system with battery and thermal storage. (7 refs.)

5.5(13) AN INTRODUCTION TO NEW ENERGY SOURCES FOR DESALINATION.
Buros, O.K. Desalination, 39 (1-3) December 1981 pp. 37-41.
Discusses potential energy sources for commercial desalination including solar, wind, ocean, and waste heat. (23 refs.)

5.5(14) PASSIVE COOLING BY NATURAL VENTILATION: A REVIEW AND RESEARCH PLAN.
Chandra, S. (Florida Solar Energy Centre, Cape Canaveral, USA) Paper presented at International Solar Energy Society - American Section/Et Al 1981 Annual Conference, Philadelphia 26-30 May 1981 V. 2 pp. 911-915.
Presents a literature review concerning wind-driven natural ventilation for passive cooling applications. Past and current research and patent literature on the subject is surveyed and design concepts for enhancing natural ventilation in humid climates explained. (23 refs.)

5.5(15) SEAFARERS RETHINK TRADITIONAL WAYS OF HARNESSING THE WIND FOR COMMERCE.
Marx, W. Smithsonian, 12 (9) December 1981 pp. 50-59.
Fishing vessels, tankers, and freight ships are being converted for sail working. Several examples of sail-assisted schooner and tanker design are given.

5.5(16) WIND AND SOLAR POWERED REVERSE OSMOSIS DESALINATION UNITS DESIGN, START UP, OPERATING EXPERIENCE.
Petersen, G. and others. (GKSS Research Centre, W. Germany) Desalination, 39 (1-3) December 1981 pp. 125-135.
Describes a photovoltaic solar generator and wind energy converter. Both projects demonstrate the operation of reverse osmosis units with defined plate modules coupled with unconventional energy systems. An extensive measurement programme is planned for the 6 kW wind energy converter. (5 refs.)

5.5(17) WIND ENERGY FOR REMOTE TELE-COMMUNICATIONS POWER.
Thorn, W.R. (TERA Corp., Berkeley, California, USA) In 3rd INTELEC International Telecommunications Energy Conference, London, U.K. 19-21 May 1981. London, Inst. Electrical Engineers 1981 IEE Conference Publ. no: 196 pp. 186-190.
Small wind energy conversion systems are commercially available and, for select applications, are cost competitive. They have been used to provide electric power in remote areas where access to an electric distribution grid is too difficult or expensive. As microwave repeater stations expand into underpopulated regions they will require their own power sources and small wind energy conversion systems are presented as a technology appropriate for this application. (8 refs.)

5.5(18) SOLAR AND WIND ENERGY FOR ELECTRIC SUPPLY OF AN AUTOMATIC LEVEL CROSSING IN SARDINIA.
Cesetti, G. and Pavone, F. Ingegneria Ferrov., 37 (4) April 1982 pp. 172-180 (In Italian)
Describes the equipment consisting of a wind generator-alternator driven by a propeller and of a photovoltaic solar panel generator, and the changes made to the electric circuits of the automatic level crossing. These changes allow for the transition to the reserve supply and the preservation of existing safety conditions.

5.5(19) UNIVERSAL SOLAR ENERGY DESALINATION SYSTEM.
Fusco, V.S. (Catalytic Inc.) In proc. IECEC 1982, 17th Intersociety Energy Conversion Engineering Conference, Los Angeles, USA 8-12 August 1982 V. 3. New York, USA, Institute Electrical & Electronic Engineers 1982 Paper no: 829252 pp. 1535-1537. (IEEE Catalogue no: B2CH1789-7)
Describes a flexible design for a solar powered water desalination plant in which a combination of solar, thermal and wind energy operates the reverse osmosis desalination unit.

5.5(20) WIND GENERATOR SYSTEM FOR MICROWAVE RADIO RELAY STATION.
Eguchi, N. (NTT, Engineering Bur, Japan) Japan Telecommunications Review, 25 (3) July 1983 pp. 216-220.
Describes an economical and reliable wind generator system, consisting of a wind turbine generator as main power source and three diesel driven generators as back up. The wind turbine generator adopts a Darrieus system and generates 8 kW maximum power with strong winds blowing over 15 m/s.

5.5(21) THE AUTOGYRO FOR SHIP PROPULSION.
Bose, N. Glasgow University. Int. Shipbuild. Prog., 30 (348) August 1983 pp. 179-186.
A strip theory approach is described which is designed to obtain the lift and drag coefficients of a wind turbine rotor operating as an autogyro. The theory was extended to include the effects of varying blade planform, blade section and blade twist, necessary in order to evaluate the performance of the variable pitch rotors used for the propulsion of marine craft when operating in beam winds in the autogyro mode. Results were obtained for two autogyro rotors and these are presented to give an idea of the typical performance of a wind turbine rotor when operating in the autogyro mode. The characteristics of autogyros are discussed in relation to their use as propulsion systems for wind propelled marine vehicles. (From author's abstract.)

5.5(22) LIVINGSTON MUNICIPAL WIND FARM THE FIRST
YEAR AND ONE HALF OF OPERATION AND PERFORMANCE.
Stern, E.D. (Livingston Community Development
Director, MT.) In American Wind Energy Assn.
Wind Energy Expo National Conference, San
Francisco 17-19 October 1983 pp. 283-293.

The wind farm is comprised of five 25 kW
wind generators, each of which has an individual
control and electrical interface system. They
are connected as a unit to the Montana Power Co.
Electrical Grid. Power failures, generator
replacements, lightning strikes, voltage surges,
and other operating and equipment problems are
chronicled. Although the wind farm turbines
and components are not yet fully reliable, the
project is anticipated to be a successful, viable
venture.

5.5(23) USE OF WIND POWER TO ASSIST IN STRIPPER
(OIL) WELL PUMPING.
Gilmore, E.H. Canyon, USA, West Texas State
Univ. Report no: TENRAC/EDF-111 31 August 1983
47 pp.

The site and its characteristics are described
as are important details on the assembly and
erection of the tower with the turbine on it.
An operating log covering the period from
1 January 1982 through to May 1983 is provided
in Appendix B and an energy production log in
Appendix C. Appendix D provides considerable
information on data collected in the periods
when the turbine was operational. The test
results are then presented which show operating
costs that are presently too high to compete
with local utility rates, but which will become
more competitive if prices for electricity
continue to escalate. (From author's abstract.)

5.5(24) ANALYSIS OF THE PERFORMANCE AND COST
EFFECTIVENESS OF NINE SMALL WIND ENERGY
CONVERSION SYSTEMS FUNDED BY THE DOE SMALL
GRANTS PROGRAM.
Kay, J. (Berkeley, USA, California Univ. at
Berkeley) Lawrence Berkeley Lab. Report no:
LBL-15998 April 1982 133 pp.

Analyses the technical performance and
cost effectiveness of nine small wind energy
conversion systems. An analytic framework
with which to evaluate the systems is given,
and each of the nine projects is reviewed,
including project technical overviews,
estimates of energy savings, and results
of economic analysis. Technical, economic,
and institutional barriers are summarised
that are likely to inhibit widespread dis-
semination of SWECS technology. Six systems
generate electricity, two pump water, and
one generates heat by hydraulic friction.

5.5(25) AN ANALYTICAL STUDY OF A HYBRID
WIND-PASSIVE SOLAR SYSTEM.
Bell, B.F. and McGowan, J.G.(Massachusetts
University at Amherst) In Intl. Wind Energy
Symposium 5th Annual Energy-Sources Technology
Conference, New Orleans, USA 7-10 March 1982
pp. 213-238. New York, Amer. Soc. Mech. Engrs.
1982.

Presents the results of a performance and
economic evaluation of a combined passive solar
and wind powered residential heating and electrical energy system. Simulated in a New
England wind and weather environment, the
modelled system is based on the coupling of a
synchronous generator vertical axis wind
machine with a highly insulated passive solar
house. The analytical model is composed of
four major sections: residential heating and
energy load; wind turbine generator system;
energy system performance model; and life-cycle
costing economic analysis.

5.5(26) DEMONSTRATION PROJECT: WIND TURBINES
FOR MUNICIPAL WATER PUMPING.
Nelson, V. (Canyon, USA, West Texas State Univ.)
U.S. Dept. of Energy Report no: DOE/R6-12060-TI
August 1982 25 pp.

Describes the trial wind power system used in
operation of the municipal water system for
Canyon, Texas, USA. Gives details of design and
operation of these machines. Presents data on
energy produced and wind speeds. Results include
energy feed back to the utility, predicted energy
from measured wind speeds and wind turbine reliability. Economic analysis indicates that the
wind turbines are not cost effective at present
for this well pumping purpose.

5.5(27) SMALL WIND ENERGY CONVERSION SYSTEMS,
APPLICATION AND PERFORMANCE.
Norton, J.H. (North Wind Power Co. Inc.) In
Intl. Wind Energy Symposium 5th Annual Energy-
Sources Technology Conference, New Orleans, USA
7-10 March 1982 pp. 45-50. New York, Amer. Soc.
Mech. Engrs. 1982.

Examines the two basic applications of SWECS-
remote and utility interface. First, the discussion of remote power briefly reviews the
history of this type of application, describes
the specifications of the HR2 high reliability
2 kW wind system, and examines in detail two
commercial applications in the petroleum
industry. Second, the discussion of utility
interface briefly reviews the history of the
L16 line interface 6 kW wind system, and
discusses its potential application in small
commercial windfarms, remote community power
systems and dispersed industrial, residential
and agricultural applications.

5.5(28) CARE AND FEEDING OF WIND FARMS.
Kahn, R. (Robert D. Kahn & Co.) Solar Age,
9 (9) September 1984 pp. 39-41.

A computer-based information system has been
integrated into the 26-turbine wind farm
owned by American Diversification Corp. in
Altamont Pass, California. Windnet automatically records the performance of wind
generators and gives wind farmers the
monitoring data they need. Power production
at specific wind speeds is documented, and
managers are alerted when malfunctions occur.

5.5(29) ROTOR POWER OUSTS THE AEROFOIL.
Pike, D. New Sci., 103 (1419) 30 August 1984
pp. 31-33.

Describes the principles of the Magnus effect, the lift observed on rotating cylinders and discusses its applications in design of wind turbines. Developments include the sailing vessel Tracker system incorporating a Magnus effect 8.4 m rotor, developed by the U.S. Windship Company in collaboration with Windfree. Also the Borg/Luther Group is investigating applications for the U.S. Navy for steering and underwater propulsion.

5.5(30) WIND POWER FOR SHIP PROPULSION.
Nanos, C.T. (Medina Yacht Co. Ltd.) In Wind Energy Conversion, proc. 5th BWEA Wind Energy Conference, Reading, U.K. 23-25 March 1983. P. Musgrove (ed.) pp. 318-325. Cambridge, U.K., Cambr dge Univ. Press 1984 375 pp. (ISBN: 0-521-26250-X)

Soft sail square rigs include the Windrise Ships Ltd. and the Dynaship design. Fore and aft rigged ships have many variants and developments from Indonesia and Greece are typical. Outlines some design concepts for rigid and semi-rigid aerofoils and Magnus effect (rotating cylinder) devices. Describes a Department of Industry (U.K.) project to study wind turbine propulsion for a ship of about 4,000 tons. Suggests that airborne sails are very effective on small sailing boats.

5.5(31) WIND-POWERED ELECTRIC GENERATION RUNWAY LIGHTING SYSTEM DEMONSTRATION PROJECT. INTERIM FINAL TECHNICAL REPORT; 15 SEPTEMBER, 1981-15 DECEMBER, 1983.
Mesa, D. (California State Dept. of Transportation, Sacramento, Div. of Aeronautics) U.S. Dept. of Energy Report no: DOE/SF/11611-T1 January 1984 58 pp. (DE84 014962/XAD)

Describes a small scale demonstration project to determine the feasibility of using wind-powered generation to operate the runway lighting system at Half Moon Bay Airport. This project seeks to determine if wind power has practical application to an airport environment as a cost-effective means of providing an alternate source of energy.

6 Bibliographies

6(1) ALTERNATIVE ENERGY SOURCES.
Leicestershire Libraries and Information
Service. LLIS October 1979 24 pp.
   Bibliography and survey of solar energy,
wind power, tidal power, wave power and
geothermal energy.

6(2) ALTERNATIVE SOURCES OF ENERGY -
3RD REVISED EDITION.
Architectural Association Library.
AA Library February 1979 22 pp. (AA Library
Bibliography New Series no: 51)
   Controlled terms: energy/ water/ wind/
heat/ solar energy/ bibliography/ alternative
energy/ methane/ heat pump/ design/ recycling.

6(3) DOE RESEARCH ON WIND ENERGY (A BIBLIO-
GRAPHY).
Oak Ridge, USA, Department of Energy May 1980
59 pp. Report no: DOE/TIC/SDI-2-2.
   Contains 126 abstracts.

6(4) SELECTED REFERENCES ON SMALL WIND ENERGY
CONVERSION SYSTEMS.
Golden, USA, Rockwell International 15 April
1980 11 pp. (DOE Rocky Flats Wind Syst.
Programme)
   Lists selected bibliographies, books,
conference proceedings and periodicals covering wind energy conversion systems, their
design, siting and aerodynamic performance.
Information on small wind systems with under
100 kW design output predominates.

6(5) WIND ENERGY LITERATURE IN THE DENMARK
TECHNICAL LIBRARY. 3RD EDITION.
Anderson, C.E. Lyngby, Denmark, National
Technol. Libr. Denmark 1980 94 pp. (Publi-
cation no: 59) (In Danish)
   Comprises 47 pages of journal references
and 32 pages of book references covering the
period 1878 to 1980.

6(6) WIND POWER BIBLIOGRAPHY.
Randall, P. London, U.K., Institution of
Electrical Engineers May 1980 111 pp.
   Abstracts, covering the years 1969-1979, are
drawn largely from the INSPEC database.
Sections include: equipment; small and large
scale systems; unconventional wind energy
systems; energy storage; site evaluation/
wind characteristics; modelling/simulation;
windpower and the energy crisis; vertical
windmills; siting and economics.

6(7) CATALOGUE OF WIND ENERGY LITERATURE IN
THE RISOE LIBRARY.
Risoe National Lab., Roskilde, Denmark. Report
no: RISO-M-2297 July 1981 120 pp. (In Danish)
(DOE Report no: DE82 902190)
   Presents a special collection of wind energy
literature established in connection with the
Test Plant for Small Windmills at Risoe National
Laboratory. About 900 publications, including
conference papers, reports, journal articles
and books, are listed in 24 subject categories.

6(8) SMALL WIND ENERGY CONVERSION SYSTEMS
(SWECS). 1976 - SEPTEMBER 1982. (CITATIONS
FROM THE ENERGY DATA BASE).
National Technical Information Service, Springfield, VA, USA. Report for 1976 - September
1982, September 1982 120 pp. (Prepared in cooperation with the Department of Energy,
Washington, DC) (PB 82-876376).
   Contains citations concerning the technology,
design and performance of small wind energy
conversion systems. Tower dynamics and rotor/
wind dynamic interactions are considered.
Federal and state regulations, economic analyses,
and marketing are discussed. Safety and
reliability are also covered.   (Contains
125 citations.)

6(9) WIND POWER: WIND ENERGY ENGINEERING. 1976 -
DECEMBER 1982. (CITATIONS FROM THE ENERGY DATA
BASE).
National Technical Information Service, Springfield, VA, USA. Report for 1976 - December 1982,
December 1982 193 pp. (Prepared in cooperation
with the Department of Energy, Washington, DC)
(PB 83-857425).
   Contains citations concerning the engineering
aspects of wind energy development and utilisation. Topics include electrical power systems,
energy storage techniques, electric grid matching and interconnections, the dynamics of wind
power stations, clustering of wind generators
into arrays, hybrid solar and wind energy
systems, and economic analyses. (Contains 208
citations.)

6(10) THE WINDMILL: ARCHITECTURE AND ENERGY.
Cable, C. Vance Bibliographies, Architecture
Series A-891 1983 8 pp.

100 references dating from 1920 cover new
designs and construction, and the energy production potential of windmills.

6(11) WINDMILL DESIGN, DEVELOPMENT, CONSTRUCTION
AND PERFORMANCE, 1973 - OCTOBER 1982: CITATIONS
FOR THE BHRA FLUID ENGINEERING (FLUIDEX) DATA-
BASE.
Springfield, USA. NTIS October 1982 113 pp.
(PB83-850685)

Gives references on the design, development,
construction and performance of windmills and
associated systems for wind energy conversion.

7 Author Index

161

Griffith, S.K.	5.1(23)	Hughes, W.L.	1.2.3(4);2.2.2(4); 3.1(29);3.2(35)
Guerrero, J.V.	4.1(6)	Hugosson, S.	1.3.3(5);3.2.2(16)
Guild, D.H.	5.2.2(9)	Hunnicutt, W.	3.2.2(10)
Gunkel. W.W.	5.4(15)	Hunt, V.D.	3.1(34)
Guscott, J.B.	1.3.1.1(17)	Hunter, G.H.	2.1(35)
		Hurst, J.	1.2.3(22)
		Hutton, M.	1.3.1.1(2)

H

Haack, B.N.	5.2.2(38);5.2.3(11)		
Haas, S.M.	1.2.3(31);4.1(1)	Ibrahim-Said, M.A.	1.1(5)
Hadley, D.L.	1.2.3(54)	Ide, H.	3.2.3(10)
Hagen, L.J.	5.4(4)	Igra, O.	3.2(33)
Halberg, N.	5.2.3(17)	Inall, E.K.	1.7(9);3.2.1(13)
Halliday, J.A.	2.2.2(18)(19);5.2.3 (44);5.4(30)	Infield, O.G.	5.2.3(46)
		Inhaber, H.	1.2.1(10)(31)
Halpern, D.A.	2.1.1(22)	Ireland, F.E.	4.2(20)
Ham, N.D.	3.2.1(15)(16)(25)	Isaacs, N.P.	5.1(1)
Hanlon, J.	1.3.1.2(3)		
Hansen, A.C.	1.2.3(21);3.2(8); 3.2.5(19)(44)		
		Jacob, A.	5.2.1(3)
Harb, A.	3.2.5(51)	Jacobs, E.W.	1.2.3(57);3.2.3(13) 5.1(5)(11)
Hardy, A.C.	1.3.1.1(4)		
Hardy, D.M.	1.2.3(7);2.1(3)	Jagadeesh, A.	1.6(3)(6)(12)
Hardy, W.E.	3.1.1(9)	Jagadish,B.S.	5.4(34)
Harley, R.G.	5.5(5)	Jakubowski, G.S.	3.2.5(31)
Harper, J.R.	5.2.2.1(4)(6)(8)(10 (11)(13)	James, P.	1.2.2(25)
		Jamieson, P.H.	3.2.2(26)
Harper, M.R.	5.4(9)	Janetzke, D.C.	3.2.2(32)
Harris, R.I.	3.1.1(8)	Janna, W.S.	2.2.1(30);5.2.3(22)
Haskins, H.J.	5.2.1.1(13)	Janssen, A.	5.2.2(39)
Haslett, J.	5.2.2(22)	Jaran, C.	3.2.2(12)
Hassan, U.	3.2.5(52)	Jarass, L.	1.3.4(20);4.2(19)
Hau, E.	1.3.4(25)	Javid, S.H.	5.2.3(39)
He, D.	1.6(2)	Jayader, T.S.	5.2.1(1)
Healy, T.J.	3.1(36)	Jeffs, L.H.	5.1(1)
Heiko, L.K.	5.2.2(30)	Jeng, D.R.	3.2.5(37)
Hellickson, M.A.	5.4(8)	Jensen, J.	5.2.1.1(9)
Helm, S.	3.2(16)	Jensen, S.A.	3.1.2(19);3.2(58); 3.2.5(19)
Hendrick, P.L.	3.1.2(25)		
Henry, B.	3.2.1(32)	Johansen, O.S.	1.3.3(25)
Hershbery, G.	4.1(12)	Johanson, E.E.	5.1(6)(19)
Hibbs, B.H.	3.2.5(21)	Johansson, M.	1.3.3(11)
Hickey, M.	4.3(7)	Johansson, T.B.	1.3.3(9)
Hicks, B.B.	2.1.1(9)	Johnson, B.	1.2.1(23)
Higashi, K.K.	3.2.1(27)	Johnson, R.F.	2.1.1(30)
Hilbert, C.B.	2.2.1(30);5.2.3(22)	Johnson, R.T.	4.2(26)
Hill, L.M.	1.3.1.1(12)	Johnston, S.F.	3.2.1(18)
Hinrichsen, D.	1.3.3(2)	Jones, B.W.	5.2.2(37)
Hipp, K.	3.2.5(41)	Jong, M .T.	5.2.2(41);5.2.3(12) 5.3(24)
Hock, L.J.	5.1(25)		
Hocking, J.D.	5.1(1)	Jufer, M.	3.1(52)
Hodges, D.C.	5.3(14)	Justus, C.G.	2.1(15);3.1.2(1)
Hodgkins, P.	2.2.1(26)	Juvet, P.J.D.	3.2.5(29)
Hoffer, T.	2.1(28)		
Hoffman, L.	5.1(12)		
Hohenhemser,K.H.	3.2.2(37);3.2.5(22)	Kadlec, E.G.	3.2.1(3)
Hollandsworth, R.P.	5.2.1.1(12)	Kahawita, R.	2.2.2(11)
Holmes, B.A.	1.3.1.1(21);3.2.1 (24)	Kahn, E.	5.2.2(4)(40)
		Kahn, R.	5.5(28)
Holmes, P.G.	5.2.1(32)	Kaier, U.	5.3(29)
Holmgren, B.	2.1(20)	Kaminsky, F.C.	3.1(54)
Hoock, S.	5.2.2.1(12)	Kant, M.	5.2.1(4)
Hori, A.M.	2.2.1.1(5)(11)	Kareem, A.	3.2(38)
Hormander, O.	1.3.3(8)	Kasmarik, M.D.	3.2.3(23)
Horvath, E.	3.1(14)(45)	Kassakian, J.G.	5.2.1.1(2)
Hottin, F.	4.2(29)	Kato, Y.	3.2.1(19)
Howard, S.M.	2.1(4)	Katz, M.J.	1.2.1(11)
Howe, J.W.	1.1(3)	Kaufman, J.W.	3.2(21)
Howes, H.E.	3.2.2(9)	Kaupang, B.M.	5.2.2.1(16)
Hsu, C.T.	3.2.3(10)(11)	Kay, J.	5.5(24)
Huang, C.H.	2.1.1(29)	Karza, R.V.	3.2.2(32)
Hubbard, H.H.	4.2(21)(28)	Kear, E.B.	5.4(13)

I

J

K